AIR POLLUTION – NEW DEVELOPMENTS

Edited by **Anca Maria Moldoveanu**

Air Pollution – New Developments
Edited by Anca Maria Moldoveanu

Published by InTech
Janeza Trdine 9, 51000 Rijeka, Croatia

Publishing Process Manager Ivana Lorković

Technical Editor Teodora Smiljanic

Cover Designer InTech Design Team

Additional hard copies can be obtained from orders@intechopen.com

Air Pollution – New Developments, Edited by Anca Maria Moldoveanu
p. cm.
ISBN 978-953-307-527-3

Meet the editor

Dr. Anca Maria Moldoveanu
Details of Education: Medical Doctor Degree, University of Medicine and Pharmacy Timisoara 1981, PhD in Medicine, University of Medicine and Pharmacy Bucharest, 1997, Certificates from London School of Hygiene and Tropical Medicine (1996, 1997), Michigan State University (2005, 2006), World Health Organization (1995, 1996).
Career to date: Specialist in Hygiene and Environmental Medicine since 1988. Specialist with the highest degree in Hygiene and Environmental Medicine since 1993. Appointments held: Clinical Hospital "Coltea" Bucharest, Romania, 1981 -1984 as medical doctor; Institute of Hygiene and Public Health Bucharest, Romania, 1985 -1991 with full time job as specialist in Hygiene and Environmental Medicine; National Institute of Public Health, Bucharest, part time job 1991 onward; University of Medicine and Pharmacy "Carol Davila" Bucharest, chair of hygiene and medical ecology – full time job since 1991: assistant professor (1991-1997), lecturer (1997 – 2005), associate professor (2005 onward), head of the chair of hygiene and medical ecology.
Publications: 4 book, 3 chapters in books, 66 articles and 38 communications in scientific events in Romania and abroad.

Contents

Preface

Today, an important issue is environmental pollution, especially air pollution. Due to pollutants present in air, human health as well as animal health and vegetation may suffer.

The book can be divided in two parts. The first half presents how the environmental modifications induced by air pollution can have an impact on human health by inducing modifications in different organs and systems and leading to human pathology (the influence of air pollution with particles on respiratory health, the molecular markers associated with biological response to aromatic hydrocarbons from urban air in humans, primordial prevention of non-communicable diseases and air pollution). This part also presents how environmental modifications induced by air pollution can influence human health during pregnancy.

The second half of the book presents the influence of environmental pollution on animal health and vegetation and how this impact can be assessed (the use of the micronucleus tests on TRADESCANTIA to evaluate the genotoxic effects of air pollution, the use of transplanted lichen PSEUDEVERNIA FURFURACEA for biomonitoring the presence of heavy metals, the monitoring of epiphytic lichen biodiversity to detect environmental quality and air pollution, the assessment of the genotoxicity indicator capacity of lichens exposed to air pollution, the assessment of the decline of silver fir in the forests in Slovenia due to air pollution, the interaction of urban vegetation and air pollutants from ambient air environment, the influence of greenhouse gases on *Populus tremuloides*). One of the chapters presents a detailed modeling study of the removal mechanisms of gaseous pollutants in a tropical boundary layer and the quantification of air pollutants removal rates around a heavily afforested power plant.

The book is the result of the effort of many experts. I would like to acknowledge the authors, who are from different countries, for their contribution to the book.

I wish to offer special thanks to the team from InTech Open Access Publisher for their exceptional assistance.

Anca Maria Moldoveanu
University of Medicine and Pharmacy "Carol Davila" Bucharest
Romania

Part 1

Air Pollution and Pregnancy

1

Air Pollution Exposure During Pregnancy and Reproductive Outcomes

Ferran Ballester and Carmen Iñiguez
University of Valencia, Nursing School, Valencia,
Center for Public Health Research (CSISP), Valencia,
Spanish Consortium for Research on Epidemiology and Public Health (CIBERESP)
Spain

1. Introduction

Air pollution is largely recognized as a risk factor for several outcomes including increased mortality, increased hospital admissions and emergency visits for both respiratory and cardiovascular diseases, and impairment of respiratory function, including reduced lung function, exacerbation of asthma and chronic obstructive pulmonary disease (Brunekreef & Holgate, 2002). Some populations have shown to be more susceptible to these effects, and among them including those people in the tails of the age distribution, i.e. the elderly and infants (Laumbach, 2010).

The study of fetal growth and birth outcomes has become an important emerging field of environmental epidemiology (Sram et al., 2005). Birth outcomes are important indicators of pregnancy and infant care as well as newborn and infant health. Besides that, reduction in fetal growth has been associated with health problems and developmental delays during childhood, from an increase in hospitalizations (Morris et al., 1998) to poor cognitive and neurological development (Richards et al., 2002) as well as with an increased risk of chronic diseases later in life (Barker, 2007).

Fetuses, like infants, present a special vulnerability, compared to adults, regarding environmental toxicants due to differences in exposure, physiological immaturity, and longer life expectancy after exposure (Perera et al., 2002; Schwartz, 2004). Results from epidemiological and experimental studies show that fetuses and infants are especially susceptible to the toxic effects of pollutants such as suspended particles, polycyclic aromatic hydrocarbons (PAH), and tobacco smoke (Perera et al., 2003). In recent years, a growing body of epidemiological research has focused on the potential impact of prenatal exposure to air pollution on birth outcomes.

In recent years a number of authors have reviewed the evidence on the relation between prenatal exposure to air pollutants and reproductive outcomes (Glinianaia et al., 2004; Maisonet et al., 2004; Lacasaña et al., 2005; Sram et al., 2005; Wang & Pinkerton, 2007; Hackley et al., 2007; Wigle et al., 2008; Bosetti et al., 2010; Vrijheid et al., 2011; Shah & Balkhair, 2011). In these reviews several outcomes have been related to exposure to air pollution during pregnancy, including low birth weight, reduced birth size, fetal growth retardation, pre-term birth, stillbirth, congenital malformations, and infant mortality.

A limited number of studies have linked different birth defects to routinely measured air pollutants, but results thus far are equivocal at best (Ritz, 2009). Practically the same could be said concerning stillbirth. Very few epidemiologic studies have elaborated on the impact of air pollution on the risk for stillbirth; results were inconsistent and the studies did not elaborate on the susceptible time of pregnancy. As an example, a recent study conducted in northern England found no association between black smoke air pollution and the risk of stillbirth during pregnancy (Pearce et al., 2010).
Some methodological problems have been argued in the epidemiological studies assessing the impact of air pollution on reproductive outcomes. Authors of recent methodological reviews (Ritz & Wilhelm, 2008; Slama et al., 2008; Woodruff et al., 2009) agree that new prospective studies should allow adequate evaluation of fetal growth (i.e. through ultrasound measurements), valid assessment of air pollution exposure, consider different time windows of exposure, and collect sufficient information on confounding variables.
Our objective in this chapter is to examine and summarize the evidence on the relation between ambient air pollution during pregnancy and birth outcomes. Additionally, we aim to describe some methodological problems and make recommendations for future studies.

2. Methodological aspects

In the past decade, there has been a sharp increase in the number of scientific studies describing possible effects of air pollution on perinatal health. These studies resulted in a considerable amount of research articles published on this topic. Reviews of the preceding literature have been published generally concluding that the evidence, although suggestive of an adverse effect, was difficult to synthesize. Variability existed of course in the nature of the pollutants and outcomes investigated but also there were important differences in the availability of data, study design and statistical analysis.
To discuss the contribution of methodological aspects on findings, heterogeneity is the basis to identify priorities and make recommendations for future research. This aim led to several international workshops and published reviews covering the main challenges and the main methodological aspects of perinatal air pollution epidemiology (Slama et al., 2008). Also trying to discern how differences in research methods contributed to variations in findings, the datasets from 20 different studies along six countries have been analyzed using a standardised protocol (Parker et al., 2011), within a recent initiative by international collaborators (Woodruff et al., 2010).
A synopsis of key methodological issues surrounding the study of air pollution effects on perinatal health is highlighted bellow.

2.1 Study design

Time-series is the typical design in air pollution epidemiology, and it has been used to study preterm birth and fetal death (Pereira et al., 1998, Sagiv et al., 2005). Under this design, population-aggregated daily counts of health events relating to daily levels of pollutants are analyzed. The time series approach, by design, controls for confounders constant over time. In consequence, there is much less concern for personal characteristics than for weather-related confounders. However, this approach only addresses short-term effects and relies on temporal variations. It appears that taking into consideration geographical variations and also medium-term temporal contrasts could be more efficient in reducing residual confounding. Furthermore, traditional time-series design assumes that the population at risk remains stable across time, a hypothesis that may be not satisfied dealing with adverse birth outcomes because of seasonality of birth influenced by socio-demographic factors (Boback et al., 2001).

Linkage of **registry data** such as birth certificates, with exposure measures of ambient air quality, typically from outdoor stationary monitors has been applied much more in the last decade, resulting in a fast-growing body of evidence on adverse effects of air pollution on fetal development (Salam et al., 2005). This approach has the advantage of allowing large-size studies at a very low cost because it relies on routinely collected data. Its limitation is that relevant information at the individual level is unavailable from birth registers and its absence in analyses may clearly lead to misclassification and confounding.

Potential confounders in this context are socio-economic and occupational status, adverse behaviours such as alcohol or tobacco use, or poor diet. Apart from confusion, effect modification can occur as certain subpopulations of women and fetuses may be especially vulnerable to air pollutants. To this respect, several studies suggest a stronger effect for males that for females (Jedrychowski et al., 2009; Ghosh et al., 2007).

For these reasons, prospective **cohort studies** with recruitment of women at earlier stages of pregnancy are a promising alternative: They collect detailed information on potential confounders and allow the personal assessment of exposure and the use of biomarkers (Estarlich et al., 2011). As an added value, under some conditions, causality may be inferred from them. Its main limitation is the associated high cost, leading to reduced sample sizes.

The trade-off between small studies rich in individual information and big studies based on registries with scarce information about confounders and risk factors should be taken in consideration (van den Hooven et al., 2009). A possibility is to combine the strengths of these two designs by conducting **case–control studies** (Hansen et al., 2009) with collection of additional information at the individual level for a sample nested within a big cohort constituted from birth records.

2.2 Exposure assessment

Various approaches may be used to estimate air pollution exposure, from the use of biomarkers of exposure to environmental models. Obviously, the most accurate information of individual exposure should be derived from short-term **personal monitoring** and **biomarkers**. However, both are scarcely used because of their extreme cost and inability to discern the appropriate timing of measurement or source contributions. Studies using personal measurements are relatively small in size and commonly used as validation studies of other modelling approaches (Nethery et al., 2009).

As was said before, there are a number of studies (Xu et al., 1995; Boback 2000; Ritz et al., 2000) with exposure assessment based on data from **monitoring networks.** These studies use data from the monitoring station closest to the subject's home address or interpolating data for neighbouring monitors, for which measurements are averaged over the entire pregnancy or over each trimester of pregnancy. Advantages of this approach are the use of readily available exposure data, simple implementation and, as pollutants are assessed on an hourly or daily basis, high flexibility in terms of the temporal exposure window considered (Lepeule et al., 2010). Nevertheless, the individual exposure assignment based exclusively on them is prone to inaccuracy because the number of sampling locations is often scarce failing to capture the spatial variability. Furthermore sampling locations could be biased towards specific sources of pollution (i.e. traffic, background, industry) and not always provide continuously measured data (i.e. Particles ≤ 2.5 μm diameter ($PM_{2.5}$) in the US is often measured every 3–6 days).

Nitrogen dioxide (NO_2) is one air pollutant frequently used as a **surrogate** for traffic related pollution in prospective studies, both in adults and in children (Jerrett et al., 2008; Brunekreef, 2007). This is the case because outdoor NO_2 levels correlate well with traffic generated pollutants. NO_2 concentrations may be easily measured using passive samplers, and are routinely measured by air quality networks allowing for correction for temporal variations. Several epidemiologic studies have examined associations between maternal exposure to nitrogen dioxide during pregnancy and reproductive outcomes such as prematurity, fetal growth retardation and birth weight (e.g., Aguilera et al., 2009; Ballester et al., 2010; Bell et al., 2007; Liu et al., 2007; Ritz & Wilhelm 2008; Slama et al., 2008; Brauer et al., 2008). Some of them have introduced other geostatistical models based on measurement campaigns with fine spatial resolution and geographic information systems (GIS) plus temporal adjustment based on background monitoring stations (Lepeule et al., 2010).

Recently, GIS-based approaches have become more commonly used to capture small area variations in pollution. The great development experienced by the GIS managing and displaying spatial data (Vine et al., 1997; Bellander et al., 2001; Briggs, 2005, 2007) has been decisive in this respect. Within GIS-based approaches, air dispersion models (Malvqvist 2011), traffic proximity models and Land-use regression are among the most used. Exposure assessments using **traffic proximity** (Jerrett et al., 2008; Brunekreef, 2007) or distance-weighted traffic densities as a proxy for individual exposure to air pollution relies on the premise that traffic is a major source of local air pollution. A potential problem with this method is the underlying assumption that pollution spreads uniformly away from its source. **Land-use regression** (LUR) methods apply regression to map air pollution using geographical variables such as land use, traffic intensity and population density as predictors. (Ghering et al., 2011; Iñiguez et al., 2009; Henderson et al.,2007; Aguilera et al., 2009; Slama et al., 2008; Wheeler et al., 2008). LUR models are an attractive option because they are able to capture fine spatial variability at a very low cost.

Whatever the method used for mapping outdoor levels of air pollution, in most studies personal exposure is the estimated concentration at a home address derived from the map. A large proportion of personal exposure is commuting time, workplace exposure or come from indoor sources, therefore personal exposure assessments should be improved by considering **time-activity patterns** and indoor sources such as tobacco smoke and cooking. In the case of studies dilated in time, taking into account a possible change of address could be also an important point. Obviously, improving exposure assessment and reducing exposure misclassification would lead to strengthen effect estimates.

Another methodological challenge recently highlighted is the convenience of exploring **critical windows of exposure** shorter than the whole pregnancy. Pregnancy is a period of big change and the timing of exposure could be as important as the intensity. Studies suggest that exposures at the periconceptional period or first trimester may lead to suboptimal placental or fetal development in earliest stages of pregnancy with ulterior adverse consequences on the duration of pregnancy and fetal growth. At this time, there is a lack of toxicological information to help guide selection of relevant exposure periods for most reproductive outcomes, but an appropriate scale, searching for critical windows may be months or trimesters. For low birth weigth (LBW) and preterm birth, first and third trimester air pollution exposures appears as the most relevant, while for birth defects, the development time of the specific organ has to be considered (Ritz, 2008).

Although trimesters is the most common scale, offering comparability with existing studies, exposure windows shorter than trimesters have also been proposed. In this case it is important to consider if that accuracy is possible given the exposure metrics. (Woodruf, 2009). LUR models are meant to characterize spatial rather than temporal variability in air pollution levels. Aiming to explore critical exposure windows within pregnancy, fine temporal resolution added to fine spatial resolution is required. Temporal patterns observed at ambient monitor stations (measured most days) could be used to adjust LUR model estimates as done by Slama et al. (2007). This assumes the spatial surface (variability) is stable over time and additional monitoring may be needed to verify this assumption.

Another point is the statistical treatment of exposure. The usual way is to consider a **linear** effect but this assumption widely depends on the examined toxicant. High exposures can lead to outcomes such as spontaneous abortions or stillbirth, removing fetuses at risk for and adverse birth outcome such as LBW or preterm birth, translating to non-linearity in dose effects. Therefore, to explore the shape of relationships by means of non-linear models may be advisable.

Finally, it is also advisable the use of **multipollutant models** to try to disentangle individual effects. However, multipollutant models may be unfeasible due to high correlation in time and space between air pollutants sharing sources. Given that people breath a mixture of air pollutants, to study synergistic effects may also be important.

2.3 Outcomes

The majority of published air pollution and perinatal health studies have evaluated the effect of exposure on fetal growth (usually assessed on the basis of attained weight at birth) and the effect on the duration of gestation. Other important but less studied outcomes related to pregnancy are infertility, fetal loss (including stillbirth and spontaneous abortion) and congenital malformations. These outcomes are briefly defined below in order of its eventual occurrence.

Infertility is the failure of a couple to reproduce after a reasonable time with regular and unprotected sexual intercourse to obtain a pregnancy. One year is usually considered as a deadline, then fertility is treated as a dichotomy, assessing the proportion of couples that fail to conceive within a 12-month period. Because there is not a discontinuity in the probability of conception after 12 cycles, a common alternative approach is to examine the monthly probability of conception in a life-table approach. Both approaches can be derived by assessing time to pregnancy.

Fetal loss is defined as a spontaneous end of pregnancy, without living birth, occurring between conception and the end of labor and excluding induced abortions and medical termination of pregnancy and ectopic pregnancies. Since the clinical, physical and psychological consequences of a fetal loss increase with the time of occurrence, fetal losses are divided in three categories according gestational age (GA): early fetal losses, occurring before than 6 weeks (usual time of detection of pregnancy), spontaneous abortions, occurring between 6 weeks and the time when the fetus is viable, and stillbirth, occurring after this point. The gestational age cut-point between spontaneous abortions and stillbirths could vary between 20 and 28 weeks according local legal definitions. However in epidemiological studies, the 20 or 22 gestational weeks cut-points are most often used.

Congenital malformations: The definition of congenital malformations classically includes structural malformations, syndromes and chromosomal anomalies present in the fetus or the newborn as defined by ICD-10 Q00.0 to Q89.9. It is recognized that a number of congenital malformations are detected during childhood or even adulthood. Since some of them are a major cause of abortion (Eurocat, 2005), these functional defects are identified only in a subset of all fetuses affected. The prevalence of (detected) congenital malformations ranges from 2% (when only major malformations are selected) to 7% when minor forms are included.

Prematurity is defined by the duration of gestation in completed weeks or days. The most common cut-point is 37 weeks (259 days). Accordingly, a preterm birth is defined as an infant born before the completion of the 37th week of gestation. Other dichotomous outcome measures with stringent cut-points aiming to define more extreme deviations from the normal gestational duration are: Extremely premature birth (< 28 weeks or 196 days) and Very preterm birth (< 32 weeks or < 224 days). Prematurity is an important indicator of perinatal and future health. In fact, preterm babies account for 75% of perinatal mortality and more than half of the long- term morbidity (Skalkidou A, et al., 2010).

Fetal Growth is also an important predictor of postnatal health, often based on the attained weight at birth. Birth weight is the standard measure of fetal growth because it is a sensible endpoint and data are greatly available by medical registers. Other measures of neonatal anthropometry such as length, head, abdominal circumference, etc. might provide additional information as well as longitudinal information about fetal parameters (ultrasounds), but they are rarely used in epidemiological studies because of their scarce availability. Different metrics based on birth weight have been used to identify suboptimal fetal growth including a reduction in birth weight (as a continuous variable), low birth weight (LBW), very low birth weight (VLBW), small for gestational age (SGA), intrauterine growth restriction (IUGR) and fetal growth restriction (FGR).

Low birth weight (LBW) is defined by the World Health Organization (WHO) as a birth weight less than 2500 grams, while very low birth weight is defined as a birth weight less than 1500 grams. As birth weight greatly depends of gestational duration, often both outcomes are combined to examine fetal growth, as is the case for small for gestational age.

A newborn is defined as small for gestational age (SGA) if its birth weight is lower than the 10th percentile of a suitable gestational age-specific weight reference. Therefore, identification of SGA births requires relevant reference curves.

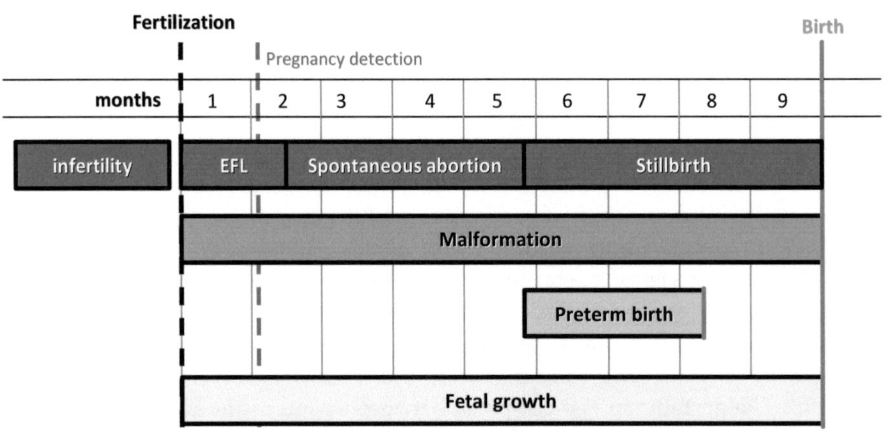

EFL: Early fetal loss

Fig. 1. Overview of the main pregnancy-related outcomes covered

The term *intra-uterine growth restriction* supposes that the fetus was retarded in its growth by a pathological process during fetal life. However, there is a lack of a clinical definition for IUGR and many studies use the same statistical limits to identify SGA or IUGR infants. In other studies an infant is defined as IUGR if it is a full-term infant with a low birth weight.

SGA fails at distinguishing "constitutionally small" infants, according to anthropometric characteristics, to those who are "growth retarded," that is, smaller than what their parental characteristics, sex, gestational duration would predict.
A newborn is defined as *fetal-growth retarded* if he fails to achieve its genetic growth potential in utero. The classification is done by adjusting birth weight by gestational age, infant sex and maternal characteristics.

3. Results

As introduced above, during the last years different authors have reviewed the epidemiological evidences about the potential effects of air pollution exposure during pregnancy and reproductive outcomes. In the table 1 we summarized the pollutant and outcomes included in the main systematic reviews on these topic.

Review	Period of study	Pollutants included	Outcomes (and # of studies)						Total # of studies
			Stillbirth	Preterm birth	BW, LBW, VLBW	SGA, IUGR	Perinatal & Infant mortality	Others	
Glinianaia et al., 2004	1996-2001	TSP, PM_{10}	3	3	3	3	-	-	12
Maisonet et al., 2004	1996-2001	TSP, PM_{10}, $PM_{2.5}$, CO, SO_2, NO_2, O_3	-	5	9	3	-	-	12
Sram et al., 2005	1966-2001	TSP, PM_{10}, $PM_{2.5}$, CO, SO_2, NO_2, O_3, PAH, POM	2	4	9	4	7	-	23
Lacasaña et al., 2005	1994-2003	TSP, PM_{10}, $PM_{2.5}$, CO, SO_2, NO_2, NOx, O_3, PAH	2	10	17	3	8	Congenital malformation (#: 2)	31
Bosseti et al., 2010	1966-june 2010	TSP, PM_{10}, $PM_{2.5}$	-	13	17	4	-	-	30
Shah & Balkhair, 2011		TSP, PM_{10}, $PM_{2.5}$, CO, SO_2, NO_2, NO, O3, HAP, BETX	-	11	25	10	-	-	41

BW: birth weight; LBW: low birth weight (<2500 g), VLBW: very low birth weight (<1500 g)
SGA: small for gestational age, IUGR: intrauterine growth restriction
TSP: total suspended particulate; PM_{10}: particulate matter <10 μm; $PM_{2.5}$: particulate matter <2.5 μm, PAH: polycyclic aromatic hydrocarbon, POM: polycyclic organic matter; BTEX: benzene, toluene, ethyl-benzene, xylene

Table 1. Systematic reviews of the epidemiological studies assessing the relation between prenatal exposure to air pollution and reproductive outcomes

3.1 Infertility

Throughout the last decades the crude human birth rate (live births per 1000 population) declined, suggesting a potential decline in fecundity (the potential to conceive). Detection of environmental contaminants in human tissues, together with reports of a global decline in semen quality, promoted the hypothesis that environmental toxicants could be important causal agents of such decline (Foster et al., 2008).
Epidemiological studies demonstrated exposure to ambient levels of air pollutants, especially airborne particulate matter, has been linked in particular to DNA fragmentation (Selevan et al., 2000) (Evenson et al., 2005) and in general to abnormal sperm morphology and reduced sperm performance in men. (Somers 2011; Hammoud et al., 2010). Recent experimental animal data indicated that female fertility is also affected by air pollutants (Veras et al., 2010).

3.2 Fetal loss

A previous review of environmental toxicants and reproductive outcomes only suggested as related to fetal loss the exposure to dichlorodiphenyltrichloroethane and bisphenol-A (Stillerman et al., 2008). Apart from a proven adverse effect of second-hand smoking exposure (Peppone et al., 2009; George et al., 2006), the evidence of an impact of air pollution on this reproductive outcome remains scarce. Only exposure to airborne particulate matter has been related to fetal loss in a study performed in mice (Veras et al., 2009). The authors found the mean post-implantation loss rate increased by 70% among those exposed to particulate matter before and during pregnancy versus no exposed.

3.3 Prematurity

Besides prenatal growth (birth weight and growth restriction), prematurity is the outcome for which there exist more published articles. In their recent review, Bosetti et al (2010) identified 13 studies assessing the relationship between exposure to **particulate matter** and preterm birth. Most of them (n: 9) considered PM_{10} as the fraction of interest (Ritz et al., 2000; Sagiv et al., 2005; Wilhelm and Ritz 2005; Leem et al., 2006; Hansen et al., 2006; Kim et al., 2007; Jiang et al., 2007; Lee et al., 2008; Brauer et al., 2008). Four, generally more recent studies, included fine particulates ($PM_{2.5}$) (Wilhelm and Ritz, 2005; Huynh et al., 2006; Ritz et al., 2007; Brauer et al., 2008); and the oldest ones investigated the effects of total suspended particulates (TSP) (Xu et al., 1995; Bobak, 2000). Overall, 9 studies found significant associations between PM exposure during pregnancy and risk of preterm birth. Four out the 13 studies were time series in design, i.e. capturing effects related with short-term exposures. The rest had designs more appropriate for evaluating more chronic exposure: 8 registry-based cohorts and one case-control study. Where a significant relation was found estimates of the association for an increase of 10 $\mu g/m^3$ for PM_{10} or TSP in time series studies ranked from an increase in preterm risk of 1% (Xu et al., 1995) to 4.4% (Jiang et al., 2007). Among the studies assessing chronic exposure, significant estimates for a PM_{10} increase of 10 $\mu g/m^3$ were around 2-4 % (Ritz et al., 2000; Hansen et al., 2006; Leem et al 2006; Jiang et al., 2007), and of 1% in a case control study in California, USA (Huynh et al., 2006) for a $PM_{2.5}$ increase of 10 $\mu g/m^3$ in the last 2 weeks of pregnancy. In another study in California (Ritz et al., 2007), a 29% increase in risk of preterm birth for exposure to $PM_{2.5}$ above 21.36 $\mu g/m^3$ during the first trimester was found. Most of the studies examined the relation taking into account air pollution exposure by trimesters. First and third trimester exposures were those most frequently associated with the risk of preterm delivery. Also, in a study in Taiwan using the extended Cox model with time-dependent PM_{10} exposures for each trimester during pregnancy (Suh et al., 2009), estimates for first and third trimester were slightly higher than those of the second trimester. A recent time series study in the city of Guangzhou (China) (Zhao et al., 2011) using the Generalized Additive Model (GAM) extended Poisson regression model and Principal Component Analysis described a relation between daily PM_{10} increases of 10 $\mu g/m^3$ and significant increases in preterm risk less than 1%, suggesting that, at least for short-term exposures, the likely impact of PM on preterm risk is small.

Besides particulate matter, **nitrogen oxides** (NOx) have also been frequently assessed as the exposure of interest related with preterm risk. In their review up to 2003, Lacasaña et al. (2005) identified 8 articles examining the relation between pregnancy exposures to NO_X and risk of preterm delivery, with nitrogen dioxide (NO_2) being the most studied pollutant. However, the overall results were inconsistent. In the past six years, nine more studies have been published analyzing the relationship between NO_2 exposure during pregnancy and the risk of preterm birth (Hansen et al., 2006; Leem et al., 2006; Jalaludin et al., 2007; Ritz et al., 2007; Brauer et al., 2008; Darrow et al., 2009; Llop et al., 2010; Gehring et al., 2011; Zhao et al., 2011). Three of the studies found no significant association (Hansen et al., 2006; Brauer et al., 2008; Gehring et al., 2011). Jalaludin et al. (2007) observed an unexplained protective effect for NO_2 exposure during the first month and first trimester of pregnancy, while five other studies found a significant relationship between preterm births and NO_2 exposure at both the beginning and/or the end of pregnancy (Leem et al., 2006; Ritz et al., 2007; Darrow et al., 2009; Llop et al., 2010; Zhao et al., 2011). Another recent study conducted in Los Angeles (USA) analyzed the relationship between NOx levels and preterm birth (Wu et al., 2009) and found this association was significant when exposure occurred throughout the pregnancy, even at relatively low air pollution levels. On the contrary, the Brauer et al. (2008) study conducted in Vancouver found a significant relation between NO exposure obtained with the inverse of the distance method and preterm birth, OR: 1.26 (95% Confidence interval, 95% CI: 1.08–1.47). Llop et al. (2010) assessed the concentration-response function. The results suggest a threshold level around 46 $\mu g/m^3$ of NO_2 concentration, after which the risk of preterm birth increases.

Following another recent review (Shah & Balkhair, 2011), 5 studies fulfilling methodological criteria have assessed the relation between **sulfur dioxide (SO_2)** exposure during pregnancy and preterm delivery (Xu et al., 1995; Landgren et al., 1996; Bobak 2000; Liu et al., 2003; Sagiv et al., 2005). All but one (Landgren et al., 1996) reported significant associations with preterm delivery. Three of the studies describing significant associations considered the relationship linear, and reported a increased risk for preterm birth ranging from 3.5% (Bobak et al., 2001; Sagiv et al., 2005) in the Czech Republic and Pennsylvania to 6.4% in Vancouver (Liu et al., 2003) for a 10 $\mu g/m^3$ increase in SO_2 levels. The Xu et al. study (1995) in Beijing, China examined a log linear relationship and found a 21% increase in the risk of preterm delivery for each ln $\mu g/m^3$ increase in SO_2 levels. In general, associations were higher for exposures at late pregnancy, i.e. third trimester or last weeks before birth. On the contrary, in a study in Croatia in the vicinity of a coal power plant, Mohorovic (2004) found the first two months of pregnancy as the critical time for preterm delivery in relation with SO_2 exposure.

Carbon monoxide (CO) exposure has also been an object of analysis in relation with preterm birth. Ritz et al (2000; 2007) and Wilhelm and Ritz (2005) have focused on this pollutant in their birth certificate based study in southern California. They first reported carbon monoxide and PM associations for preterm birth in the South Coast Air Basin using data from 1989 to 1993 (Ritz et al., 2000). The study continued from 1994 to 2000 (Wilhelm & Ritz, 2005), and reported an 8-24% increase in risk of preterm birth per 1-ppm (1.145 mg/m^3) increase in CO during the first trimester among women who lived close to stations measuring carbon monoxide. Depending on the distance to the monitoring station, they also observed a 9-30% increase in the risk of preterm birth when average CO concentrations were high (≥1.9 ppm) 6 weeks before birth. Finally, they conducted a case-control survey nested within their birth cohort and collected detailed risk factor information to assess the extent to which residual confounding and exposure misclassification may impact air pollution effect estimates (Ritz et al., 2007). For the first trimester, the odds of preterm birth consistently increased with increasing CO exposures, regardless of the type of data (cohort/sample) or covariate adjustment. . Women exposed to carbon monoxide above 0.91 ppm during the last 6 weeks of pregnancy experienced increased odds of preterm birth. The results did not change substantially after further adjustment, except for time-activity patterns, which strengthened the observed associations. They considered CO may be a better marker of vehicle exhaust toxins than $PM_{2.5}$, since the latter includes both particles directly emitted in vehicular exhaust and those created secondarily through atmospheric reactions (Ritz et al., 2007).

A few studies included other specific air pollutants. Llop et al (2010) assessed the relation between prenatal exposure to **benzene**, as a marker for both traffic and industrial sources. The researchers found an association with preterm risk at benzene levels above 2.7 $\mu g/m^3$ throughout the entire pregnancy. The authors acknowledged, however, this was an isolated finding with no precedent in the literature, and therefore, it should be regarded with caution.

Instead of examining specific air pollutants, researchers have investigated the role of other indicators of **traffic or industrial air pollution**. Wilhelm & Ritz (2003) used a distance-weighted traffic density measure to take into account residential proximity to and level of traffic on roadways surrounding homes. They obtained a clear exposure-response pattern with preterm birth, with a RR of 1.08 (95% CI: 1.01-1.15) for the highest exposure. In Taiwan, Yang et al. (2003) investigated the association between traffic-related air pollution and preterm deliveries in a zone along the Zhong-Shan Freeway in Taiwan. The prevalence of deliveries of preterm birth infants was significantly higher among mothers who lived within 500 m of the freeway than among mothers who resided 500-1,500 m from the freeway. The adjusted odds ratio was 1.30 (95% CI:1.03-1.65) for delivery of preterm infants born to mothers who lived within 500 m of the freeway. Brauer et al. (2008) used a geostatistical method to assess exposure to traffic air pollution. In this study, however, no associations were observed between the simple road proximity measures and preterm birth < 37 weeks, and there were no cases of births < 30 weeks that were within 50 m of a highway. Similarly, no association was found when traffic proximity was assessed in the Generation R study in Holland (van de Hooven et al., 2009). In a recent study in Shizuoka, Japan, Yorifuji et al. (2011) classified 14,226 liveborn single births from 1997 to 2008, according to proximity to major roads. They found positive associations between proximity to major roads and preterm birth at all considered gestational ages. Living within 200 m increased the risk of birth before 37 weeks by 1.5 times (95% CI : 1.2-1.8), birth before 32 weeks by 1.6 times (1.1-2.4), and birth before 28 weeks by 1.8 times (1.0-3.2). Proximity specifically increased the risk of preterm births with preterm premature rupture of the membranes and with pregnancy hypertension.

Lastly, regarding industrial pollution some studies in Taiwan revealed associations between preterm birth and maternal residence within 3km of a major oil refinery (OR: 1.41, 95% CI 1.08-1.82) (Lin et al., 2001), near three oil refineries (OR: 1.14, 95% CI=1.01-1.28) (Yang et al 2004), within 2 km of a cement plant (OR: 1.30, 95% CI 1.09–1.54) (Yang et al., 2003), within 2 km of an industrial complex, including petrochemical, petroleum, steel, and shipbuilding industries (OR: 1.11, 95% CI 1.02-1.21) (Tsai et al., 2003) or within 3 km of coal-based electricity-generating stations (OR: 1.14, 95% CI 1.01-1.30) (Tsai et al., 2004).

3.4 Fetal growth (crude neonatal anthropometry (birth weight, LBW), GA adjusted neonatal anthropometry (SGA, IUGR), ultrasounds

Fetal growth deficit, measured in different ways and with different indicators, is with no doubt the most studied reproductive outcome in relation to air pollution exposure during pregnancy. Because of its availability from medical records, weight at birth and low birth weight (LBW, i.e. weight at birth < 2500 g) have been the main reproductive outcomes assessed in the first studies examining the impact of air pollution on fetal development. Birth weight may be considered as a fetal growth measure if the analyses are adjusted by gestational age or if LBW studies are restricted to term newborn. Also, a considerable number of studies have also evaluated the impact of air pollution exposure during pregnancy on SGA or IUGR. Particulate matter (either as TSP, PM_{10} or as $PM_{2.5}$) and NO_2 have been the pollutants most frequently examined in studies on air pollution and birth outcomes. Two of the six reviews mentioned previously have included only particulate matter as exposure (Glinianaia et al., 2004, Bosetti et al., 2010).
The use of passive sensors has allowed for a better spatial resolution when assessing individual exposure to air pollution (Raaschou-Nielsen et al., 2000; Gauderman et al., 2005). For these reasons and because studies involving more particles have been subjected to frequent literature reviews, this chapter addresses more detailed studies evaluating the relationship between **NO_2** exposure and birth weight, as well as weight in grams including low birth weight, or small for gestational age (SGA).
In Table 2, we present a summary of the 17 articles studying the effect of NO_2 exposure on birth weight (Lacasana et al., 2005). The number is itself an indication of the growing interest in this issue. Ten out the 17 studies assessed the relationship of NO_2 with birth weight, also 10 examined the relationship with LBW, and 9 the relationship with SGA. Among the 10 studies analyzing birth weight, a significant association was found in four studies (Gouveia et al., 2004; Mannes et al; 2005; Bell et al., 2007; and Morello-Frosch et al., 2010). In two other studies, an association was clearer when some sub-analysis was made trying to improve exposure assessment by restricting either spatially, i.e. women living <2km away from a monitoring station in the French study in Nancy and Poitiers (Lepeule et al., 2010), or taking into account time activity, i.e. women spending <2hr/day in non-residential outdoor in a cohort of the Spanish INMA study (Aguilera et al., 2009). Of the 10 studies that considered LBW, only 4 found an association with NO_2 (Lee et al., 2003; Bell et al., 2007; Brauer et al., 2008, Morello-Frosch et al., 2010). On the other hand, six of the nine articles (Liu et al., 2003; Salam et al., 2005; Mannes et al., 2005; Liu et al., 2007; Brauer et al., 2008; Ballester et al., 2010) studying SGA found an association with NO_2. This discrepancy may be due to the number of cases of SGA is greater than that of LBW term. Moreover, the use of SGA, calculated for each week of gestation, enables the effect of gestational length to be more effectively controlled than LBW, which is done by simply selecting births that take place after a certain period of gestation (between weeks 37- 44 weeks). On the other hand, it

should be noted that all these studies found an association between LBW and exposure to other pollutants, mainly particulate matter, CO and SO_2.

Regarding studies focusing on **PM exposure**, the paper by Glinianaia et al. (2004) was the first work systematically reviewing the evidence on the association between air pollution and fetal health outcomes. Twelve articles published between 1996 and 2001 were identified that included PM as the exposure of interest. Limitations in design and lack of confounding factors were defined. Authors concluded that currently available evidence was compatible with either a small adverse effect of particulate air pollution on fetal growth or with no effect. In their recently published review, Bosetti et al. (2010) identified 17 articles including information on PM exposure and LBW or very low birth weight (i.e. weight at birth < 1500 g). In most of the reviewed studies (14/17), LBW was the outcome studied; 2 studies considered VLBW infants, and another one considered infants with a birth weight between 2,500 and 3,000 g (Slama et al., 2007). With reference to the pollutant investigated (Bosetti et al., 2010), 5 studies (the older ones) considered exposure to TSP, 12 exposure to PM_{10} , and only 3 considered exposure to $PM_{2.5}$ (Bell et al., 2007; Brauer et al., 2008; Slama et al., 2007). Besides limitations described in the Glinianaia et al. (2004) article, Bosetti et al. (2010) underscored poor exposure assessment and inconsistent reporting of findings, especially in relation to time windows of exposure-effect. The authors concluded in the same way of Glinianaia 6 years before (2004): the excess risk associated with exposure to particulates during pregnancy, if any, is small and it is unclear if it is causal or due to misclassification or residual confounding (Bosetti et al., 2010).

For **CO** exposure, in their recent review, Shah & Balkhair (2011) identified 13 studies reporting results for LBW and four SGA. Among them only three reported an increased risk of LBW, two associated with exposures during the third trimester (Maisonet et al., 2001; Ritz & Yu, 1999) and the other at first and second trimesters (Lee et al., 2003). Only one of the studies assessing SGA reported a significant higher risk (Liu et al., 2003).

In the same review (Shah & Balkhair 2011), 14 studies were summarized describing results from studies including **SO_2** and LBW, but only one examined the impact of SO_2 on SGA. Five studies reported increased odds of LBW births following SO_2 exposure and one reported significant association with Very LBW (Rogers et al., 2000).

Other pollutants such ozone (O_3) and benzene have also been included in studies evaluating the effects of air pollution in fetal growth According the review by Shah & Balkhair (2011), no significant effect of **O_3** on LBW or SGA was reported. Regarding **benzene** a significant association was found in the Spanish INMA cohort in Sabadell, especially among women spending more time in their residence (Aguilera et al., 2009). However, a recent combined analysis of four cohorts in the INMA study, including Sabadell, Valencia, Asturias and Gipuzkoa cohorts (Estarlich et al., 2011) confirmed the association between NO_2 exposure and reductions in both length and weight at birth but not as a result of benzene exposure.

To try to overcome the difficulties of synthesizing previous findings due to differences in study design, several multicenter international initiatives have been launched recently. One of them, previously cited, ICAPPO has recently published its first results (Parker et al., 2011). Another large multicenter international study is the European Study of Cohorts for Air Pollution Effects (ESCAPE) (http://www.escapeproject.eu), is currently working to develop uniform criteria for assessing air pollution effects on reproductive outcomes.. ESCAPE will include information on more than 70,000 mother-child pairs and air-pollution. The results from this project can provide strong evidence on existence of a risk in fetal growth retardation related with exposure to air pollution during pregnancy.

Study	Location (time period)	Design N° of births	Outcome(s)	Exposure: Assessment: -data source; -individual assignment	Exposure: Mean (SD) NO_2 levels, in μg/m³	Exposure: Pregnancy periods examined	Exposure: NO_2 increase assessed	Results β (95% CI) grams OR (95% CI)	Adjusted for[a]	Other outcomes/ air pollutants studied / comments
Lepeule et al., 2010	Poitiers and Nancy, France (2003-2006)	Cohort of pregnant women 776	Birth weight	2 methods: 1. AQMS: - 3 and 6 monitoring stations for each city, respectively - nearest station 2. TAG.:- 61, and 89 passive samplers in 9 and 10 campaigns for each city, respectively -residential prediction using LUR and temporal adjustment	AQMS: Poitiers: 24.9 Nancy: 31.2 TAG: Poitiers: 20.3 Nancy: 31.2	Trimester; Entire pregnancy	10 μg/m³ (5.3 ppb) increase among women living <1 ,<3, <5km from a monitoring station	-β for birth weight 1st trimester, for women living < 2 km away from a monitoring station -AQMS: –37 (–75 to 1) -TAG: –51 (–128 to 26)	1,3,5,8,13,15,18,19,25	Despite some pattern related with the association presented higher *p*-value for women living within 5 or 1 km of a monitoring station and for the others studied periods.
Gehring et al., 2011	Amsterdam, Holland (2003-2004)	Cohort of pregnant women 7610	Birth weight among born ≥37 <43 weeks SGA	-13 4-week measurements using passive samplers at 62 sites, and GIS; -residential prediction using LUR and-temporal adjustment using background monitors	P50 38.7 (IQR:35.3; 43.3)	Trimester; Entire pregnancy	Quintiles	No association between NO_2 and birth weight or SGA.	1,2,7,8,16,18,19,27,28,36,37	
Morello-Frosch et al., 2010	California, USA (1996-2006)	Birth register-based study 3,545,177	Birth weight LBW among born between 37-44 weeks	-residential estimation (at ZIP code) using timely geocoded levels by air quality networks	45.5 (17.9)	Trimester; Entire pregnancy	18.8 μg/m³ (10 ppb) increase among women living <3, <5, <10km from a monitoring station	-β for birth weight Entire pregnancy <3 km: -8.3 (-9.6 to -7.0) <5 km: -9.7 (-10.6 to -8.8) <10 km: -9.0 (-9.6 to -8.4) 1st trimester, small significant associations (~-3) 2nd trimester: no significant association 3rd trimester <3 km: -8.1 (-10.2 to -6.1) <5 km: -7.9 (-9.2 to -6.5) <10 km: -7.0 (-7.9 to -6.0) -OR for LBW Entire pregnancy <3 km: 1.03 (1.01-1.05) <5 km: 1.04 (1.03-1.05) <10 km: 1.03 (1.02-1.04)	1,2,7,8,12,18,19,21,23,27,30	Reductions in birth weight also associated with CO, O_3, PM_{10}, $PM_{2.5}$, PM_{coarse} Unexpected positive association with SO_2 Increased odds of LBW also associated with CO, and $PM_{2.5}$

Study	Location (time period)	Design N° of births	Outcome(s)	Exposure				Results β (95% CI) grams OR (95% CI)	Adjusted for[a]	Other outcomes/ air pollutants studied / comments
				Assessment: -data source; -individual assignment	Mean (SD) NO_2 levels, in μg/m³	Pregnancy periods examined	NO_2 increase assessed			
Madsen et al., 2010	Oslo, Norway (1999-2002)	Birth register-based study 25,229	Birth weight LBW excluded <37 wg SGA excluded <37 wg	-dispersion model based on GIS and background air pollution levels -monitoring station	29.8 (11.2) Q1 Q4 35.6 (4.2)	Entire pregnancy	Q4 (>38.0) vs Q1 (<20.3) μg/m³	No association between NO_2 and birth weight, LBW or SGA among all population. Higher effect among multiparious mothers: -8.6 (-30.0 to 12.7)	1,2,7,8,13,18	$PM_{2.5}$ and PM_{10}. Also only small but non significant associations among multiparious mothers
Ballester et al., 2010	Valencia; Spain (2004-2006)	Cohort of pregnant women 785	Birth weight SGA	-4 campaigns using passive samplers at 93 sites, monitoring network, and GIS; -residential prediction using Kriging + LUR and temporal adjustment	36.9 (11.1)	Trimester; Entire pregnancy	10 μg/m³ (5.3 ppb)	-β for birth weight 1st trimester: -12.8 (-34.5 to 9.0) Entire pregnancy: -9.7(-33.2 to 13.8) -OR for SGA 2nd trimester: 1.37 (1.01-1.85) Entire pregnancy: 1.28 (0.94-1.74)	1-19	Association with birth length and birth head circumference
Aguilera et al., 2009	Sabadell, Spain (2004-2006)	Cohort of pregnant women 570	Birth weight	-3 campaigns using passive samplers at 57 sites, monitoring network, and GIS; -residential prediction using LUR	32.17 (8.89)	Trimester; Entire pregnancy	IQR: 12.0 μg/m³	No association between NO_2 and birth weight among all population. Only significant effect among women spending <2hr/d in non-residential outdoor: β for birth weight 2nd trimester: -74.7 (-140.4 to -9.0)	1,3,5,7,8,13,17-19	Significant birth weight reductions associated with BTEX exposure
Brauer et al., 2008 [34]	Vancouver, Canada (1999-2002)	Birth register-based study 70,249	SGA LBW excluded <37 wg	-Monitoring network and 2 campaigns using passive samplers at 116 sites; -nearest and inverse-distance weighting (IDW) area monitors, LUR temporally adjusted	32.5 (range:15.3; 53.6)	Month; Entire pregnancy	10 μg/m³ (5.3 ppb)	Entire pregnancy:, NO_2 IDW OR for SGA :1.14 (1.09-1.18) OR for LBW:1.11 (1.01-1.23)	1,2,7,8,13,18-22	SGA also associated with CO, SO_2, and $PM_{2.5}$ but not with O_3.
Slama et al., 2007 [12]	Munich, Germany (1998,1999)	Cohort of pregnant women 1,016	BW<3000 g among births>2500g and >37 <44 weeks	- 2 campaigns with passive samplers at 40 sites, and GIS; -residential prediction using LUR.	35.8 (P5th:28.3; P95th: 42.5)	Trimester; Entire pregnancy	10 μg/m³ (5.3 ppb) , Quartiles	PR 1st trimester: 0.96 (0.73- 1.20) 2nd trimester: 1.18 (0.95- 1.44) 3rd trimester: 1.13 (0.91- 1.35) Entire pregnancy: 1.21 (0.86-1.68) No significant associations by quartiles	1,3,5-7,13,14,18,19	Significant associations with $PM_{2.5}$ and $PM_{2.5}$ absorbance

Study	Location (time period)	Design N° of births	Outcome(s)	Exposure: Assessment: -data source; -individual assignment	Exposure: Mean (SD) NO_2 levels, in µg/m³	Exposure: Pregnancy periods examined	Exposure: NO_2 increase assessed	Results β (95% CI) grams OR (95% CI)	Adjusted for[a]	Other outcomes/ air pollutants studied / comments
Bell et al., 2007 [10]	Massachusetts and Connecticut , USA (1999-2002)	Birth register-based study 358,504	Birth weight LBW excluded <37 wg	Average county-level concentration from monitoring networks	32.7 (9.4)	Trimester; Entire pregnancy	9.0 µg/m³ (4.8 ppb)	β for birth weight 1st trimester: nr (-9.6 to -8.8) Entire pregnancy: -8.9(-10.8 to -7.0) OR for LBW Entire pregnancy: 1.027 (1.002-1.051) Associations for other trimesters were less consistent	1,2,8,13,18, 21,23-28	Exposures to CO, $PM_{2.5}$ and PM_{10} also lowered birth weight. SGA was also associated with CO, $PM_{2.5}$, PM_{10}, and SO_2.
Hansen et al., 2007 [35]	Brisbane; Australia (2000-2003)	Birth register-based study 21,432	SGA	-4 monitoring stations; -average of measurements	16.5 (7.7)	Trimester ; Month	IQR: 11.1 µg/m³ (5.9 ppb) Quartiles	No association between NO_2 and SGA	1,2,10,18,19,20,23,24, 29	Effect of NO_2 (IQR:11.1 mg/m³) third trimester on crow-heel length : -0.15cm (95%CI: -0.25 to -0.05). No effects for PM_{10}, BSP, O_3
Liu et al., 2007 [32]	Calgary, Edmonton, and Montreal; Canada (1985-2000)	Birth register-based study 386,202	SGA among born between wg37-42	-2, 4, 8 monitoring stations in each city, respectively; -mean of measurements in the residential area	45.1 (IQR:32.9; 55.5)	Trimester ; Month	37.6 µg/m³ (20 ppb)	OR for Trimester 1st : 1.16 (1.09-1.24) 2nd: 1.14 (1.06- 1.21) 3rd : 1.16 (1.09- 1.24)	1,2,7,18,19, 21,30	Also associated with CO and $PM_{2.5}$. In multipollutant models only CO showed robustness while effects of NO_2 and $PM_{2.5}$ where no longer observed. No effects for SO_2 and O_3.
Mannes et al., 2005 [31]	Sydney Australia(1998-2000)	Birth register-based study 138,056	Birth weight SGA	Average of the monitoring stations in the city	43.6 (13.9)	Trimester; one month before birth	1.88 µg/m³ (1 ppb) increase among women living <5km from a monitoring station	β for birth weight 1st trimester: -26.2(-35.4 to -17.0) 2nd trimester: -33.9(-43.4 to -24.3) 3rd trimester: -32.2(-41.8 to -22.5) Month before pregnancy: -19.7(-27.8 to -11.5) OR for SGA 1st trimester: 1.06 (0.99- 1.14) 2nd trimester: 1.14 (1.07- 1.21) 3rd trimester: 1.13 (1.05- 1.21) Month before pregnancy: 1.07 (1.00-1.14)	1,2,10,13,18,19,20	Also associated with CO and PM_{10}. In multipollutant models NO_2 appeared as the most important pollutant.

Study	Location (time period)	Design N° of births	Outcome(s)	Exposure				Results β (95% CI) grams OR (95% CI)	Adjusted for[a]	Other outcomes/ air pollutants studied / comments
				Assessment: -data source; -individual assignment	Mean (SD) NO_2 levels, in μg/m³	Pregnancy periods examined	NO_2 increase assessed			
Salam et al., 2005 [30]	California, USA (1975-1987)	Birth register-based study 3,901	Birth weight SGA <P15 born between wg 37-44 LBW born between wg 37-44	Spatial interpolation from the 3 nearest monitoring stations (in <50 km) When there are available stations in <5 km the nearest station data are assigned	67.9 (29.0)	Trimester Entire pregnancy	47 μg/m³ (25 ppb)	β for birth weight Entire pregnancy: -7.2 g (-34.7 to 20.4) OR for SGA 1stTrimester : 1.2 (1.0-1.4) Entire pregnancy: 1.1 (0.9-1.3) No association for other trimesters or LBW	1,2,10,12,13,18,27,31	O3 exposure during 2nd and 3rd trimester and CO exposure during 1st trimester were associated with reduced birth weight.
Wilhelm and Ritz, 2005 [29]	Los Angeles County, CA, USA (1994-2000)	Birth register-based study 106,483	LBW at term Born between 90-320 days; excluded weight <500g or >5000g	15 monitoring stations for the county-level analysis 11 stations for the address-level analysis	73.5 (range:38.7-116.6	First month Trimester 6 weeks before birth Entire pregnancy	nr	Results from one pollutant models for NO_2 are not provided. No association between NO_2 and LBW after adjusting for CO and/or PM10.	1,2,7,8,18,19,23,27,32,33	Association of LBW with CO and PM_{10} but not with O_3. Clearer association for women residing 1 mile from a station
Gouveia et al., 2004 [28]	Sao Paulo, Brasil (1997)	Birth register-based study 179,460	Birth weight LBW at term; excluded <37wg or weight <1000g >5500g	Mean of the hourly maximum from between 4-12 NO_2 monitoring stations in the city	117.9 (51.2)	Trimester	10 μg/m³ (5.3 ppb)	β for birth weight 1st trimester: -7.0 g (-14.3 to 0.3) No association between NO_2 and LBW	1,2,7,10,18,23,24	An association between LBW and CO and $PM_{2.5}$ exposure during 1st trimester was found
Lee et al.,2003 [27]	Seul, Korea (1996-1998)	Birth register-based study 388,105	LBW born between wg37-44	Daily mean of the 20 monitoring stations in the city	61.1 (19.2)	Month Trimester Entire pregnancy	IQR: 27.6 μg/m³ (14.7 ppb)	OR for LBW 1st trimester: 1.02 (0.99-1.04) 2nd trimester: 1.03 (1.01-1.06) 3rd trimester: 0.98 (0.96-1.00) Entire pregnancy: 1.04 (1.00-1.08)	1,8,18,19,21,25,34	An association between LBW and CO, PM_{10} and SO_2 exposure was found except tor 3rd trimester for all pollutants, and for 1st trimester for SO_2.
Liu et al., 2003 [26]	Vancouver (Canada) (1986-1998)	Birth register-based study 229,085	SGA born between wg 37-44 LBW excluded <500g or <22wg	Mean of the monitoring stations in the residential area of each mother	36.5(P5th:21.6; P95th: 60.0)	First Month Last month Trimester	18.8 μg/m³ (10 ppb)	OR for SGA First month: 1.05 (1.01–1.10) Last month: 0.98 (0.92–1.03) 1st trimester: 1.03 (0.98–1.10) 2nd trimester: 0.94 (0.88–1.00) 3rd trimester: 0.98 (0.92–1.06) OR for LBW First month: 0.98 (0.90–1.07) Last month: 0.94 (0.85–1.04)	1,2,7,18,19	An association between SGA and CO and SO_2 first month exposure was also found

Table 2. Results from studies assessing NO_2 effect on birth weight published in the period 2003- 2010

Table abbreviations: β (95% CI): regression coefficient for birth weight (in grams) and 95% confidence interval; OR: odds ratio; PR: prevalence ratio; SGA: small for gestational age: <10th percentile from population charts unless otherwise indicated in the table; LBW: birth weight <2500gr unless otherwise indicated in the table. LBW at term: LBW among those born ≥37 weeks of gestation; nr: not reported; wg: weeks of gestation; BSP: Black smoke particles; ppb: parts per billion. P5th: Percentile 5th. P95th: Percentile 95th ; IQR: Interquartile range; Q: quartile (i.e. Q1: first quartile); BTEX: benzene, toluene, ethylbenzene, xylene

[a] Covariates considered: 1: Gestational age 2: Maternal age; 3: Maternal pre-pregnancy weight; 4: Gestational weight gain; 5: Maternal height; 6: Maternal body mass index; 7: Parity; 8: Maternal education; 9: Maternal working status; 10: Maternal socio economic status; 11: Mother's country of origin; 12:Living with partner; 13: Maternal smoking during pregnancy; 14: Maternal environmental Tobacco Exposure at home; 15: Maternal environmental general tobacco exposure; 16: Maternal alcohol consumption; 17: Paternal height; 18: Infant sex; 19: Season; 20: Country of origin ; 21: Year of birth; 22: Income; 23: Prenatal care; 24: Type of delivery; 25: Birth order; 26: Weather; 27: Mother's ethnicity; 28: Marital status; 29: Previous abortions; 30: Location of residence; 31: Diabetes; 32:Time since last delivery; 33: Previous preterm; 34: Paternal education; 35: Neighbourhood SES measures; 36: Work stress during pregnancy; 37:Depression during pregnancy

Few studies have examined the relation between air pollution exposure during pregnancy and other anthropometric indicators at birth such as head circumference (HC). Studies in two cohorts of pregnant women have assessed the relationship between prenatal exposure to airborne polycyclic aromatic hydrocarbons (PAH) and fetal growth (Choi et al., 2006). One of these cohorts was from Cracow (Poland) and the other was made up of predominantly African-American and Dominican women from New York City (US). PAH exposure was related to a reduction in birth weight, length and HC among babies from Cracow, where exposure levels were higher. Among the participants from New York, results indicated a greater effect on African-American women, with higher reductions in both birth weight and HC, showing some kind of susceptibility to PAH exposure. In another study Hansen et al. (2007) assessed birth length and HC among 26,617 term births in Brisbane, Australia. Exposures to the four pollutants studied (i.e. PM_{10}, black smoke, ozone and NO_2) were not significantly associated with a reduction in HC. However an interquartile range increase in NO_2, but not in other pollutants, during the third trimester was associated with a reduction in length: -0.15 cm (95%CI -0.25 to -0.05) for a 10 $\mu g/m^3$ increase in NO_2. In the French Eden cohort (Slama et al., 2006) a reduction of -0.31 cm (95%CI -0.73 to 0.10) in HC at birth was found when comparing NO_2 exposure in the highest tertile (>31.4 $\mu g/m^3$) to that in the lowest tertile. In a study in the Spanish INMA cohort in Valencia (Ballester et al., 2010) a reduction in birth length of -0.07 cm (95%CI -0.15 to 0.02) was associated with a 10 $\mu g/m^3$ increase in NO_2 during the first trimester. When exposure above 40 $\mu g/m^3$ was compared with exposure equal to or less than 40 $\mu g/m^3$, a reduction of -0.27 cm (95%CI -0.51 to -0.03) in birth length was found. In the combined analysis of four cohorts within the INMA study a (Estarlich et al., 2011) an increase of 10 $\mu g/m^3$ in NO_2 exposure during pregnancy was associated with a change in birth length of -0.09 cm (95% CI: -0.18 to -0.01). For the subset of women who spent ≥15 hr/day at home, the change in birth length was twice that.

Finally, regarding studies assessing fetal growth using ultrasound data instead of neonatal outcomes, we found only three studies. The first one was an Australian study that examined 15,623 scans at mid pregnancy and assigned air pollution data to each woman's residential postal code from the closest monitoring station (Hansen et al., 2008). The authors reported a negative impact of several air pollutants, but no impact of NO_2. They indicated as a possible cause for inconsistent results the increase in the exposure measurement error with increasing distance from a monitor. The second study (Slama et al., 2009) examined three scans for 271 women at second and third trimesters and a personal assessment of maternal benzene exposure in a French pregnancy cohort. The authors found this exposure associated with a reduction of head dimensions during the second trimester. In the third one, Aguilera et al (2010) conducted a study with 1692 ultrasounds for 562 women within the Sabadell INMA-cohort. An adverse effect of NO_2 exposure on fetal growth measures during the second and third trimester was found, after restricting analyses to women with less mobility.

3.5 Congenital malformations

Among the environmental factors related to the risk of congenital anomalies, air pollution is among the most studied (Ritz 2010). Thus, studies conducted in California (Ritz et al., 2002) and Texas (Gilboa et al., 2005) showed a positive association between specific cardiac defects and various environmental pollutants measured at stations near the residence of mothers between weeks 3 and 8 of gestation, but the evidence was insufficient for other AC. However, Hwang et al (2008) identified associations between oral abnormalities and exposure to ozone during the first two months of pregnancy, but not with other air pollutants. In Atlanta, Strickland et al. (2009) studied exposure to PM_{10} during 3-7 weeks of gestation and found a small increased risk for one of 12 specific cardiac abnormalities. Hansen et al. (2009) in Brisbane, showed an association with exposure to ozone and SO_2 during 3-8 weeks of gestation and two different subgroups of heart defects (pulmonary artery and aorta) and oral defects. In the north of England, Rankin et al. (2009) described a weak positive association between airborne particles measured as black smoke and nervous system abnormalities, and a negative association between exposure to black smoke and SO_2 during the first trimester of pregnancy and cardiac abnormalities. Also in England, Dummer et al. (2003) found an increased risk of lethal congenital anomalies (spina bifida and heart defects) in relation to living near incinerators, and an increased risk of fetal death and anencephaly in relation to living near incinerators and crematoriums. Exposure to air pollutants from combustion (SO_2, NO_2 and PM_{10}) has been associated with tetralogy of Fallot (Dolk et al., 2010).

A systematic review and meta-analysis of epidemiological studies on ambient air pollution and congenital anomalies was recently published (Vrijheid et al., 2011). The authors conducted meta-analyses if at least four studies published risk estimates for the same pollutant and anomaly group. Summary risk estimates were calculated for risk at high versus low exposure level in each study, and risk per unit increase in continuous pollutant concentration.

An analysis of the data indicated that each individual study reported statistically significant increased risks for some combinations of air pollutants and congenital anomalies, among many combinations tested. In meta-analyses, NO_2 and SO_2 exposures were related to increased risk of coarctation of the aorta (OR per 10 ppb NO_2=1.17, 95%CI: 1.00-1.36; OR per 1 ppb SO_2=1.07, 95%CI: 1.01- 1.13) and tetralogy of Fallot (OR per 10ppb NO_2=1.20, 95%CI :1.02-1.42; OR per 1 ppb SO_2=1.03, 95%CI: 1.01-1.05), and PM_{10} exposure to an increased risk of atrial septal defects (OR per 10 $\mu g/m^3$=1.14, 95%CI: 1.01-1.28). Meta-analyses found no statistically significant increase in risk of other cardiac anomalies or oral clefts. The authors conclude that they found some evidence for an effect of ambient air pollutants on congenital cardiac anomaly risk. Similar to reviews on other reproductive outcomes, the authors suggest that improvements in the areas of exposure assessment, outcome harmonization, assessment of other congenital anomalies, and mechanistic knowledge are needed to advance this field.

4. Discussion and conclusions

4.1 General overview

One general conclusion of such studies was that although there is evidence prenatal exposure to air pollutants has a detrimental effect on fetal development (Sram et al., 2005) and infant mortality (Lacasana et al., 2005), results are less consistent for other outcomes. Specifically, in studies that have examined the relationship between exposure to pollutants and preterm birth or congenital anomalies, the evidence is insufficient to suggest causality (Hackley et al., 2007) (Vrijheid et al., 2010). Even for findings related with fetal development, recent reviews point out substantial heterogeneity in the results, making it difficult to draw robust conclusions (Shah & Balkhair, 2011). In any case, associations are of small magnitude. Among the issues that have been identified as crucial in comparing and interpreting the results on prenatal air pollution exposure and fetal development, the following have been highlighted (Slama et al., 2008; Ritz & Wilhelm, 2008; Woodruff et al., 2009; Shah & Balkhair, 2011): study design, exposure assessment (time activity and critical windows of exposure) and outcomes definition.

4.2 Biological mechanisms

Numerous biologic pathways have been identified whereby particulate air pollutants might impact the placenta and fetus. Kannan et al. (2006) performed a robust study on plausible fisiopathological mechanisms by which exposure to particulate matter may lead to adverse perinatal outcomes. They identified five mechanistic pathways including a) oxidative stress; b) pulmonary and placental inflammation; c) blood coagulation; d) alteration of endothelial function; and e) hemodynamic responses to particulate exposure. As Ritz & Wilhelm given attention (2008), these pathways may not act independently. For example, an increase in maternal blood pressure and an impaired trophoblast invasion of the spiral arteries may induce uteroplacental hypoperfusion and a state of relative hypoxia surrounding the trophoblast. Some experimental studies have added evidence, as in the study by Veras et al. (2009) where the researchers exposed female mice during pre-gestational and gestational periods to filtered or non-filtered air in exposure chambers in a garden near high density traffic. Then placentas were collected from near-term pregnancies and prepared for microscopical examination and alterations on placental functional morphology were found in placentas from those exposed to non-filtered air. Besides that, fetal weight declined in exposed group.

Polycyclic aromatic hydrocarbons (PAH), an important component of fine particulates from combustion sources, have been proposed as having a role in adverse reproductive outcomes. Experimental evidence showed that prenatal exposure of rats to maternal inhalation of benzo(a)pyrene (a PAH known for its potential toxicity) significantly compromised fetal survival rate and birth weight (Archibong et al., 2002). PAH exposure effects on fetal growth have been described in epidemiological studies in humans (Choi et al., 2006). Induction of apoptosis after DNA damage from PAH, endocrine disruption, and binding to the aryl hydrocarbon receptor for placental growth factors, resulting in decreased exchange of oxygen and nutrients have been proposed as the mechanisms for PAH toxicity (Choi et al., 2006; Veras et al 2010; Dejmek et al., 2000). On the other hand, adult male exposure to inhaled PAH has also been found to affect fertility in male rates (Ramesh et al., 2008).

On the other hand, NO_2 is a potent oxidant and increased lipid peroxidation in the maternal and/or fetal compartment has been found in preterm birth (Moison et al., 1993). Tabacova et al. (1997) investigated the relationship between exposure to nitrogen-oxidizing species and pregnancy complications in an area in Bulgaria highly polluted by oxidized nitrogen compounds. Methemoglobin, a biomarker of individual exposure, was determined, and glutathione balance and lipid peroxide levels were measures of oxidant/antioxidant status. A high percentage of women suffered from pregnancy complications. The most common ones were anemia (67%), threatened abortion/premature labor (33%), and signs of preeclampsia (23%). Methemoglobin was significantly elevated in all three conditions, compared with normal pregnancies. Reduced total glutathione, an indicator of maternal antioxidant reserves, decreased, whereas cell-damaging lipid peroxide levels increased. Mohoroviz found similar results for methemoglobin in a polluted area of Croatia (Mohorovic, 2004). These results suggest that maternal exposure to environmental oxidants can increase the risk of pregnancy complications through stimulation of the formation methemoglobin, which may lead to hypoxia and hypoxemia in pregnant women and has an important influence on maternal health as well as placental and fetal development. The study in Croatia also described the impact of early exposures (first two months), as this is the time of greatest susceptibility in human gestation. It remains unclear if NO_2 is just a marker of air pollution from traffic or other combustions (i.e. PM, PAH or volatile organic compounds VOC), or it is a pollutant playing an important role on .

Another studied pollutant, carbon monoxide (CO) is known to induce fetal hypoxia by forming carboxyhemoglobin at the expense of oxyhemoglobin, thus resulting in an increased risk of fetal underdevelopment and neonatal mortality (Veras et al., 2010). These effects, however, have been described in relation to high CO concentrations related with tobacco exposure, some times higher than those present in ambient air. This fact raises the issue of whether CO may be reflective of the action of other toxins as PAH, metals or VOC (Ritz & Wilhelm, 2008)

Finally, the role of genetic polymorphisms, mainly as effect modifiers, on the relation between maternal air pollution exposure and reproductive outcomes should be considered. Some preliminary studies have reported how some genotypes as GSTM1 and CYP1A1*2A modified the effect of environmental exposure on birth weight and prematurity (Sram et al., 2006).

4.3 Public health implications

Air pollution is still an important public health problem. Exposure to elevated levels of air pollution can cause a variety of adverse health outcomes, including reproductive outcomes, yet being respiratory and cardiovascular diseases the two groups of causes where higher burden is borne. Air quality in developed countries has been generally improved over the last three decades. However, many recent epidemiological studies have consistently shown positive associations between low-level exposure to air pollution and adverse health outcomes. Thus, adverse health effects of air pollution, even at relatively low levels, remain a public concern and strategies to improve air quality around the world are being defined (EU 2005).

Fetuses are very susceptible to environmental exposures and disruption of their development may have an impact on child and adult health. Exposure to ambient air pollution is ubiquitous and even if increased risks of adverse reproductive outcomes are small, they can have a big impact measured as number of attributable cases at the population level (Slama et al., 2008). Two examples can illustrate this. One cost-benefit analysis of air pollution regulations in the USA (Wong et al., 2004), examined the child-specific health impacts derived from the U.S. Clean Air Act (CAA). It was estimated that from 1990-2010, CAA regulations would prevent 10,000 fewer low birth weight infants, which represented an estimated savings of 230 million U.S. dollars in health care costs. A recent health impact assessment in Korea (Seo et al., 2010) estimated that population-attributable risk of low birthweight related to PM_{10} pollution ranged between 5% and 19% in seven Korean cities, indicating that a large proportion of LBW could be prevented if air pollution was reduced.

Fetal and child health are a clearly identifiable topic adding argues to need to reinforce international, national and local efforts to improve air quality around the world, as well as providing health professional and the population, especially pregnant women indications and recommendations to prevent hazardous exposures during pregnancy

4.4 Next steps: future research

Fairly all the reviews and methodological papers on this topic published recently agree that current evidence suggests that air pollution may play some role in adverse pregnancy outcomes (Slama et al., 2008; Ritz & Wilhellm, 2008; Woodroof et al., 2009). Also, it is widely shared that owing the importance on the study and prevention of potential environmental insults during pregnancy this is a clear developing field in epidemiology and biomedical research. Besides that a number of limitations in former studies and some opportunities for better advance in the knowledge on the issue have been identified, leading to proposal of a series of recommendations including the following:

- Develop collaborations to establish large international cohorts with high availability and quality of exposure and potentially confounding factors, with enough sample size and avoiding as much as possible selection bias.
- Expand, clearly defined, new outcomes to be considered in studies, including subfertility, fetal loss, pregnancy complications (i.e. preeclampsia, hypertension), as well as measured characteristics after birth (i.e. placental weight, sexual differentiation, perinatal neurological scores).
- Improve exposure assessment by using models allowing for spatial and temporal resolution and also taking into account time-activity patterns, allowing for identification time-windows of susceptibility during pregnancy.
- Include biomarkers of exposure to air pollution as well as those of susceptibility (i.e. genetic polymorphisms).
- Report extensively cohort characteristics as well as information on outcomes, exposure and covariables.
- Develop experimental studies to help identify relevant biological mechanisms (Slama et al., 2008)

5. Acknowledgements

We are very grateful to Melanie Larson for her invaluable help in editing the manuscript.

6. References

Aguilera, I; Guxens, M; Garcia-Esteban, R; Corbella, T; Nieuwenhuijsen, MJ; Foradada, CM & Sunyer, J. (2009). Association between GIS-based exposure to urban air pollution during pregnancy and birth weight in the INMA Sabadell Cohort. *Environmental Health Perspectives*. Vol.117, N°.8, pp. 1322-7, ISSN: 0091/6765.

Aguilera, I; Garcia-Esteban, R; Iñiguez, C; Nieuwenhuijsen, MJ; Rodríguez, A, Paez, M; Ballester, F & Sunyer, J. (2010). Prenatal exposure to traffic-related air pollution and ultrasound measures of fetal growth in the INMA Sabadell cohort. *Environmental Health Perspectives*. Vol.118, N°.5, pp.705-11. ISSN: 0091/6765.

Archibong, AE; Inyang, F; Ramesh, A; Greenwood, M; Nayyar, T; Kopsombut, P; Hood, DB & Nyanda, AM. (2002). Alteration of pregnancy related hormones and fetal survival in F-344 rats exposed by inhalation to benzo(a)pyrene. *Reproductive Toxicology*. Vol.16, N°6, pp.801-8. ISSN: 0890-6238.

Ballester, F; Estarlich, M; Iñiguez, C; Llop, S; Ramón, R; Esplugues, A; Lacasaña, M & Rebagliato, M. (2010). Air pollution exposure during pregnancy and reduced birth size: a prospective birth cohort study in Valencia, Spain. *Environmental Health*. 9:6. ISSN: 1476-069X.

Barker, DJ. (2007). The origins of the developmental origins theory. *Journal of Internal Medicine*. Vol.261, N°.5,pp.412-7. ISSN: 0954-6820.

Bell, ML; Ebisu, K & Belanger, K. (2007). Ambient air pollution and low birth weight in Connecticut and Massachusetts. *Environmental Health Perspect*. Vol.115, N°.7, pp.1118-1124, ISSN: 0091/6765.

Bobak, M. (2000). Outdoor air pollution, low birth weight, and prematurity. *Environmental Health Perspect*. Vol.108, N°.2,pp.173-6. ISSN: 0091/6765.

Bobak, M & Gjonca, A. (2001). The seasonality of live birth is strongly influenced by socio-demographic factors. *Human Reproduction*. Vol.16, N°.7, pp.1512-7. ISSN: 1028-7825.

Bosetti, C; Nieuwenhuijsen, MJ; Gallus, S; Cipriani, S; La Vecchia, C & Parazzini, F. (2010). Ambient particulate matter and preterm birth or birth weight: a review of the literature. *Archives of Toxicology*. Vol.84, N°.6, pp.447-60. ISSN:0171-9750.

Brauer, M; Lencar, C; Tamburic, L; Koehoorn, M; Demers, P & Karr, C. (2008). A cohort study of traffic-related air pollution impacts on birth outcomes. *Environmental Health Perspect*. Vol.116, N°.5, pp.680-6. ISSN: 0091/6765.

Brunekreef, B & Holgate, ST. (2002). Air pollution and health. *The Lancet*. Vol.360, N°. 9341, pp.1233–42. ISSN: 0140-6736.

Brunekreef, B. (2007). Health effects of air pollution observed in cohort studies in Europe. *Journal of Exposure Science and Environmental Epidemiology*. Vol.17, Suppl.2, pp.S61-5. ISSN:1559-0631.

Choi, H; Jedrychowski, W; Spengler, J; Camann, DE; Whyatt, RM; Rauh, V; Tsai, WY & Perera, FP. (2006). International studies of prenatal exposure to polycyclic aromatic hydrocarbons and fetal growth. *Environmental Health Perspect*. Vol.114, N°.11, pp.1744-50. ISSN: 0091/6765.

Darrow, LA; Klein, M; Flanders, WD; Waller, LA; Correa, A; Marcus, M; Mulholland, JA; Russell, AG & Tolbert, PE. (2009). Ambient air pollution and preterm birth: a time-series analysis. *Epidemiology*. Vol.20, N°.5, pp.689-98. ISSN: 1044-3983.

Dejmek, J; Solanský, I; Benes, I; Lenícek, J & Srám, RJ. (2000). The impact of polycyclic aromatic hydrocarbons and fine particles on pregnancy outcome. *Environmental Health Perspect*. Vol.108, N°.12, pp.1159-64. ISSN: 0091/6765.

Dolk, H; Armstrong, B; Lachowycz, K; Vrijheid, M; Rankin, J; Abramsky, L; Boyd, PA & Wellesley, D. (2010). Ambient air pollution and risk of congenital anomalies in England, 1991-1999. *Occupational and Environmental Medicine*. Vol.67, N°.4, pp.223-7. ISSN:1076-2752.

Dummer, TJ; Dickinson, HO & Parker, L. (2003). Adverse pregnancy outcomes around incinerators and crematoriums in Cumbria, north west England, 1956-93. *Journal of Epidemiology and Community Health*. Vol.57, N°.6, pp.456-61. ISSN:0143-005X.

ESCAPE—European Study of Cohorts for Air Pollution Effects. Available online: http://www.escapeproject.eu/ (accessed on 18 August 2009).

Estarlich, M; Ballester, F; Aguilera, I; Fernández-Somoano, A; Lertxundi, A; Llop, S; Freire, C; Tardón, A; Basterrechea, M; Sunyer, J & Iñiguez, C. (2011). Residential Exposure to Outdoor Air Pollution during Pregnancy and Anthropometric Measures at Birth in a Multicenter Cohort in Spain. *Environmental Health Perspect*. [Epub ahead of print]. ISSN: 0091/6765.

EU legislation. Communication of 21 September 2005 from the Commission to the Council and the European Parliament - Thematic Strategy on Air Pollution. Available at: http://europa.eu/legislation_summaries/environment/air_pollution/l28159_en.htm [accessed at 15 may 2011]

EUROCAT. EUROCAT Guide 1.3 and reference documents Instructions for the Registration and Surveillance of Congenital Anomalies. EUROCAT, 2005.

Evenson, DP & Wixon, R. (2005). Environmental toxicants cause sperm DNA fragmentation as detected by the Sperm Chromatin Structure Assay (SCSA). *Toxicology and Applied Pharmacology*. Vol.207. Suppl.2, pp.532-7. ISSN:0041-008X.

Foster, WG; Neal, MS; Han, MS & Dominguez, MM. (2008). Environmental contaminants and human infertility: hypothesis or cause for concern? *Journal of Toxicology and Environmental Health. Part B: Critical reviews*. Vol.11, N°.3-4, pp.162-76. ISSN:1521-6950.

George, L; Granath, F; Johansson, AL; Annerén, G & Cnattingius, S. (2006). Environmental tobacco smoke and risk of spontaneous abortion. *Epidemiology*. Vol.17, N°.5, pp.500-5. ISSN: 1044-3983.

Gehring, U; Wijga, AH; Fischer, P; de Jongste, JC; Kerkhof, M; Koppelman, GH; Smit, HA & Brunekreef, B. (2011). Traffic-related air pollution, preterm birth and term birth weight in the PIAMA birth cohort study. *Environmental Research*. Vol. 111, N°.1, pp.125-35. ISSN:0013-9351.

Ghosh, R; Rankin, J; Pless-Mulloli, T & Glinianaia, S. (2007). Does the effect of air pollution on pregnancy outcomes differ by gender? A systematic review. *Environmental Research*. Vol.105, N°.3, pp.400-8. ISSN:0013-9351.

Glinianaia, SV; Rankin, J; Bell, R; Pless-Mulloli, T & Howel, D. (2004). Particulate air pollution and fetal health: a systematic review of the epidemiologic evidence. *Epidemiology*. Vol.15, N°.1, pp.36-45. ISSN: 1044-3983.

Gouveia, N; Bremner, SA & Novaes, HM. (2004). Association between ambient air pollution and birth weight in Sao Paulo, Brazil. *Journal of Epidemiology and Community Health*. Vol.58; N°.1, pp.11-7. ISSN:0143-005X.

Hackley, B; Feinstein, A & Dixon, J. (2007). Air pollution: impact on maternal and perinatal health. *Journal of Midwifery and Women's Health.* Vol.52, N°.5, pp:435-43. ISSN:1526-9523.

Hammoud, A; Carrell, DT; Gibson, M; Sanderson, M; Parker-Jones, K & Peterson, CM. (2010). Decreased sperm motility is associated with air pollution in Salt Lake City. *Fertility and Sterility.* Vol.93, N°.6, pp.1875-9. ISSN:0015-0282.

Hansen, C; Neller, A; Williams, G & Simpson, R. (2006). Maternal exposure to low levels of ambient air pollution and preterm birth in Brisbane, Australia. *British Journal of Obstetrics and Gynaecology.* Vol.113, N°.8, pp.935-41. ISSN: 1470-0328.

Hansen, C; Neller, A; Williams, G & Simpson, R. (2007). Low levels of ambient air pollution during pregnancy and fetal growth among term neonates in Brisbane, Australia. *Environmental Research.* Vol.103, N°.13, pp.383-9. ISSN:0013-9351.

Hansen, CA; Barnett, AG & Pritchard, G. (2008). The effect of ambient air pollution during early pregnancy on fetal ultrasonic measurements during mid-pregnancy. *Environmental Health Perspectives.* Vol.116, N°.3, pp.362-9. ISSN: 0091/6765.

Hansen, CA; Barnett, AG; Jalaludin, BB & Morgan, GG. (2009). Ambient air pollution and birth defects in Brisbane, Australia. *PLoS One.* Vol.4, N°.4, pp.e5408. ISSN:1932-6203.

Huynh, M; Woodruff, TJ; Parker, JD & Schoendorf, KC. (2006). Relationships between air pollution and preterm birth in California. *Paediatric and Perinatal Epidemiology.* Vol.20, N°.6, pp.454-61. ISSN:0269-5022.

Hwang, B-F & Jaakkola, JJK. (2008). Ozone and other air pollutants and risk of oral clefts. *Environmental Health Perspectives.* Vol.116, N°.10, pp.1411-5. ISSN: 0091-6765.

Iñiguez, C; Ballester, F; Estarlich, M; Llop, S; Fernandez-Patier, R; Aguirre-Alfaro, A; Esplugues A & INMA Study group, Valencia. (2009). Estimation of personal NO2 exposure in a cohort of pregnant women. *Sci Total Environ.* Vol.407, N°.23, pp.6093-9. ISSN: 0048-9697

Gauderman, WJ; Avol, E; Lurmann, F; Kuenzli, N; Gilliland, F; Peters, J & McConnell; R. (2005). Childhood asthma and exposure to traffic and nitrogen dioxide. *Epidemiology.* Vol.16, N°.6, pp.737-43. ISSN: 1044-3983.

Gilboa, SM; Mendola, P; Olshan, AF; Langlois, PH; Savitz, DA; Loomis, D; Herring, AH & Fixler, DE. (2005). Relation between ambient air quality and selected birth defects, seven county study, Texas, 1997-2000. *American Journal of Epidemiology.* Vol.162, N°.3, pp.238-52. ISSN: 0002-9262.

Jalaludin, B; Mannes, T; Morgan, G; Lincoln, D; Sheppeard, V & Corbett, S. (2007). Impact of ambient air pollution on gestational age is modified by season in Sydney, Australia. *Environmental Health.* 6:16. ISSN:1476-069X.

Jedrychowski, W; Perera, F; Mrozek-Budzyn, D; Mroz, E; Flak, E; Spengler, JD; Edwards, S; Jacek, R; Kaim, I & Skolicki, Z. (2009). Gender differences in fetal growth of newborns exposed prenatally to airborne fine particulate matter. *Environmental Research.* Vol.109, N°.4, pp.447-56. ISSN:0013-9351.

Jerrett, M ; Shankardass, K ; Berhane, K ; Gauderman, WJ ; Künzli, N ; Avol, E ; Gilliland, F ; Lurmann, F ; Molitor, JN ; Molitor, JT ; Thomas, DC ; Peters, J & McConnell, R. (2008). Traffic-related air pollution and asthma onset in children: a prospective cohort study with individual exposure measurement. *Environmental Health Perspectives.* Vol.116, N°.10, pp.1433-8. ISSN: 0091/6765.

Jiang, LL ; Zhang, YH ; Song, GX ; Chen, GH ; Chen, BH ; Zhao, NQ & Kan, HD. (2007). A time series analysis of outdoor air pollution and preterm birth in Shanghai, China. *Biomedical and Environmental Sciences*. Vol.20, N°.5, pp.426-31. ISSN: 0895-3988

Kannan, S; Misra, DP; Dvonch, JT & Krishnakumar, A. (2006). Exposures to airborne particulate matter and adverse perinatal outcomes: a biologically plausible mechanistic framework for exploring potential effect modification by nutrition. *Environmental Health Perspectives*. Vol.114, N°.11, pp.1636-42. ISSN: 0091/6765.

Kim, OJ; Ha, EH; Kim, BM; Seo, JH; Park, HS; Jung, WJ; Lee, BE; Suh, YJ; Kim, YJ; Lee, JT; Kim, H; Hong, YC. (2007). PM_{10} and pregnancy outcomes: a hospital-based cohort study of pregnant women in Seoul. *Journal of Occupational and Environmental Medicine*. Vol.49, N°.12, pp.1394-402. ISSN:1076-2752.

Lacasana, M; Esplugues, A & Ballester, F. (2005). Exposure to ambient air pollution and prenatal and early childhood health effects. *European Journal of Epidemiology*. Vol.20, N°.2, pp.183–99. ISSN: 0393-2990.

Landgren, O. (1996). Environmental pollution and delivery outcome in southern Sweden: a study with central registries. *Acta Paedriatrica*.Vol.85, N°.11, pp.1361-4. ISSN: 0803-5253.

Laumbach RJ. (2010). Outdoor air pollutants and patient health. *American Family Physician*. Vol.81, N°.2, pp.175-80. ISSN: 0002-838X.

Lee, BE; Ha, EH; Park, HS; Kim, YJ; Hong, YC; Kim, H & Lee, JT. (2003). Exposure to air pollution during different gestational phases contributes to risks of low birth weight. *Human Reproduction*. Vol. 18, N°.3, pp.638–43. ISSN: 1460-2350.

Lee, SJ; Hajat, S; Steer, PJ & Filippi, V. (2008). A time-series analysis of any short-term effects of meteorological and air pollution factors on preterm births in London, UK. *Environmental Research*. Vol.106, N°.2, pp.185-94. ISSN:0013-9351.

Leem, JH; Kaplan, BM; Shim, YK; Pohl, HR; Gotway, CA; Bullard, SM; Rogers, JF, Smith, MM &, Tylenda, CA. (2006). Exposures to air pollutants during pregnancy and preterm delivery. *Environmental Health Perspectives* Vol.114, N°.6, pp.905-10. ISSN: 0091/6765.

Lepeule, J; Caïni, F; Bottagisi, S; Galineau, J; Hulin, A; Marquis, N; Bohet, A; Siroux, V; Kaminski, M; Charles, MA & Slama, R. (2010). EDEN Mother–Child Cohort Study Group. Maternal exposure to nitrogen dioxide during pregnancy and offspring birth weight: comparison of two exposure models. *Environmental Health Perspectives*. Vol.118, N°.10, pp.1483-9. ISSN: 0091/6765.

Lin, MC; Chiu, HF; Yu, HS; Tsai, SS; Cheng, BH; Wu, TN; Sung, FC & Yang, CY. (2001). Increased risk of preterm delivery in areas with air pollution from a petroleum refinery plant in Taiwan. *Journal of Toxicology and Environmental Health. Vol.*A64, pp.637–44. ISSN: 0013-273X.

Liu, S; Krewski, D; Shi, Y; Chen, Y & Burnett, RT. (2003). Association between gaseous ambient air pollutants and adverse pregnancy outcomes in Vancouver, Canada. *Environmental Health Perspectives*. Vol.111, N°.14, pp.1773-8. ISSN: 0091/6765.

Liu, S; Krewski, D; Shi, Y; Chen, Y & Burnett, RT. (2007). Association between maternal exposure to ambient air pollutants during pregnancy and fetal growth restriction. *Journal of Exposure Science and Environmental Epidemiology*. Vol.17, N°.5, pp.426-32. ISSN:1559-0631.

Llop, S; Ballester, F; Estarlich, M; Esplugues, A; Rebagliato, M; Iñiguez, C. (2010). Preterm birth and exposure to air pollutants during pregnancy. *Environmental Research.* Vol.110, N°.8, pp.778-85. ISSN:0013-9351.

Madsen, C; Gehring, U; Walker, SE; Brunekreef, B; Stigum, H; Naess, O & Nafstad, P. (2010). Ambient air pollution exposure, residential mobility and term birth weight in Oslo, Norway. *Environmental Research* . Vol.110, N°.4, pp.363-71. ISSN:0013-9351.

Maisonet, M; Bush, TJ; Correa, A & Jaakkola, JJ. (2001). Relation between ambient air pollution and low birth weight in the Northeastern United States. *Environmental Health Perspectives.* Vol.109, Suppl 3, pp.351-6. ISSN: 0091/6765.

Maisonet, M; Correa, A; Misra, D & Jaakkola, JJ. (2004). A review of the literature on the effects of ambient air pollution on fetal growth. *Environmental Research.* Vol.95, N°.1, pp.106–15. ISSN:0013-9351.

Mannes, T; Jalaludin, B; Morgan, G; Lincoln, D; Sheppeard, V & Corbett, S. (2005). Impact of ambient air pollution on birth weight in Sydney, Australia. *Occupational and Environmental Medicine.* Vol.62, N°.2, pp.524-30. ISSN:1076-2752.

Mohorovic, L. (2004). First two months of pregnancy-critical time for preterm delivery and low birthweight caused by adverse effects of coal combustion toxics. *Early Human Development.* Vol.80, N°.2, pp.115-23. ISSN:0378-3782.

Moison, RM; Palinckx, JJ; Roest, M; Houdkamp, E & Berger, HM. (1993). Induction of lipid peroxidation of pulmonary surfactant by plasma of preterm babies. *The Lancet.* Vol. 341, N°.8837, pp.79-82. ISSN:0140-6736.

Morello-Frosch, R; Jesdale, BM; Sadd, JL & Pastor, M. (2010). Ambient air pollution exposure and full-term birth weight in California. *Environmental Health.* 9:44. ISSN: 1476-069X.

Morris, SS; Victora, CG; Barros, FC; Halpern, R; Menezes, AM; César, JA; Horta, BL & Tomasi, E. (1998). Length and ponderal index at birth: associations with mortality, hospitalizations, development and post-natal growth in Brazilian infants. *International Journal of Epidemiology.* Vol.27, N°.2, pp.242-7. ISSN:0300-5771.

Nethery, E; Brauer, M & Janssen, P. (2009). Time-activity patterns of pregnant women and changes during the course of pregnancy. *Journal of Exposure Science and Environmental Epidemiology.* Vol.19, N°.3, pp.317-24. ISSN:1559-0631.

Parker, J; Rich, DQ; Glinianaia, SV; Leem, JH; Wartenberg, D; Bell, ML; Bonzini, M; Brauer, M; Darrow L; Gehring, U; Gouveia, N; Grillo, P; Ha, E; van den Hooven, EH; Jalaludin, B; Jesdale, BM; Lepeule, J; Morello-Frosch, R; Morgan, GG; Slama, R; Pierik, FH; Pesatori, AC; Sathyanarayana, S; Seo, J; Strickland, M; Tamburic, L & Woodruff, TJ. (2011). The International Collaboration on Air Pollution and Pregnancy Outcomes: Initial Results. *Environmental Health Perspectives.* [Epub ahead of print]

Pearce, MS; Glinianaia, SV; Rankin, J; Rushton, S; Charlton, M; Parker, L & Pless-Mulloli, T. (2010). No association between ambient particulate matter exposure during pregnancy and stillbirth risk in the north of England, 1962-1992. *Environmental Research.* Vol.110, N°.1, pp.118-22. ISSN:0013-9351.

Peppone, LJ; Piazza, KM; Mahoney, MC; Morrow, GR; Mustian, KM; Palesh, OG & Hyland, A. (2009). Associations between adult and childhood secondhand smoke exposures and fecundity and fetal loss among women who visited a cancer hospital. *Tobacco Control.* Vol.18, N°.2, pp.115-20. ISSN:0964-4563.

Pereira, LA; Loomis, D; Conceição, GM; Braga, AL; Arcas, RM; Kishi, HS; Singer, JM; Böhm, GM & Saldiva, PH. (1998). Association between air pollution and intrauterine mortality in São Paulo, Brazil. *Environmental Health Perspectives*. Vol.106, N°.6, pp.325-9. ISSN: 0091/6765.

Perera, FP ; Illman, SM ; Kinney, PL ; Whyatt, RM ; Kelvin, EA ; Shepard, P; Evans, D ; Fullilove, M; Ford, J ; Miller, RL ; Meyer, IH & Rauh, VA. (2002). The challenge of preventing environmentally related disease in young children: community-based research in New York City. *Environmental Health Perspectives*. Vol.110, N°.2, pp.197-204. ISSN: 0091/6765.

Perera, FP; Rauh, V; Tsai, WY; Kinney, P; Camann, D; Barr, D; Bernert, T; Garfinkel, R; Tu, YH; Diaz, D; Dietrich, J & Whyatt, RM. (2003). Effects of transplacental exposure to environmental pollutants on birth outcomes in a multiethnic population. *Environmental Health Perspectives*. Vol.111, N°.2, pp.201-6. ISSN: 0091/6765.

Raaschou-Nielsen, O; Hertel, O; Vignati, E; Berkowicz, R; Jensen, SS; Larsen, VB; Lohse, C & Olsen, JH. (2000). An air pollution model for use in epidemiological studies: evaluation with measured levels of nitrogen dioxide and benzene. *Journal of Exposure Analysis and Environmental Epidemiology*. Vol.10, N°.1, pp.4-14. ISSN:1053-4245.

Ramesh, A; Inyang, F; Lunstra, DD; Niaz, MS; Kopsombut, P; Jones, KM; Hood, DB; Hills, ER & Archibong, AE. (2008). Alteration of fertility endpoints in adult male F-344 rats by subchronic exposure to inhaled benzo(a)pyrene. (2008). *Experimental and Toxicologic Pathology*. Vol.60, N°.4-5, pp.269-80. ISSN:0940-2993.

Rankin, J; Chadwick, T; Natarajan, M; Howel, D; Pearce, MS & Pless-Mulloli, T. (2009). Maternal exposure to ambient air pollutants and risk of congenital anomalies. *Environmental Research*. Vol.109, N°.2, pp.181-7. ISSN:0013-9351.

Richards, M; Hardy, R; Kuh, D & Wadsworth, ME. (2002). Birthweight, postnatal growth and cognitive function in a national UK birth cohort. *International Journal of Epidemiology*. Vol.31, N°.2, pp.342-8. ISSN:0300-5771.

Ritz, B & Yu, F. (1999). The effect of ambient carbon monoxide on low birth weight among children born in southern California between 1989 and 1993. *Environmental Health Perspectives*. Vol.107, N°.1, pp.17-25. ISSN: 0091/6765.

Ritz, B; Yu, F; Chapa, G & Fruin, S. (2000). Effect of air pollution on preterm birth among children born in Southern California between 1989 and 1993. *Epidemiology*. Vol.11, N°.5, pp.502-11. ISSN: 1044-3983.

Ritz, B; Yu, F; Fruin, S; Chapa, G; Shaw, GM & Harris, JA. (2002). Ambient air pollution and risk of birth defects in Southern California. *American Journal of Epidemiology*. Vol.155, N°.1, pp.17-25. ISSN: 0002-9262.

Ritz, B; Wilhelm, M; Hoggatt, KJ & Ghosh, JK. (2007). Ambient air pollution and preterm birth in the environment and pregnancy outcomes study at the University of California, Los Angeles. *American Journal of Epidemiology*. Vol.166, N°.9, pp.1045–52. ISSN: 0002-9262.

Ritz, B & Wilhelm, M. (2008). Ambient air pollution and adverse birth outcomes: methodologic issues in an emerging field. *Basic and Clinic Pharmacology and Toxicology*. Vol.102, N°.2, pp.182–90. ISSN:1742-7835.

Ritz, B. (2010). Air-pollution and congenital anomalies. *Occupational and Environmental Medicine*. Vol.67, N°.4, pp.221-2. ISSN:1076-2752.

Rogers, JF; Thompson, SJ; Addy, CL; McKeown, RE; Cowen, DJ & Decoufle, P. (2000). Association of very low birth weight with exposures to environmental sulfur dioxide and total suspended particulates. *American Journal of Epidemiology*. Vol.151, N°.6, pp.602–13. ISSN: 0002-9262.

Sagiv, SK; Mendola, P; Loomis, D; Herring, AH; Neas, LM; Savitz, DA & Poole, C. (2005). A time-series analysis of air pollution and preterm birth in Pennsylvania, 1997-2001. *Environmental Health Perspectives*. Vol.113, N°.5, pp.602-6. ISSN: 0091/6765.

Salam, MT; Millstein, J; Li, YF; Lurmann, FW; Margolis, HG & Gilliland, FD. (2005). Birth outcomes and prenatal exposure to ozone, carbon monoxide, and particulate matter: results from the Children's Health Study. *Environmental Health Perspectives*. Vol.113, N°.11, pp.1638-44. ISSN: 0091/6765.

Schwartz, J. (2004). Air pollution and children´s health. *Pedriatrics*. Vol.113, Suppl.4, pp.1037-43. ISSN:0918-5739.

Selevan, SG; Borkovec, L; Slott, VL; Zudová, Z; Rubes, J; Evenson, DP & Perreault, SD. (2000). Semen quality and reproductive health of young Czech men exposed to seasonal air pollution. *Environmental Health Perspectives*. Vol.108, N°.9, pp.887-94. ISSN: 0091/6765.

Seo, JH; Leem, JH; Ha, EH; Kim, OJ; Kim, BM; Lee, JY; Park, HS; Kim, HC; Hong, YC & Kim, YJ. (2010). Population-attributable risk of low birthweight related to PM_{10} pollution in seven Korean cities. *Paediatric and Perinatal Epidemiology*. Vol.24, N°.2, pp.140-8. ISSN:0269-5022.

Shah, PS & Balkhair, T. (2011). Knowledge Synthesis Group on Determinants of Preterm/LBW births. Air pollution and birth outcomes: a systematic review. *Environment International*. Vol.37, N°.2, pp.498-516. ISSN: 0160-4120.

Skalkidou, A; Kieler, H; Stephansson, O; Roos, N; Cnattingius, S & Haglund, B. (2010). Ultrasound pregnancy dating leads to biased perinatal morbidity and neonatal mortality among post-term-born girls. *Epidemiology*. Vol.21, N°.6, pp.791-6. ISSN: 1044-3983.

Slama, R; Sinno-Teller, S; Thiébaugeorges, O; Goua, V; Forhan, A; Ducot, B; Annesi-Maesano, I; Heinrich, J; Schweitzer, M; Magnin, G; Bouyer, J; Kaminski, M & Charles, MA. (2006). Eden Study Group. Relation between atmospheric pollutants and head circumference in utero and at birth: a cohort study relying on ultrasound imaging during pregnancy. *Epidemiology*. Vol.17, pp.S129-S130. ISSN: 1044-3983.

Slama, R; Morgenstern, V; Cyrys, J; Zutavern, A; Herbarth, O; Wichmann, HE & Heinrich, J. (2007). LISA Study Group. Traffic-related atmospheric pollutants levels during pregnancy and offspring's term birth weight: a study relying on a land-use regression exposure model. *Environmental Health Perspectives*. Vol.115, N°.9, pp.1283-92. ISSN: 0091/6765.

Slama, R; Darrow, L; Parker, J; Woodruff, TJ; Strickland, M; Nieuwenhuijsen, M; Glinianaia, S; Hoggatt ,KJ; Kannan, S; Hurley, F, Kalinka, J; Srám, R; Brauer, M; Wilhelm, M; Heinrich, J & Ritz, B. (2008). Meeting report: atmospheric pollution and human reproduction. *Environmental Health Perspectives*. Vol.116, N°.6, pp.791-8. ISSN: 0091/6765.

Slama, R; Thiebaugeorges, O; Goua, V; Aussel, L; Sacco, P; Bohet, A; Forhan, A; Ducot, B; Annesi-Maesano, I; Heinrich, J; Magnin, G; Schweitzer, M; Kaminski, M & Charles, MA. (2009). EDEN Mother-Child Cohort Study Group. Maternal personal exposure to airborne benzene and intrauterine growth. *Environmental Health Perspectives*. Vol.117, N°.8, pp.1313-21. ISSN: 0091/6765.

Somers, CM. (2011). Ambient air pollution exposure and damage to male gametes: human studies and in situ 'sentinel' animal experiments. *Systems biology in reproductive medicine*. Vol.57, N°.1-2, pp.63-71. ISSN:1939-6368.

Sram, RJ; Binkova, B; Dejmek, J & Bobak M. (2005). Ambient air pollution and pregnancy outcomes: a review of the literature. *Environmental Health Perspectives*. Vol.113, N°.4, pp.375–82. ISSN: 0091/6765.

Sram, RJ; Binkova, B; Dejmek, J; Chvatalova, I ; Solansky, I & Topinka, J. (2006). Association of DNA adducts and genotypes with birth weight. *Mutaiot Research*. Vol.608, N°.2, pp.121-8. ISSN: 0027-5107.

Stillerman, KP; Mattison, DR; Giudice, LC & Woodruff, TJ. (2008). Environmental exposures and adverse pregnancy outcomes: a review of the science. *Reproductive Sciences*. Vol.15, N°.7, pp.631-50. ISSN: 1933-7191.

Strickland, MJ; Klein, M; Correa, A; Reller, MDL; Mahle, WT; Riehle-Colarusso, TJ; Botto, LD; Flanders, WD; Mulholland, JA; Siffel, C; Marcus, M & Tolbert, PE. (2009). Ambient air pollution and cardiovascular malformations in Atlanta, Georgia, 1986-2003. *American Journal of Epidemiology*. Vol.169, N°.8, pp.1004-14. ISSN: 0002-9262.

Suh, YJ; Kim, H; Seo, JH; Park, H; Kim, YJ; Hong, YC & Ha, EH. (2009). Different effects of PM_{10} exposure on preterm birth by gestational period estimated from time-dependent survival analyses. *International Archives of Occupational and Environmental Health*. Vol.82, N°.5, pp.613-21. ISSN: 0340-0131.

Tabacova, S; Balabaeva, L & Little, RE. (1997). Maternal exposure to exogenous nitrogen compounds and complications of pregnancy. *Archives of Environmental Health*. Vol.52, N°.2, pp.341-7. ISSN: 0003-9896.

Tsai, SS; Yu, HS; Liu, CC & Yang, CY. (2003). Increased incidence of preterm delivery in mothers residing in an industrialized area in Taiwan. *Journal of Toxicology and Environmental Health*. Vol.66, N°.11, pp.987-94. ISSN: 0013-273X.

Tsai, SS; Yu, HS; Chang, CC; Chuang, HY & Yang, CY. (2004). Increased risk of preterm delivery in women residing near thermal power plants in Taiwan. *Archives of Environmental Health*. Vol.59, N°.9, pp.478–83. ISSN: 0003-9896.

Van den Hooven, EH; Jaddoe, VW; de Kluizenaar, Y; Hofman, A; Mackenbach, JP; Steegers, EA; Miedema, HM & Pierik, FH. (2009). Residential traffic exposure and pregnancy-related outcomes: a prospective birth cohort study. *Environmental Health*. 8:59. ISSN: 1476-069X.

Veras, MM ; Damaceno-Rodrigues, NR ; Guimarães-Silva, RM ; Scoriza, JN ; Saldiva, PH ; Caldini, EG & Dolhnikoff, M. (2009). Chronic exposure to fine particulate matter emitted by traffic affects reproductive and fetal outcomes in mice. *Environmental Research*. Vol.109, N°.5, pp.536-43. ISSN:0013-9351.

Veras, MM ; Caldini, EG ; Dolhnikoff, M & Saldiva, PH. (2010). Air pollution and effects on reproductive-system functions globally with particular emphasis on the Brazilian population. *Journal of Toxicology and Environmental Health. Part B: Critical reviews*. Vol.13, N°.1, pp.1-15. ISSN:1521-6950.

Vrijheid, M; Martinez, D; Manzanares, S; Dadvand, P; Schembari, A; Rankin, J & Nieuwenhuijsen, M. (2011). Ambient Air Pollution and Risk of Congenital

Anomalies: A Systematic Review and Meta-Analysis. *Environmental Health Perspectives*. Vol.119, N°.5, pp.598-606. ISSN: 0091/6765.

Wang, L & Pinkerton, KE. (2007). Air pollutant effects on fetal and early postnatal development. *Birth Defects Research Part C: Embryo Today*. Vol.81, N°.3, pp.144-54. ISSN: 1542-975X.

Wigle, DT; Arbuckle, TE; Turner, MC; Bérubé, A; Yang, Q; Liu, S & Krewski, D. (2008). Epidemiologic evidence of relationships between reproductive and child health outcomes and environmental chemical contaminants. *Journal of Toxicology and Environmental Health. Part B: Critical reviews*. Vol.11, N°.5-6, pp.373-517. ISSN:1521-6950.

Wilhelm, M & Ritz, B. (2003). Residential proximity to traffic and adverse birth outcomes in Los Angeles County, California, 1994-1996. *Environmental Health Perspectives*. Vol.111, N°.2, pp.207-16. ISSN: 0091/6765.

Wilhelm, M & Ritz, B. (2005). Local variations in CO and particulate air pollution and adverse birth outcomes in Los Angeles County, California, USA. *Environmental Health Perspectives*. Vol.113, N°.9, pp.1212-21. ISSN: 0091/6765.

Wong, EY; Gohlke, J; Griffith, WC; Farrow, S & Faustman EM. (2004). Assessing the health benefits of air pollution reduction for children. *Environmental Health Perspectives*. Vol.112, N°.2, pp.226-32. ISSN: 0091/6765.

Woodruff, TJ; Parker, JD; Darrow, LA; Slama, R; Bell, ML; Choi, H; Glinianaia, S; Hoggatt, KJ; Karr, CJ; Lobdell, DT & Wilhelm, M. (2009). Methodological issues in studies of air pollution and reproductive health. *Environmental Research*. Vol.109, N°.3, pp.311-20. ISSN:0013-9351.

Woodruff, TJ; Parker, JD; Adams, K; Bell, ML; Gehring, U; Glinianaia, S; Ha, EH; Jalaludin, B & Slama, R. (2010). International Collaboration on Air Pollution and Pregnancy Outcomes (ICAPPO). *International Journal of Environmental Research and Public Health*. Vol.7, N°.6, pp.2638-52. ISSN 1660-4601.

Wu, J; Ren, C; Delfino, RJ; Chung, J; Wilhelm, M & Ritz, B. (2009). Association between local traffic-generated air pollution and preeclampsia and preterm delivery in the south coast air basin of California. *Environmental Health Perspectives*. Vol.117, N°.11, pp.1773-9. ISSN: 0091/6765.

Xu, X; Ding, H & Wang, X. (1995). Acute effects of total suspended particles and sulfur dioxides on preterm delivery: a community-based cohort study. *Archives on Environmental Health*. Vol.50, N°.6, pp.407-15. ISSN: 0003-9896.

Yang, CY ; Chang, CC ; Chuang, HY; Ho, CK ; Wu, TN & Tsai, SS. (2003). Evidence for increased risks of preterm delivery in a population residing near a freeway in Taiwan. *Archives on Environmental Health*. Vol.58, N°.10, pp.649–54. ISSN: 0003-9896.

Yang, CY; Chang, CC; Chuang, HY; Ho, CK; Wu, TN & Chang, PY. (2004). Increased risk of preterm delivery among people living near the three oil refineries in Taiwan. *Environment International*. Vol.30, N°.3, pp.337-42. ISSN: 0160-4120.

Yorifuji, T; Naruse, H; Kashima, S; Ohki, S; Murakoshi, T; Takao, S; Tsuda, T & Doi, H. (2011). Residential proximity to major roads and preterm births. *Epidemiology*. Vol.22, N°.1, pp.74-80. ISSN: 1044-3983.

Zhao, Q; Liang, Z; Tao, S; Zhu, J & Du, Y. (2011). Effects of air pollution on neonatal prematurity in Guangzhou of China: a time-series study. *Environmental Health*. 10:2. ISSN: 0022-0892.

2

Air Pollution and Adverse Pregnancy Outcome

Jong-Han Leem[1] and Eun-Hee Ha[2]
[1]*Department of Occupational & Environmental Medicine, School of Medicine, Inha University,*
[2]*Department of Preventive Medicine, School of Medicine, Ewha Womans University*
Korea

1. Introduction

Air pollution is known to be associated with increased total mortality, including cardiovascular and respiratory mortality. People with chronic disease in adulthood, such as cardiovascular disease, metabolic syndrome and respiratory diseases, are very susceptible with air pollutants (Kwon HJ et al., 2003). However, air pollution adversely impacts not only adults and the elderly but also fetuses and children. In fact, fetuses are the most vulnerable group to air pollution because vulnerability and susceptibility to air pollution are formed at early ages. Low birth weight (LBW), pre-term delivery (PTB), intrauterine growth restriction, and post-neonatal infant mortality are such undesirable outcomes.

LBW affects 20 million infants worldwide (UNICEF, 2004). LBW is comprised of two overlapping etiologies: PTB and intrauterine growth retardation (IUGR). In particular, LBW is associated with a higher risk of infant and childhood mortality, coronary heart disease, and other health problems. LBW additionally has a well-established association with early-onset insulin resistance and a later risk of adulthood diseases, including all aspects of the metabolic syndrome.

PTB remains the leading cause of perinatal mortality and occurs in approximately 4-10% of pregnancies (Reagan and Salsberry 2005). Known risk factors for PTB include lower social class, less education, single marital status, low income, younger maternal age, low body weight, ethnicity, smoking, and poor housing, along with medical factors such as induction, premature rupture of membranes, infection, multiple pregnancy, intrauterine death, fetal and uterine abnormalities and chorioamnionitis (Bibby and Stewart 2004).

LBW and PTB are both significantly associated with infant mortality and an array of infant morbidities that range from pulmonary to neurologic outcomes. These associations form the basis for the "fetal origins" or the "Barker hypothesis" which postulates that "fetal growth retardation consequent to malnutrition has long-term structural and physiologic impacts that predispose an individual to chronic diseases in adulthood" (Barker, 2007).

Is there the association between air pollution and adverse pregnancy outcomes, such LBW, and PTB? In this chapter, we will review the association between air pollution and adverse pregnancy outcomes(APO), such as LBW and PTB. We also will estimate the disease burden of LBW and PTB caused by air pollution, and discuss how to decrease these APOs.

2. Air pollution and adverse pregnancy outcomes

2.1 Air pollution and low birth weight

Fifteen (Alderman et al., 1987; Bell et al., 2007; Bobak and Leon, 1999; Bobak, 2000; Dugandzic et al., 2006; Gouveia et al., 2004; Ha et al., 2001; Lee et al., 2003; Lin et al., 2004; Liu et al., 2003; Maisonet et al., 2001; Morello-Frosch et al., 2010; Ritz and Yu, 1999; Rogers et al., 2000; Salam et al., 2005) studies reported data on birth outcomes following SO_2 exposure (Table1): ten reported increased odds of LBW births following SO_2 exposure, five reported no association with LBW, and one reported significant association with extreme LBW (Rogers et al., 2000).

Little is known about the association between nitrous oxide (NO) and APO. But in natural condition, NO is converted into nitrogen dioxide (NO_2), which is 5~10 times more toxic than NO. NO_2 is known to be associated with APO by several studies. The association between NO_2 exposure and LBW was explored in 11 studies (Bell et al., 2007; Bobak, 2000; Gouveia et al., 2004; Ha et al., 2001; Lee et al., 2003; Lin et al., 2004; Liu et al., 2003; Madsen et al., 2010; Maroziene and Grazuleviciene, 2002; Morello-Frosch et al., 2010; Salam et al., 2005) (Table 1). Increased LBW risk with increased NO_2 exposure was reported by Ha et al. (2001) (first trimester), Lee et al. (2003) (second trimester), Bell et al. (2007) (during the entire gestation) and Morello-Frosch et al. (2010). However, other reports did not identify significant increases in LBW births.

The association between carbon monoxide(CO) exposure and birth outcomes was explored in 13 studies (Alderman et al., 1987; Bell et al., 2007; Chen et al., 2002; Gouveia et al., 2004; Ha et al., 2000; Huynh et al., 2006; Lee et al., 2003; Lin et al., 2000; Liu et al., 2003; Maisonet et al., 2001; Morello-Frosch et al., 2010; Salam et al., 2000; Ritz and Yu, 1999) (Table 1). Of these, five studies (Ritz and Yu, 1999; Ha et al., 2000; Maisonet et al., 2001; Lee et al., 2003; Morello-Frosch et al., 2010) reported an increased risk of LBW births.

Seven studies (Chen et al., 2002; Dugandzic et al., 2006; Gouveia et al., 2004; Ha et al., 2001; Lin et al., 2004; Liu et al., 2003; Morello-Frosch et al., 2010) investigated the association between exposure to ozone and LBW (Table 1). None of these studies reported a statistically significant increase in LBW with higher exposure to ozone.

The effects of $PM_{2.5}$ (Particulate matter less than 2.5 μm in aerodynamic diameter) on LBW were evaluated in four studies (Bell et al., 2007; Huynh et al., 2006; Madsen et al., 2010; Morello-Frosch et al., 2010) (Table 1). Huynh et al. (2006) reported an association between high levels of $PM_{2.5}$ and LBW when the exposure was measured at any time during the gestation, and particularly in the last 2 weeks of pregnancy and the first month of gestation. Bell et al. (2007) and Morello-Frosch et al. (2010) also reported that high levels of $PM_{2.5}$ were associated with LBW. But Madsen et al. (2010) did not report any association between exposure to high levels of $PM_{2.5}$ and LBW.

Twelve studies (Bell et al., 2007; Chen et al., 2002; Dugandzic et al., 2006; Gouveia et al., 2004; Lee et al., 2003; Maisonet et al., 2001; Lin et al., 2004; Salam et al., 2005; Seo et al., 2010; Madsen et al., 2010; Morello-Frosch et al., 2010; Xu et al., 2011) assessed the effects of PM_{10} (Particulate matter less than 10 μm in aerodynamic diameter) on LBW (Table 1). Lee et al. (2003) and Xu et al. (2011) reported on a possible association between LBW and an increase of more than an interquartile range in PM_{10} exposure during the first and second trimesters, and Gouveia et al. (2004) reported higher odds of LBW among mothers in the highest quartile of exposure during the second trimester. Seo et al. (2010) reported that among seven Korean cities, two had higher odds of LBW births with incremental exposure to PM_{10}, whereas five had no association. Other studies have reported no association between PM_{10} and LBW births.

Of five studies (Bobak, 2000; Bobak and Leon, 1999; Ha et al., 2001; Rogers et al., 2000; Wang et al., 1997) that reported on the association between total suspended particles and LBW (Table 1), three (Ha et al., 2001; Rogers et al., 2000; Wang et al., 1997) reported an increased risk of LBW births with higher concentrations. The other two (Bobak, 2000; Bobak and Leon, 1999) reported no association between TSP (total suspended particle) and LBW births.

Pollutants or exposure variable	Results	Reference
SO_2	NE*	Alderman et al ,1987
	AOR**=1.22(95% CI, 1.03–1.44) for > 5.5 ppm during the last trimester	Ritz and Yu,1999
	AOR=1.10(95% CI, 1.02–1.17) for 50 μg/m³ increase	Bobak and Leon, 1999
	NE	Bobak, 2000
	AOR=2.88(95% CI, 1.16–7.13) for > 56.75 μ/m³ in annual exposure for very LBW infant outcome	Rogers et al.,2000
	ARR=1.06(95% CI, 1.02–1.10) for IQR increase in the first trimester	Ha et al.,2001
	Second trimester exposures falling within the 25 and < 50th (AOR 1.21; CI 1.07,1.37), the 50 to < 75th (AOR 1.20; CI 1.08,1.35), and the 75 to < 95th (AOR 1.21; CI 1.03,1.43) percentiles were also at increased risk for term LBW when compared to those in the reference category (< 25th percentile).	Maisonet et al. 2001
	OR=1.14(95% CI, 1.04–1.24) for IQR increase in all trimesters OR=1.06(95% CI, 1.02–1.11) for IQR increase in the second trimester.	Lee et al. 2003
	AOR=1.11(95% CI, 1.01–1.22) for 5 ppb increase in the first month	Liu et al. 2003
	NE	Gouveia et al. 2004
	AOR=1.16(95% CI, 1.02–1.33) for 7.1-11.4 ppb increase in entire pregnancy AOR=1.26(95% CI, 1.04–1.53) for >11.4 ppb increase in entire pregnancy AOR=1.20(95% CI, 1.01–1.41) for >12.4 ppb increase in third trimester	Lin et al. 2004
	NE	Salam et al., 2005
	ARR=1.15(95% CI, 1.00–1.31) increase in the first trimester	Dugandzic et al. 2006
	NE	Bell et al. 2007
	AOR=1.01(95% CI, 1.00- 1.02) for ppb increase at 10 km monitor distance	Morello-Frosch et al. 2010
NO_2	NE	Bobak. 2000
	ARR=1.07(95% CI, 1.03–1.11) for IQR increase in the	Ha et al. 2001

	first trimester	
	NE	Maroziene and Grazuleviciene. 2002
	OR=1.04(95% CI, 1.00–1.08) for IQR increase in all trimester OR=1.03(95% CI, 1.01–1.06) for IQR increase in the second trimester	Lee et al. 2003
	NE	Liu et al. 2003
	NE	Gouveia et al. 2004
	NE	Lin et al. 2004
	NE	Salam et al. 2005
	AOR=1.03(95% CI, 1.00–1.05) for IQR increase	Bell et al. 2007
	NE	Madsen et al. 2010
	AOR=1.03(95% CI, 1.01–1.05) for ppm increase at 3 km monitor distance AOR=1.04(95% CI, 1.03–1.05) for ppm increase at 5 km monitor distance AOR=1.03(95% CI, 1.02–1.04) for ppm increase at 10 km monitor distance	Morello-Frosch et al. 2010
CO	NE	Alderman et al., 1987
	AOR=1.22(95% CI, 1.03–1.44) for > 5.5 ppm increase during the last trimester	Ritz and Yu, 1999
	AOR=1.08(95% CI, 1.04-1.12) for IQR increase in the first trimester	Ha et al. 2001
	AOR=1.31(95% CI, 1.06–1.62) for 1 ppm increase in the third trimester	Maisonet et al. 2001
	NE	Chen et al. 2002
	OR=1.05(95% CI, 1.01–1.09) increase in all trimesters OR=1.04(95% CI, 1.01–1.07) for IQR increase in the first trimester OR=1.03(95% CI, 1.00–1.08) increase in the second trimester	Lee et al. 2003
	NE	Liu et al. 2003
	NE	Gouveia et al. 2004
	NE	Lin et al. 2004
	NE	Salam et al. 2005
	NE	Huynh et al. 2006
	NE	Bell et al. 2007
	AOR=1.06(95% CI, 1.03–1.09) for ppm increase at 5 km monitor distance AOR=1.04(95% CI, 1.02–1.06) for ppm increase at 10 km monitor distance	Morello-Frosch et al. 2010
Ozone(O_3)	NE	Ha et al. 2001
	NE	Chen et al. 2002
	NE	Liu et al. 2003

	NE	Gouveia et al. 2004
	NE	Lin et al. 2004
	NE	Dugandzic et al. 2006
	NE	Morello-Frosch et al. 2010
$PM_{2.5}$	AOR=1.14(95% CI, 1.07-1.23) for 17.7-22.1 μg/m³ increase at any time during the gestation AOR=1.15(95% CI, 1.07-1.24) for > 22.1 μg/m³ increase at any time during the gestation AOR=1.09(95% CI, 1.01-1.17) for 12.5-18.2 μg/m³ increase in the first month AOR=1.14(95% CI, 1.06-1.22) for 18.2-23.0 μg/m³ increase in the first month AOR=1.21(95% CI, 1.12-1.30) for > 23.0 μg/m³ increase in the first month AOR=1.11(95% CI, 1.04-1.19) for 10.2-15.6 μg/m³ increase in the last 2 weeks AOR=1.18(95% CI, 1.10-1.19) for 15.6-23.3 μg/m³ increase in the last 2 weeks AOR=1.17(95% CI, 1.09-1.27) for > 23.3 μg/m³ increase in the last 2 weeks	Huynh et al. 2006
	OR=1.05(95% CI, 1.02-1.09) for IQR increase	Bell et al. 2007
	NE	Madsen et al. 2010
	AOR=1.05(95% CI, 1.02-1.08) for 10 μg/m³ increase at 5 km monitor distance AOR=1.04(95% CI, 1.02-1.07) for 10 μg/m³ increase at 10 km monitor distance	Morello-Frosch et al. 2010
PM_{10}	NE	Maisonet et al. 2001
	NE	Chen et al., 2002
	OR=1.06(95% CI, 1.01-1.10) for IQR increase in all trimesters OR=1.03(95% CI, 1.00-1.07) for IQR increase in the first trimester OR=1.04(95% CI, 1.00-1.08) for IQR increase in the second trimester	Lee et al. 2003
	AOR=1.25(95% CI, 1.03-1.53) for highest quartile of exposure increase in the second trimester	Gouveia et al. 2004
	NE	Lin et al. 2004
	NE	Salam et al. 2005
	NE	Dugandzic et al. 2006
	NE	Bell et al. 2007
	AOR=1.24(95% CI, 1.02-1.52) for increments (difference between the maximum and minimum concentrations) increase in Pusan AOR=1.19(95% CI, 1.04-1.37) increase in Daegu	Seo et al. 2010
	NE	Madsen et al. 2010

	NE	Morello-Frosch et al. 2010
	AOR=1.13(95% CI, 1.02-1.25) for IQR (7μg/m^3) increase in the first trimester AOR=1.10(95% CI, 1.00-1.22) for IQR (7μg/m^3) increase in the second trimester	Xu et al. 2011
TSP	AOR=1.10(95% CI, 1.05-1.14) for 100 μg/m^3 increase	Wang et al. 1997
	NE	Bobak and Leon 1999
	NE	Bobak. 2000
	AOR=2.88(95% CI, 1.16-7.13) for > 56.75 μ/m^3 increase for very LBW infant outcome	Rogers et al. 2000
	ARR=1.04(95% CI, 1.00-1.08) for IQR increase in the first trimester	Ha et al. 2001

Air pollution and LBW
*NE: No effect, **AOR: Adjusted odds ratio

Table 1. Air pollution and low birth weight

2.2 Air pollution and preterm delivery

An association of higher exposure to SO_2 and PTB was reported in seven (Bobak, 2000; Huynh et al., 2006; Leem et al, 2006; Liu et al., 2003; Mohorovic, 2004; Sagiv et al., 2005; Xu et al., 1995) of the eight studies.

The association of NO_2 exposure and PTB was explored in 10 studies (Bobak, 2000; Gehring et al., 2011; Jalaludin et al., 2007; Leem et al, 2006; Liu et al., 2003; Llop et al. 2010; Maroziene and Grazuleviciene, 2002; Ritz et al.,2000, 2007, 2011) (Table 2). Bobak (2000) (first and third trimester), Maroziene and Grazuleviciene(2002) (first trimester), Leem et al (2006), Bobak (2000) (first and third trimester), Llop et al. (2010) (second and third trimester and entire pregnancy) reported an increased risk of PTB; however, others reported no association.

The association between CO exposure and preterm birth was explored in 6 studies (Ritz et al., 2000; Huynh et al., 2006; Leem et al., 2006; Liu et al., 2003; Wilhelm and Ritz, 2005; Ritz et al., 2007). Liu et al. (2003) (last month of the pregnancy), Wilhelm and Ritz (2005) (first trimester), leem et al (2006) (first and third trimester) and Ritz et al. (2007) (first trimester) reported a higher risk of PTB with higher concentration of CO. 4 of 6 studies reported a higher risk of PTB with CO exposure around 1ppm. In particular, Leem et al's study (2006) showed that the relationships between PTB and exposures to CO was dose dependent ($p<0.001$). But Ritz et al.(2000) and Huynh et al. (2006) did not report the association between CO exposure and preterm birth There was a significant dose-dependent association between gestational age and sulfur dioxide and total suspended particulate concentrations (Xu et al, 1995).

In 3 studies (Ritz et al., 2000; Liu et al., 2003; Ritz et al., 2007, the association between exposure to ozone and PTB was investigated (Table 2). None of these studies reported a statistically significant increase in PTB with higher exposure to ozone.

The effects of $PM_{2.5}$ on PTB were evaluated in three studies (Huynh et al., 2006; Ritz et al., 2007, 2011) *(Table 2)*. Huynh et al. (2006) and Ritz et al. (2007, 2011) reported an association of high levels of $PM_{2.5}$ with PTB when the exposure was measured at any time during the gestation.

In 4 studies (Leem et al., 2006; Ritz et al., 2000; Sagiv et al, 2005; Wilhelm and Ritz,2005) , the effects of PM_{10} on PTB were assessed (Table 2). Leem et al. (2006) and Ritz et al. (2000) reported on the association between PTB and high levels in PM_{10} exposure during the first

trimester. But Wilhelm and Ritz (2005) and Sagiv et al(2005) did not report any association between exposure to PM_{10} and PTB.
Three studies (Bobak, 2000; Bobak and Leon, 1999; Xu et al., 1995) reported on total suspended particles and an association with PTB (Table 2), all of which reported an increased risk of PTB births with higher concentrations.
Some studies reported the association between living near petrochemical industrial complexes or living with 500m of freeway, or living 200m of main roads and PTB. Llop et al. reported the association between benzene exposure > 2.7 μg/m^3 and PTB.

Pollutants or exposure variable	Results	Reference
SO_2	AOR** = 1.21 (95% CI, 1.01–1.45) for 100 μg/m^3 increase	Xu et al. 1995
	NE*	Langren, 1996
	AOR = 1.27 (95% CI, 1.16–1.39) for 50 μg/m^3 increase in the 1st trimester	Bobak 2000
	AOR=1.09(95% CI, 1.01–1.19) for 5.0 ppb increase	Liu et al 2003
	Significantly shorter gestation for SO_2 exposure during the initial two months of pregnancy	Mohorovic 2004
	AOR=1.15(95% CI, 1.00, 1.32) for 15 ppb increase during 6 weeks before birth	Sagiv et al., 2005
	AOR=2.31(95% CI, 1.29-4.15) for 1 hr maximum in the first trimester	Huynh et al. 2006
	AOR=1.21 (95% CI: 1.04-1.42) in the highest quartiles in the 1st trimester	Leem et al., 2006
NO_2	AOR= 1.10(95%CI,1.00-1.21) For 50 μg/m^3 increase above mean level in the 1st trimester AOR= 1.08(95%CI,0.98-1.19) For 50 μg/m^3 increase above mean level in the 1st trimester AOR= 1.11(95%CI,1.00-1.23) For 50 μg/m^3 increase above mean level in the 1st Trimester	Bobak, 2000
	NE	Ritz et al. 2000
	AOR= 1.67(1.28-2.18) for 10 μg/m^3 increase in exposure in the first trimester	Maroziene and Grazuleviciene, 2002
	NE	Liu et al. 2003
	AOR=1.24 (95% CI: 1.09-1.41) for 0.77 1.01 ppm in the 1st trimester AOR=1.21 (95% CI: 1.07-1.37) in the highest quartiles in the 3rd trimester	Leem et al., 2006
	Decreasing PTB risk	Jalaludin et al. 2007
	NE	Ritz et al., 2007
	AOR=1.16 (95% CI: 1.07-1.26) per inter-quartile range	Ritz et al. 2011
	AOR=1.29(95% CI= 1.13- 1.46) NO_2 > 46.2 μg/m^3	Llop et al. 2010

	NE	Gehring et al., 2011
CO	NE	Ritz et al. 2000
	AOR=1.08(95% CI: 1.01-1.15) for a 1.0 ppm increase	Liu et al 2003
	AOR= 1.27 (95% CI: 1.07–1.50) for 1ppm increase in the 1st trimester	Wilhelm and Ritz 2005
	AOR= 1.25 (95%: 0.81–1.91)	Huynh et al. 2006
	AOR=1.26 (95% CI: 1.11-1.44) for 0.77 1.01 ppm in the 1st trimester AOR=1.16 (95% CI: 1.01-1.34) for 0.79 1.11 ppm in the 3rd trimester	Leem et al., 2006
	AOR= 1.25(95% CI: 1.12- 1.38) for CO more than 1.25 ppm in the 1st trimester	Ritz et al. 2007
$PM_{2.5}$	AOR = 1.15(95% CI: 1.07- 1.24).	Huynh et al. 2006
	AOR= 1.10(95% CI: 1.01- 1.20) for $PM_{2.5}$ more than 21.36 µg/m^3 in the 1st Trimester	Ritz et al. 2007
	AOR= 1.08 (1.02-1.15) per inter-quartile range	Ritz et al. 2011
PM_{10}	RR = 1.16 (95% 1.06–1.26) for 50 µg/m^3 increase in the 1st trimester	Ritz et al. 2000
	AOR=1.00(95% CI, 0.90—1.12) for 10 µg/m^3 increase in the 1st trimester	Wilhelm and Ritz 2005
	NE	Sagiv et al., 2005
	AOR=1.27 (95% CI: 1.04-1.56)) in the highest quartiles in the 1st trimester	Leem et al.,2006
TSP	AOR = 1.10 (95% CI, 1.01–1.20) for 100 µg/m^3 increase	Xu et al. 1995
	AOR= 1.11(95% CI, 1.02-1.22) for each 50 µg/m^3 increase in the 1st trimester AOR= 1.06(95% CI, 0.96-1.16) for each 50 µg/m^3 increase in the 2nd trimester AOR= 1.14(95% CI, 1.03-1.26) for each 50 µg/m^3 increase in the 3rd trimester	Bobak and Leon 1999
	AOR = 1.18 (95% CI, 1.05–1.31) for 50 µg/m^3 increase in the 1st trimester	Bobak 2000
O_3	NE	Ritz et al. 2000
	NE	Liu et al 2003
	NE	Ritz et al. 2007
SO_2 + NO_2 + PM_{10}	AOR = 1.41 (91% CI, 1.08–1.82) comparing petrochemical and control municipalities	Lin et al. 2001
SO_2 + NO_2 + PM_{10}+ CO (Air Polllution Index)	AOR=4.66(95% CI, 1.92-11.32); 95% confidence interval (95% CI), 1.92-11.32 in hispanic mothers	Woodruff et al. 2003
Benzene	AOR=1.29(95% CI= 1.13- 1.46) Benzene > 2.7 µg/m^3	Llop et al. 2010

Living near petrochemical industrial complexes	AOR=1.18 (95% CI=1.04-1.34 for mothers living near petrochemical industrial complexes	Yang et al. 2002
living near industrial districts	AOR=1.11 (95% CI, 1.02-1.21)) for mothers living near industrial districts	Tsai et al. 2003
living near oil refinery plants	AOR=1.14 (95% CI,1.01-1.28) for mothers living near oil refinery plants	Yang et al. 2004
local traffic-generated NO(x) and $PM_{(2.5)}$.	AOR= 1.42(95% CI, 1.26-1.59) For highest NO(x) and PM(2.5) quartiles	Wu et al. 2009
living within 500 m of the freeway.	AOR=1.30 (95% CI, 1.03, 1.65)	Yang et al. 2003
Living within 200 m of main roads	AOR= 1.5 (95% CI = 1.2-1.8),	Yorifuji et al. 2011

Air pollution and premature births
*NE: No effect, **AOR: Adjusted odds ratio

Table 2. Air pollution and premature births

3. Considerations to reduce bias or measurement errors

Some studies systemically reviewed published articles about air pollution and LBW or PTB (Bobak, 2005; Bonzini et al., 2010; Pope et al., 2010; Leonardi-Bee et al., 2008; Ghosh et al., 2007; Misra et al., 1999; Sram et al., 2011). Because of different exposure assessments, methods of ascertainment, measurement times and collinearity between pollutants, the results about the association showed heterogeneity and/or an absence of association. We summarized our systemic review of these research papers in tables 1 and 2. Exposure to sulphur dioxide was associated with PTB, and exposure to $PM_{2.5}$ was associated with LBW and PTB. The evidence for NO_2, ozone and carbon monoxide was inconclusive.

To investigate the association between air pollutants and adverse birth outcomes is challenging. The following are major issues in clarifying such associations.

3.1 Exposure assessment: reduction of misclassification

Epidemiological studies for APO often had limited spatial and temporal information on pollution sources and concentrations. Misclassification of exposure is often a source of bias in such environmental epidemiologic studies.

Various exposure assessment methods have been applied in different studies. Exposures were assigned to individual subjects based on residential address at delivery using the nearest ambient monitoring station data [CO, NO_2, NO, nitrogen oxides (NOx), O_3, and $PM_{2.5}$ or less than 10 μm in aerodynamic diameter (PM_{10})], both unadjusted and temporally adjusted land-use regression (LUR) model estimates, line-source air dispersion model

estimates, and a simple traffic-density measure. Proximity to main roads and photochemical industrialized zones was also applied by using geographic information system (GIS).
Reliable measurements of daily SO_2, NO_2, CO, and PM_{10} concentrations were available from several air monitoring stations by using various extrapolation method, such as kriging to predict the spatial distribution of the air pollutants (Pikhart et al. 2001; Mulholland et al., 1998; Jerrett et al. 2005b). The kriging method, unlike proximity models (Jerrett et al. 2005a), uses real pollution measurements in the computation of exposure estimates. In case of many monitoring stations, kriging methods are often preferred to other interpolation methods because they are fairly accurate in a variety of situations and avoid the artifacts that often result from the use of IDW, spline, or global/local polynomials (Jerret et al. 2005a; Ritz et al. 2000; Waller and Gotway, 2004).
Even though exposure models, such as kriging method, attempt to decrease misclassification of individual exposures by enhancing exposure assessment through spatially- and temporally-explicit exposure models, the potential remains for misclassification of exposure due to the use of surrogate ambient air pollution data. The only real way to avoid such potential misclassifications is to conduct personal exposure assessments which are often not feasible. LUR models often produced odds ratios somewhat larger in size than temporally adjusted models (Wu J et al. 2011).
Mobility patterns could introduce possible confounding when examining small-scale variations in exposure by using addresses. This could be of importance in future studies (Madsen C et al., 2010).

3.2 Biological mechanism

Interpretation of epidemiological studies reporting an association between air pollution and birthweight needs caution. Although a range of social and behavioural determinants of birthweight or preterm birth have been identified, the biological mechanisms leading to prematurity are not well understood (Berkowitz and Papiernik, 1993) and it is not clear which mechanisms could provide the link between air pollution and birthweight. The biological mechanisms whereby air pollution might influence birth weight remain unexplained, although several theories have been proposed. The pathways could be similar to those of maternal smoking, which can increase PTB risk through premature rupture of membranes and placental abruption and lower birth weight. Air pollution could affect fetal health either through direct effects on the fetus by exposure through the placenta or from effects on the mother's health and multiple mechanisms may occur simultaneously (Glinianaia et al., 2004).
One hypothesized pathway is that placental inflammation may play an important role in the physiological pathway between air pollution exposure and LBW (Lee BE et al., 2003). Although most published reports focus on genitourinary infections, maternal illness due to respiratory infection in pregnancy may also be involved. It is possible that air pollution during pregnancy leads to placental inflammation, which impairs placental function (Dexter et al., 2000). Salafia et al. (1995) reported that chronic inflammation brought about growth restriction, independently of placental vasculopathy. PM_{10} and SO_2 exposures from first through second trimesters appeared to have the largest effect on LBW. In terms of the biological mechanism on LBW, it is reasonable to consider PM_{10} and SO_2 together rather than separately because they represent fine particles that are believed to be a risk pollutant (Ha et al., 2001). In addition, these pollutants were correlated strongly with each other and exerted an effect on LBW within similar periods.

Coarse PM (PM ≤ 10μM) is emitted from residential heating and power plants, whereas fine PM (PM ≤ 2.5 μM) is emitted from cars, utilities or wood burning. Both types of PM are comprised of primary and secondary particles: primary particles are emitted directly from a source, such as construction work, and secondary particles are formed after the reaction of primary particles in the atmosphere with chemical pollutants such as SO_2 or NO_2. When PM enters the lungs it can be absorbed into the blood and hence dispersed into distant organs. Due to their relatively small size, PM escapes phagocytosis (Ritz et al., 2007). Particle exposure in vitro and in exposed animals causes oxidative stress (Kadiiska et al., 1997) and can increase the permeability of lung epithelium, allowing particles access to the endothelial cells and the blood (Donaldson et al., 2001). PM_{10} and gaseous pollutants such as SO_2 and NO_2 lead to pulmonary inflammation with a systemic release of cytokines (Walters et al., 2001; Nemmar et al., 2002) and increased blood viscosity (Peters et al., 1997; Prescott et al., 2000).
Air pollution may affect DNA or its transcription. DNA adducts are more common in areas with higher levels of pollution. Placental DNA adducts were more common among mothers exposed to higher levels of outdoor air pollution(Bobak, 2000). When toxic organic matter such as polycyclic aromatic hydrocarbons (PAH) is adsorbed onto the surface of PM, associated oxidative stress (Leem et al., 2005) and DNA adducts are formed (Perera et al., 1999). High levels of DNA adducts were associated with reduced gestational length (Liu et al., 2003; Perera et al., 1998; Perera et al., 1999), and a correlation has been observed between the adduct levels in the mother's and the newborn's blood (Topinka et al., 2009). Newborns with elevated PAH-DNA adducts (which are used as a proxy to measure individual biologically effective dose to PAH) were found to have significantly reduced birth weight and head circumference suggesting that transplacental exposures to PAHs in ambient air may negatively impact on fetal development. High levels of PAH can interfere with nourishment of the fetus by increasing blood viscosity, and reducing the flow of blood to the placenta and uterus (Liu et al., 2003; Shah PS et al.,2011; Ritz et al., 2000). The effects of air pollution on DNA adducts levels seem similar (although weaker) to the effects of cigarette smoking. There may also be a parallel with maternal smoking, an accepted risk factor for LBW, for which the biologic mechanisms are not well understood. Although the fetal exposures to air pollution are probably lower than to tobacco smoke, the biologic mechanisms (rheologic factors, DNA damage) may be partially similar(Bobak, 2000)
Another potential mechanisms could be related to hematologic factors. Rheologic variables, including blood viscosity, influence the blood perfusion of the placenta. It has been shown that inflammation in the lung caused by air pollutants increases the coagulability of the blood. Production of free radicals induced by pollutants might cause an inflammatory response, contributing to enhanced blood coagulation. Hematologic effects of air pollutants might occur from an initial inflammatory response resulting in increased blood coagulation, and subsequent decreased oxygen supply to the placenta. Human volunteers exposed to diesel particles at 300 $\mu g/m^3$ for an hour had increases in peripheral neutrophils and platelets as well as upregulation of endothelial adhesion molecules. Decreased oxygen supply from blood viscosity changes by increasing coagulability may cause chronic hypoxic injury to fetus. This theory is supported by evidence of the role of elevated blood viscosity for impaired efficiency of maternal blood flow(Ha et al., 2001). Increased blood viscosity is associated with decreased oxygen diffusion (Zondervan et al., 1988) and may interfere with the supply of oxygen and nutrients to the fetus. In addition, some toxicants from air pollutants could cross the placenta with direct effects on fetal development (Dejmek et al., 1999).

Alternatively, placental insufficiency may be an important pathway. Basic concepts of the pathophysiology of IUGR are based on different levels of maternal supply, fetoplacental competition, and fetal adaptations. In recent studies knowledge about placental development and function has been increased. Abnormalities in placental development may occur during its formation, and cellular and molecular functions may be changed leading to inadequate implantation and growth. Abnormalities in placental transport may also develop later on, because of problems in the uteroplacental circulation, exchange at the intervillous space, and umbilical and fetal circulation. All these factors lead to problems with fetal adaptation mechanisms, most importantly decreased fetal growth rate and fetal activities. Biologic mechanisms that have been suggested to support the hypothesis of an effect associated with early pregnancy exposures are related to the etiology of IUGR. Although likely multifactorial, one suggested mechanism for IUGR is abnormal placental development in early pregnancy. Placental insufficiency reduces the oxygen and nourishment supplies to the fetus and leads to growth retardation.

Exposure to air pollution in early pregnancy could cause insufficient trophoblast formation, and lead to insufficient placental vascularization (Duvekot et al., 1995). Chronic reductions of uteroplacental circulation due to the effects of air pollution could result in fetal hypoxia and IUGR (Werler et al., 1985).

CO is well known as a reproductive toxicant that can interfere with oxygen delivery to the fetus. CO shifts the oxyhemoglobin dissociation equilibrium and displaces oxygen from hemoglobin for a given partial pressure of oxygen. CO can also cause oxidative injury due to its effects on the endothelium (Hardy and Thom, 1994). CO has also been shown to cross the placental barrier (Sangalli et al., 2003) and the fetus is particularly vulnerable to CO poisoning because of 10–15% higher accumulation in fetal blood than maternal levels. Its elimination is slower in fetal blood than in maternal circulation. Fetal hemoglobin has greater affinity for binding CO than does adult hemoglobin (Longo, 1977). O_2 delivery to fetal tissues is further compromised. The resultant tissue hypoxia has the potential to reduce fetal growth (Bosley et al., 1981; Gabrielli et al., 1995; Ritz and Yu, 1999; Salam et al., 2005). Another possible toxic mechanism of CO is that it can also affect leukocytes, platelets, and the endothelium, inducing a cascade of effects resulting in oxidative injury that contributes to the toxicity of other air pollutants (Ha et al., 2001).

Gaseous pollutants such as SO_2 and NO_2 lead to pulmonary inflammation with a systemic release of cytokines (Walters et al., 2001; Nemmar et al., 2002) and increase blood viscosity (Peters et al., 1997; Prescott et al., 2000). Prenatal exposure to SO_2 can lead to developmental and functional toxicities (Singh, 1989). NO_2 suppresses antioxidant defense systems of the human body (Tabacova et al., 1998). Exposure of experimental animal models to NO_2 during pregnancy induces lipid peroxidation in the placenta and disturbs postnatal development (Tabacova et al., 1985). Exposure to any gas pollutants leads to inflammatory reactions in the lung, leading to systemic release of cytokines that may trigger PTB (Walters et al., 2001). NO_2 may also have direct toxic effects on the fetus (Maroziene and Grazuleviciene, 2002). Particle and NO_2 were correlated strongly with each other and exerted an effect on LBW within similar periods. But considering many published data, the evidence for the association between NO_2 and LBW is inconclusive. Exposure to ozone may have negative effects on birth weight (BW) and neurodevelopment (Dell'Omo et al., 1995), although the mechanism through which ozone can affect pregnancy outcomes is unclear.

Several hypotheses have been postulated to explain the mechanism of triggering PTB. One hypothesis suggests causality between uterine inflammation and PTB. The direct evidence that infection provokes preterm labor was first shown in an animal study. When Group B streptococci were injected into the amnionic fluid in preterm rhesus monkeys, amnionic fluid cytokine concentrations increased, followed by production of the prostaglandins E_2 and $F_{2\alpha}$ and, finally, uterine contractions (Gravett et al., 1994). Similarly, in humans, preterm labor due to infection is thought to be initiated by cytokines, including interleukin-1 (IL-1), tumor necrosis factor, and interleukin-6, produced by macrophages (Cram et al., 2002; Narahara and Johnston 1993: Mitreski and Radeka 2002).

Additionally, entry of PM into the body by this method may lead to oxidative inflammation in lungs and other organs, including the placenta, thereby increasing the susceptibility of the mother to begin preterm labor (Liu et al., 2003).

Because IL-1β is not present in the membranes of term-laboring patients, it may be the unique mediator by which intrauterine infection induces preterm labor (Cunningham and William, 1997). Antenatal infection can trigger intrauterine inflammation which then promotes preterm labor. In addition, periodontal disease may be an independent risk factor for preterm labor: postulated mechanisms include translocation of periodontal pathogens to the fetoplacental unit and action of a periodontal reservoir of lipopolysaccharides or inflammatory mediators (McGaw, 2002). Our inability to determine the periodontal status of the mother is a potential confounding factor. Cyclooxygenase-2 inhibitor, developed as an anti-inflammatory drug, also has tocolytic effects (Sakai et al., 2001). A similar inflammatory mechanism has been suggested for the effect of smoking on IUGR, PTB, and perinatal mortality (Klesges et al., 2001). There are reports of increased blood viscosity and plasma fibrinogen during air pollution (Peters et al., 1997). It has been speculated that chronic exposure to high pollution levels may influence placental function (Petruzzelli et al., 1998). The placental dysfunction may lead to IUGR. The effects of air pollution on pregnancy outcomes may differ according to the timing of exposure, with early exposures likely to be important for pregnancy end points such as spontaneous abortion, IUGR and birth defects (Antipenko and Kogut 1993; Dejmek et al., 1999; Dejmek et al., 2000; Hansteen et al., 1987). Intrauterine infection during pregnancy could also lead to brain damage of the developing fetus (Huleihel et al., 2004).

Recent studies suggest that antenatal infection and inflammation can increase the preterm infant's susceptibility to develop chronic lung disease. It may be that exposure of the fetal lung to high concentrations of pro-inflammatory cytokines is the cause of this increased susceptibility (Miralles et al., 2002). Photochemically produced gaseous products influence the toxic responses of cells, such as the production of cytokine, in the absence of particles (Sexton et al., 2004). PM_{10} is responsible for the production and release of inflammatory cytokines by the respiratory tract epithelium, as well as for activation of the transcription factor NfkappaB (Baeza-Sqiban et al., 1999; Bonvallot et al., 2001). Although fetal exposures to air pollution are probably much lower than exposure to the constituents of cigarette smoke, the biological mechanism of PTB could be through increased prostaglandin levels that are triggered by inflammatory mediators during exposure periods.

The pathophysiology of carbon monoxide may be more complex, involving hypoxic stress on the basis of interference with oxygen transport to the cells and possibly impairment of electron transport. Carbon monoxide can also affect leukocytes, platelets and the endothelium, inducing a cascade of effects resulting in oxidative injury (Hardy and Thom 1994). Carbon monoxide may interfere with metabolic and transport function of the placenta and, after crossing the placental barrier, concentrate more in the fetus than in the

mother (Hardy and Thom, 1994). These placenta insufficiency may be associated with preterm birth.
The causality between air pollution and risk of IUGR, LBW, short birth length, and small head circumference has been suggested through molecular epidemiologic studies where levels of DNA adducts are positively correlated with these outcomes (Sram et al., 2005). With the same biologic mechanism of the DNA damage, high levels of DNA adducts may be a cause of PTB.

3.3 Window periods

The possible biological mechanisms involved in the reduction of birth weight associated with maternal exposure to air pollution vary according to the timing of this exposure. The implantation of the fetus and the formation of the placenta occur during the first trimester while weight gain occurs predominantly during the third trimester. Therefore, exposure during both periods presents the possibility of interference with the final birth weight. In the first trimester, genetic mutations are considered to be the most important element in placental abnormalities, and in the second and third trimesters extremely complex vascular alterations are considered the main cause of placental abnormalities and consequent IUGR. Pollutants are recognized as being able to have an effect on both dimensions (Gouveia et al., 2004).
The possible biological mechanisms of air pollution on birth weight might vary according the time of pregnancy, such as the implantation of the fetus and the formation of placenta during the first trimester, as well as important weight gain during the third trimester. Placental abnormalities, DNA damage, disruption of the endocrine system and change of blood coagulability are those potential biological mechanisms, which have been reported (Dejmek et al., 2000; Maisonet et al., 2004; Perera et al., 1999, 2002; Whyatt et al., 1998).
The finding of a significant effect of PM exposure on LBW during the first trimester is consistent. Its effect is striking at window periods during the first trimester.
The highest ambient air pollution concentrations during the first trimester were significantly associated with elevated relative risks of PTB. These results are generally consistent with the findings from China, South Korea, the United States, Canada, and the Czech Republic (Bobak 2000; Liu et al., 2003; Mohorovic 2004; Ritz et al., 2000; Tsai et al., 2003; Woodruff et al., 2003; Xu et al., 1995; Yang et al., 2002a; Yang et al., 2002b, Yang et al., 2003; Yang et al., 2004; Leem et al., 2006). These studies reported significant associations between air pollution and PTB during early pregnancy (i.e., first or second month, first trimester) (Mohorovic 2004; Ritz et al., 2000), late pregnancy (i.e., last month, last trimester, 7 days or 6 weeks before birth) (Liu et al., 2003; Xu et al., 1995), or during both early and late pregnancies (Bobak 2000).

3.4 Disease burden from air pollution and smoking

Population-attributable risk (PAR) is used to determine by what percentage the incidence of a disease in a population would be reduced if exposure were eliminated. PAR measures the potential impact of control measures on a population, and is relevant to decisions on public health. PAR is a very important concept in guiding policy decisions regarding the preventive approaches to APO, such as LBW, and PTB (Seo et al., 2010).
PAR measures the potential impact of control measures on a population, and is relevant to decisions on public health. PAR is a very important concept in guiding policy decisions regarding the preventive approaches to many diseases, such as cancer, hypertension, diabetes mellitus, and stroke.

Some studies have reported the PAR levels for LBW attributable to environmental factors, such as smoking (Levi F, 1999; Matsubara et al., 2000; Suzuki et al., 2008) and indoor pollution (Boy et al., 2002). Cigarette smoking during pregnancy is a strong dose-dependent risk factor for LBW (Chiolero et al. 2005; Windham et al. 2000). Women exposed to prenatal secondhand smoke were more at risk for preterm birth (odds ratio [OR]=2.3; 95% Confidence Interval [CI] [.96, 5.96]), and their infants were more likely to have immediate newborn complications (OR=2.4; 95% CI [1.09, 5.33]) than non-exposed women. Infants of passive smoking mothers were at increased risk for respiratory distress syndrome (OR=4.9; 95% CI [1.45, 10.5]) and admission to a Neonatal Intensive Care Unit ((OR=6,5; 95% CI[1.29, 9.7]) when compared to infants of smoking mothers (OR=3.9; 95% CI [1.61, 14.9]; OR=3.5; 95% CI [2.09, 20.4], respectively). Passive smokers and/or women with hair nicotine levels greater than .35 ng/ml were more likely to deliver earlier (1 week), give birth to infants weighing less (decrease of 200-300 g), and deliver shorter infants (decrease of 1.1-1.7 cm) (Ashford et al., 2010). Environmental tobacco smoke (ETS) and traffic-related air pollution share a few characteristics. They are widespread exposures in both developed and developing countries, and they have several chemical components in common. Mothers who smoke during pregnancy are twice as likely to give birth to a LBW newborn. In high-income countries, the mean PAR for tobacco smoking in both genders combined is estimated to be 25-30% of the total cancer mortality. Some studies reported PAR for LBW attributable to environmental factors, such as smoking (Chiolero et al., 2005; Windham et al., 2000; Matsubara et al., 2000; Suzuki et al., 2008) and indoor pollution (Boy et al., 2002). Chiolero et al., (2005) reported that maternal smoking during pregnancy was closely associated with LBW, small-for-gestational age (SGA), and pre-term birth. Comparing smokers to non-smokers, the adjusted odds ratios (AOR) were 2.7 (2.1-3.5) for LBW, 2.1 (1.7-2.5) for SGA, and 1.4 (1.1-1.9) for preterm birth. Past smoking was not associated with the outcomes. In that study, maternal smoking during pregnancy accounted for 22% (15-29%) of all LBW babies in the population, 14% (10-18%) of SGA babies, and 7% (1-12%) of preterm babies. Ojima et al. (2004) reported on the population-attributable proportion of active and passive smoking for LBW. These results showed the population-attributable proportion of smoking among mothers without preeclampsia during pregnancy was 7.0% for active smoking and 15.6% for passive smoking. Leonardi-Bee et al. (2008) reported that exposure of non-smoking pregnant women to ETS reduces mean birth weight by 33 g or more, and increases the risk of birth weight below 2500 g by 22%, but has no clear effect on gestation or SGA risk.

Misra and Nguyen (1999) suggested that there is consistent evidence to relate maternal ETS exposure to an increased APO risk and that this association may be generalized to the work environment. In studies with positive findings, infants exposed to ETS antenatally were 1.5-4 times more likely to be born with LBW, but few studies examined LBW. Most studies looked at measures of IUGR. ETS was associated with reductions in birth weight (adjusted for gestational age) ranging from 25 to 90 g. Infants born to women exposed to ETS were generally 2-4 times more likely to be born SGA. ETS exposure in the workplace can and should be minimized to protect pregnant women from its adverse effects.

Such research is urgently needed so as to calculate the etiologic fractions of the PAR that contribute directly to PTB. This will enable preventive strategies to be established to protect fetuses against air pollutants.

Most studies have reported an association between exposure to air pollution and PTB, with risk ratios from 1.03-1.36. Especially, PM_{10} air pollution was found to be significantly associated with PTB.

Some studies reported PAR for PTB attributable to environmental factors, such as smoking, outdoor air pollution, and indoor pollution. Maternal smoking during pregnancy was closely associated with LBW, SGA, and PTB. In a study of seven Korean cities, air pollution accounted for 7~18% of all LBW babies (Seo et al., 2010).

PAR to PM_{10} pollution for LBW was comparable to the figure derived from maternal smoking for PTB. Because air pollution is an important risk factor for PTB, a large proportion of PTB could be prevented if air pollution is reduced.

PAR depends on the strength of the relative risk, but also on the prevalence of the risk factor. Causes for APO may include metals, inhalational of anesthetics, organic solvents, air pollution, radiation, stress, and physical stress. Common risk factors carry larger PARs than do rare risk factors. The PAR attributable to PM_{10} pollution for LBW was similar to that regarding smoking for LBW because every pregnant woman was exposed to air pollution. Though those who smoke are in the minority, the relative risk due to smoking is greater than air pollution.

Air pollution is an important risk factor for LBW and PTB because every pregnant woman is exposed to air pollution. Thus, a large proportion of LBW and PTB pregnancies could be prevented if air pollution were reduced.

3.5 Gender as effect modifier

Does the effect of air pollution on pregnancy outcomes differ by gender? Gender is known to influence pregnancy outcomes. Recent studies have reported an association between air pollution exposure and APO, but gender differences have not been considered. In order to assess the current evidence of the interactive effects between gender and air pollution on pregnancy outcomes, Ghosh R et al.(2007) undertook a systematic literature review. In total 11 studies were included. Of the 11 studies, four evaluated LBW, one each evaluated very LBW and fetal growth and six examined PTB. Females were at higher LBW risk: AOR ranged from 1.07 to 1.62. Males were at higher risk for PTB: AORs ranged from 1.11 to 1.20. In addition, there was some evidence to suggest that the effect of air pollution on LBW is gender dependent; however, the evidence was available only from four studies.

3.6 Socioeconomic status(SES): health disparity

People with low socioeconomic status (SES) are more vulnerable to air pollution than others. They are exposed to infection, nutritionally deficient, and often lived in more polluted area.

Infection in pregnancy is a predictor of premature births (Gibbs RS et al., 1992), and it could be speculated that repeated infections, possibly related to pollution, might play a part. Increased blood viscosity, found during air pollution episodes (Peters A et al., 1997) may be related to impaired placental function (Zondervan HA et al., 1987). Increased concentrations of DNA adducts have been found in the blood (Perera FP et al., 1992; Petruzzelli S et al., 1998) and placentas (Topinka J et al., 1997) of subjects living in polluted areas, and were also found to be related to birthweight (Perera FP et al., 1998). Maternal nutrition status can be acting as a effect modifier between exposures to airborne particulate matter and adverse perinatal outcomes (Kannan et al, 2006). Maternal pulmonary function has been linked to altered placental vascular function and growth retardation in asthmatic pregnancies (Bracken et al. 2003; Clifton et al. 2001; Schatz et al. 1990). Mothers with lowered pulmonary function are more likely to have increased risks of LBW and PTB. Other theories about these associations include a) altered cardiac function from changes in heart rate variability; b) inhalation by the mother of PAHs that then relate to placental exposure, potentially disrupting endocrine and nervous systems; c) changes in blood viscosity due to alveolar inflammation from PM, which in turn affects placental function; and d) binding of CO to hemoglobin binding sites, preventing the binding of oxygen and subsequent function (Glinianaia et al. 2004; Maisonet et al. 2004; Šrám et al. 2005).

Despite advances in medical care, preterm birth and its associated racial/ethnic disparities remain major public health issues. Environmental exposures may contribute to racial disparities in preterm birth (Burris HH et al., 2011). Interestingly, a study in South Korea recently demonstrated that SES modifies the association between air pollution and preterm birth (Yi O et al., 2010).

Hispanic, African-American, and Asian/Pacific Islander mothers experienced higher mean levels of air pollution and were more than twice as likely to live in the most polluted counties compared with white mothers after controlling for maternal risk factors, region, and educational status [Hispanic mothers: AOR = 4.66; 95% confidence interval (95% CI), 1.92-11.32; African-American mothers: AOR = 2.58; 95% CI, 1.00-6.62; Asian/Pacific Islander mothers: AOR = 2.82; 95% CI, 1.07-7.39](Woodruff et al. 2003).

PTB increased from 8.3% in counties with low income inequality to 10.0% in counties with high inequality. The Gini Index remained modestly associated with PTB after adjusting for individual level variables and mean county-level per capita income within the total population (AOR: 1.06; 95% CI 1.03-1.09) as well as within most of the racial/ethnic groups. PNM(post-natal mortality) increased from 1.15 deaths per 1000 live births in low inequality counties to 1.32 in high-inequality counties. However, after adjustment, income inequality was only associated with PNM within the non-Hispanic black population (AOR: 1.20; 95% CI 1.03-1.39). These findings may provide some support for the association between income inequality and PTB. Further research is required to elucidate the biological mechanisms of income inequality (Huynh M et al., 2005).

4. Conclusion

The association between air pollution, such PM and SO_2, and APO has been clearly shown in the literature although the mechanisms have not been elucidated. More work is required to fully elucidate the physiologic mechanisms by which air pollution may affect fetal growth and development and to determine if the mechanisms are pollutant specific.

The findings of prior studies of air pollution effects on adverse birth outcomes are difficult to synthesize due to differences in study design, although a few studies have included meta-analysis. Location-specific analyses of air pollution effects on birth weight need to be conducted using a common protocol and a standardized statistical approach to understand how differences in research methods contribute to variations in findings. Study groups such as The International Collaboration on Air Pollution and Pregnancy Outcome (ICPPO) have been formed to perform these kinds of collaboration study. Variability in PM_{10}-LBW relationships among study locations has remained, despite the use of a common statistical approach by a pilot study (Parker et al., 2011). A more detailed meta-analysis and use of more complex protocols for future analysis may uncover the reasons for heterogeneity across locations.

Many studies demonstrated air pollution levels critical to LBW and PTB in humans. These levels are very important because they may be a good indication on how to protect fetuses against adverse effects from air pollutants. Annual standards for air quality are certainly too high in some countries and do not prevent APO. Many studies showed that statistically significant effects of LBW and PTB are seen below the air quality standards for PM_{10} and SO_2 and potentially below the standards for CO and NO_2. The adverse effects on pregnancy are increased for smaller particles like $PM_{2.5}$. Several lines of evidence support the plausibility of a negative effect of CO exposure on birth weight. CO reduces oxygen-carrying capacity of maternal hemoglobin, which could adversely affect O2 delivery to fetal circulation. Low-concentration exposure to CO, even below 1ppm, increased PTB risk. The current air quality standard for CO is 9 ppm. The air quality standards for $PM_{2.5}$ should be established, and the air quality standards for CO should be lowered to 1ppm to protect fetuses' health against the hazardous toxicities of $PM_{2.5}$ and CO. Many studies may provide supportive evidence that reduction in the current air quality standards may increase the health of pregnancy outcomes.

We observed that exposure to sulphur dioxide was associated with PTB, and exposure to $PM_{2.5}$ was associated with LBW and PTB. The evidence for N_2O, ozone and carbon monoxide was inconclusive. However, the observed adverse effects were generally small. Possible important factors such as maternal activity pattern, diet, smoking and occupation, which are usually not reported on the birth certificate, might have led to exposure misclassification and confounding and could have hidden moderately increased risks. Additional well-conducted studies that include detailed information on maternal risk factors and use validated models for estimating maternal exposure are needed to establish the extent of the association between air pollution and birth outcomes.

In conclusion, several studies showed that relatively low concentrations of air pollution below current air quality standards during critical gestational periods may contribute to increased risk of LBW and PTB. Fetuses in the early and late stages of development are susceptible to air pollutants. Further studies are needed to validate the fetuses' susceptibility to air pollutants with more detailed information on personal exposures, confounders, and effect modifiers. Many investigators reported reductions in ETS exposure and the risk of LBW and very early preterm birth. Clues about potential mechanisms underlying the disparities in LBW and preterm birth can be gained from exploring differences in environmental exposures. Investigators should include environmental variables when studying birth outcomes. Such efforts should result in targeted interventions to decrease the incidence of LBW, preterm birth and its disparities.

5. References

Alderman BW, Baron AE, Savitz DA.(1987). Maternal exposure to neighborhood carbon monoxide and risk of low infant birth weight. *Public Health Rep* ;102(4):410-4. ISSN: 0033-3549

Antipenko YeN, Kogut NN. (1993). The experience of mutation rate quantitative evaluation in connection with environmental pollution (based on studies of congenital anomalies in human populations). *Mutat Res* 289(2):145 -155. ISSN: 0027-5107

Ashford KB, Hahn E, Hall L, Rayens MK, Noland M, Ferguson JE. (2010). The effects of prenatal secondhand smoke exposure on preterm birth and neonatal outcomes. *J Obstet Gynecol Neonatal Nurs.* 2010 Sep;39(5):525-35. . ISSN: 0884-2175

Baeza-Squiban A, Bonvallot V, Boland S, Marano F. (1999). Airborne particles evoke an inflammatory response in human airway epithelium. Activation of transcription factors. *Cell Biol Toxicol* 15(6):375-380. ISSN: 0742-2091

Barker DJ.(2007). The origins of the developmental origins theory. *J Intern Med.* 261(5):412-7. ISSN: 0954-6820

Bell ML, Ebisu K, Belanger K. (2007). Ambient air pollution and low birth weight in Connecticut and Massachusetts. *Environ Health Perspect* 115:1118–24. ISSN: 0091-6765

Berkowitz GS, Papiernik E.(1993). Epidemiology of preterm birth. *Epidemiol Res* ;15:414–43. ISSN: 0193-936X

Bibby E, Stewart A. (2004). The epidemiology of preterm birth. *Neuro Endocrinol Lett.* 25 Suppl 1:43-47. ISSN: 0172- 780X

Bobak M. (2000). Outdoor air pollution, low birth weight, and prematurity. *Environ Health Perspect* 108(2):173-176. ISSN: 0091-6765

Bobak M. (2005). Ambient Air Pollution and Pregnancy Outcomes: A Review of the Literature. *Environ Health Perspect* 113:375-382. ISSN: 0091-6765

Bobak M, Leon DA.(1999) Pregnancy outcomes and outdoor air pollution: an ecological study in districts of the Czech Republic 1986-8. *Occup Environ Med.* Aug;56(8):539-43. ISSN: 1351-0711

Bonvallot V, Baeza-Squiban A, Baulig A, Brulant S, Boland S, Muzeau F, et al. (2001). Organic compounds from diesel exhaust particles elicit a proinflammatory response in human airway epithelial cells and induce cytochrome p450 1A1 expression. *Am J Respir Cell Mol Biol* 25(4):515-521. ISSN: 1044-1549

Bonzini M, Carugno M, Grillo P, Mensi C, Bertazzi PA, Pesatori AC. (2010). Impact of ambient air pollution on birth outcomes: systematic review of the current evidences. *Med Lav* 101(5):341-63. ISSN: 0025-7818

Bosley ARJ, Sibert JR, Newcombe RG. (1981). Effects of maternal smoking on fetal growth and nutrition. *Arch Dis Child* 56:727-729 ISSN: 0003-9888

Boy E, Bruce N, Delgado H. (20002). Birth weight and exposure to kitchen wood smoke during pregnancy in rural Guatemala. *Environ Health Perspect* 110(1):109-114. ISSN: 0091-6765

Bracken MB, Triche EW, Belanger K, Saftlas A, Beckett WS, Leaderer BP. (2003). Asthma symptoms, severity, and drug therapy: a prospective study of effects on 2205 pregnancies. *Obstet Gynecol.* 102(4):739-52. ISSN:0029-7844

Burris HH, Collins JW Jr, Wright RO. (2011). Racial/ethnic disparities in preterm birth: clues from environmental exposures. *Curr Opin Pediatr. 23(2):227-32. ISSN:* 1040-8703

Chen L, Yang W, Jennison BL, Goodrich A, Omaye ST.(2002). Air pollution and birth weight in northern Nevada, 1991-1999. *Inhal Toxicol* 14(2):141-157. ISSN: 0895-8378

Chiolero A, Bovet P, Paccaud F. (2005). Association between maternal smoking and low birth weight in Switzerland: the EDEN study. *Swiss Med Wkly.* 135(35-36):525-30. ISSN: 1424-7860

Clifton VL, Giles WB, Smith R, Bisits AT, Hempenstall PA, Kessell CG, Gibson PG. (2001). Alterations of placental vascular function in asthmatic pregnancies. *Am J Respir Crit Care Med.* 164(4):546-53. ISSN: 1073-449X

Cram LF, Zapata MI, Toy EC, Baker B 3rd. (2002). Genitourinary infections and their association with preterm labor. *Am Fam Physician* 65(2):241-248. ISSN: 0002-838X

Cunningham FG, Williams JW. (1997). Parturition. In: Williams Obstetrics. 20th ed. Stamford, Conn: Appleton & Lange; 306-313 and 797-821.

Dejmek J, Selevan SG, Benes I, Solansky I, Sram RJ. (1999). Fetal growth and maternal exposure to particulate matter during pregnancy. *Environ Health Perspect.* 107(6):475-480. ISSN: 0091-6765

Dejmek J, Solansky I, Benes I, Lenicek J, Sram RJ. (2000). The impact of polycyclic aromatic hydrocarbons and fine particles on pregnancy outcome. *Environ Health Perspect* 108(12):1159-1164. ISSN: 0091-6765

Dexter SC, Pinar H, Malee MP, Hogan J, Carpenter MW, Vohr BR. (2000). Outcome of very low birth weight infants with histopathologic chorioamnionitis. *Obstet Gynecol* 96(2):172-7. ISSN: 0029-7844

Dell'Omo G, Fiore M, Petruzzi S, Alleva E, Bignami G. (1995). Neurobehavioral development of CD-1 mice after combined gestational and postnatal exposure to ozone. *Arch Toxicol.* 69(9):608-16. ISSN:0340-5761

Donaldson K, Stone V, Seaton A, MacNee W. (2001). Ambient particle inhalation and the cardiovascular system: potential mechanisms. *Environ Health Perspect.*109 Suppl 4:523-7. ISSN: 0091-6765

Dugandzic R, Dodds L, Stieb D, Smith-Doiron M.(2006). The association between low level exposures to ambient air pollution and term low birth weight: a retrospective cohort study. *Environ Health* ;5:3. ISSN: 1476-069X

Duvekot JJ, Cheriex EC, Pieters FA, Peeters LL. (1995). Severely impaired fetal growth is preceded by maternal hemodynamic maladaptation in very early pregnancy. *Acta Obstet Gynecol Scand.*74(9):693-7. ISSN:0001-6349

Gabrielli A, Layon AJ. Carbon monoxide intoxication during pregnancy: a case presentation and pathophysiologic discussion, with emphasis on molecular mechanisms. *J Clin Anesth.* 1995 Feb;7(1):82-7.ISSN:0952-8180

Gehring U, van Eijsden M, Dijkema MB, van der Wal MF, Fischer P, Brunekreef B. (2011).Traffic-related air pollution and pregnancy outcomes in the Dutch ABCD birth cohort study. *Occup Environ Med* ;68(1):36-43. ISSN: 1351-0711

Gibbs RS, Romero R, Hillier SL, Eschenbach DA, Sweet RL. (1992).A review of premature birth and subclinical infection. *Am J Obstet Gynecol* ;166:1515–28. ISSN: 0002-9378

Ghosh R, Rankin J, Pless-Mulloli T, Glinianaia S. (2007). Does the effect of air pollution on pregnancy outcomes differ by gender? A systematic review. *Environ Res.* 105(3):400-8. ISSN: 0013-9351

Glinianaia SV, Rankin J, Bell R, Pless-Mulloli T, Howel D. (2004). Particulate air pollution and fetal health: a systematic review of the epidemiologic evidence.*Epidemiology.* 15(1):36-45.ISSN: 1044-3983

Gouveia N, Bremner SA, Novaes HM. (2004). Association between ambient air pollution and birth weight in Sao Paulo, Brazil. *J Epidemiol Community Health* 58:11–7.ISSN: 0143-005X

Gravett MG, Witkin SS, Haluska GJ, Edwards JL, Cook MJ, Novy MJ. (1994). An experimental model of intraamniotic infection and preterm labor in rhesus monkeys. *Am J Obstet Gynecol* 171:1660-1667. ISSN: : 0002-9378

Ha EH, Hong YC, Lee BE, Woo BH, Schwartz J, Christiani DC.(2001). Is air pollution a risk factor for low birth weight in Seoul? *Epidemiology* 12(6):643-648. ISSN:1044-3983

Hansteen IL, Heldaas SS, Langard S, Steen-Johnsen J, Christensen A, Heldaas K.(1987). Surveillance of pregnancies as a means of detecting environmental and occupational hazards. I. Spontaneous abortions, congenital malformations and cytogenetic abnormalities in a newborn population. *Hereditas* 107(2):197-203. ISSN: 0018-0661

Hardy KR, Thom SR. (1994). Pathophysiology and treatment of carbon monoxide poisoning. *J Toxicol Clin Toxicol* 32(6):613-629. ISSN: 0731-3810

Huleihel M, Golan H, Hallak M.(2004). Intrauterine infection/inflammation during pregnancy and offspring brain damages: possible mechanisms involved. *Reprod Biol Endocrinol* 2(1):17-24. ISSN: 1477-7827

Huynh M, Parker JD, Harper S, Pamuk E, Schoendorf KC. (2005). Contextual effect of income inequality on birth outcomes. *Int J Epidemiol.* 34(4):888-95. ISSN: 0300-5771

Huynh M, Woodruff TJ, Parker JD, Schoendorf KC. (2006). Relationships between air pollution and preterm birth in California. *Paediatr Perinat Epidemiol.* 20(6):454-61. ISSN: 0269-5022

Jalaludin B, Mannes T, Morgan G, Lincoln D, Sheppeard V, Corbett S.(2007). Impact of ambient air pollution on gestational age is modified by season in Sydney, Australia. *Environ Health.* 2007 Jun 7;6:16. ISSN: 1476-069X

Jerrett M, Arain A, Kanaroglou P, Beckerman B, Potoglou D, Sahsuvaroglu T, Morrison J, Giovis C. (2005a). A review And evaluation of intraurban air pollution exposure models. : *J Expo Anal Environ Epidemiol.* 15(2):185-204. ISSN:1053-4245

Jerrett M, Buzzelli M, Burnett RT, DeLuca PF.(2005b). Particulate air pollution, social confounders, and mortality in small areas of an industrial city. *Soc Sci Med* 60(12):2845-2863. ISSN: 0277-9536

Kadiiska MB, Mason RP, Dreher KL, Costa DL, Ghio AJ. (1997). In vivo evidence of free radical formation in the rat lung after exposure to an emission source air pollution particle. *Chem Res Toxicol.* 10(10):1104-8.ISSN: 0893-228X

Kannan S, Misra DP, Dvonch JT, Krishnakumar A. (2006). Exposures to airborne particulate matter and adverse perinatal outcomes: a biologically plausible mechanistic framework for exploring potential effect modification by nutrition. *Environ Health Perspect.*114(11):1636-42. ISSN: 0091-6765

Klesges LM, Johnson KC, Ward KD, Barnard M.(2001). Smoking cessation in pregnant women. *Obstet Gynecol Clin North Am* 28(2):269-282. ISSN: 0889-8545

Kwon HJ, Cho SH, Nyberg F, Pershagen G. (2001) Effects of ambient air pollution on daily mortality in a cohort of patients with congestive heart failure. *Epidemiology.* 12(4):413-9. ISSN: 1044-3983

Lee BE, Ha EH, Park HS, Kim YJ, Hong YC, Kim H, Lee JT. (2003). Exposure to air pollution during different gestational phases contributes to risks of low birth weight. *Hum Reprod.* 18(3):638-43. ISSN:0268-1161

Leem JH, Kaplan BM, Shim YK, Pohl HR, Gotway CA, Bullard SM, Rogers JF, Smith MM, Tylenda CA. (2006). Exposures to air pollutants during pregnancy and preterm delivery. *Environ Health Perspect.* 114(6):905-10. ISSN: 0091-6765

Leem JH, Kim JH, Lee KH, Hong Y, Lee KH, Kang D, Kwon HJ.(2005). Asthma attack associated with oxidative stress by exposure to ETS and PAH. *J Asthma.* 42(6):463-7. ISSN: 0277-0903

Leonardi-Bee J, Smyth A, Britton J, Coleman T. (2008).Environmental tobacco smoke and fetal health: systematic review and meta-analysis. *Arch Dis Child Fetal Neonatal Ed*.93(5):F351-61. ISSN: 1359-2998

Levi F.(1999). Cancer prevention: epidemiology and perspectives. *Eur J Cancer* 35(7):1046-1058. ISSN: 0959-8278

Lin CM, Li CY, Yang GY, Mao IF. (2004). Association between maternal exposure to elevated ambient sulfur dioxide during pregnancy and term low birth weight. *Environ Res* 96(1):41-50. ISSN: 0013- 9351

Lin MC, Yu HS, Tsai SS, Cheng BH, Hsu TY, Wu TN, Yang CY. (2001). Adverse pregnancy outcome in a petrochemical polluted area in Taiwan. *J Toxicol Environ Health A* 63(8):565-574. ISSN: 1528-7394

Liu S, Krewski D, Shi Y, Chen Y, Burnett RT. (2003). Association between gaseous ambient air pollutants and adverse pregnancy outcomes in Vancouver, Canada. *Environ Health Perspect* 111(14):1773-1778. ISSN: 0091-6765

Llop S, Ballester F, Estarlich M, Esplugues A, Rebagliato M, Iñiguez C.(2010). Preterm birth and exposure to air pollutants during pregnancy. *Environ Res* 110(8):778-85. ISSN: 0013-9351

Longo LD.(1977). The biological effects of carbon monoxide on the pregnant woman, fetus, and newborn infant. *Am J Obstet Gynecol.* 129(1):69-103. ISSN: 0002-9378

Madsen C, Gehring U, Walker SE, Brunekreef B, Stigum H, Naess O, Nafstad P. (2010). Ambient air pollution exposure, residential mobility and term birth weight in Oslo, Norway. *Environ Res.* 110(4):363-71. ISSN: 0013- 9351

Maisonet M, Bush TJ, Correa A, Jaakkola JJ. (2001). Relation between ambient air pollution and low birth weight in the Northeastern United States. *Environ Health Perspect* 109 Suppl 3:351-356. ISSN: 0091-6765

Maisonet M, Correa A, Misra D, Jaakkola JJ.(2004). A review of the literature on the effects of ambient air pollution on fetal growth. *Environ Res* 95(1):106-115. ISSN: 0013-9351

Maroziene L, Grazuleviciene R. (2002). Maternal exposure to low-level air pollution and pregnancy outcomes: a population-based study. *Environ Health* 1(1):6-12. ISSN: 1476-069X

Matsubara F, Kida M, Tamakoshi A, Wakai K, Kawamura T, Ohno Y. (2000). Maternal active and passive smoking and fetal growth: A prospective study in Nagoya, Japan. *J Epidemiol* 10(5):335-343. ISSN: 0917-5040

McGaw T.(2002). Periodontal disease and PTB of low-birth-weight infants. *J Can Dent Assoc* 68(3):165-169. ISSN: 0008-3372

Miralles RE, Hodge R, Kotecha S. (2002). Antenatal inflammation and infection in chronic lung disease of prematurity. *Child Care Health Dev* 28 Suppl 1:11-15. ISSN: 0305-1862

Misra DP, Nguyen RH.(1999). Environmental tobacco smoke and low birth weight: a hazard in the workplace? *Environ Health Perspect.* 107 Suppl 6:897-904. . ISSN: 0091-6765

Mitreski A, Radeka G.(2002). Prostacyclin and hormone levels in patients with symptoms of miscarriage and infection. *Med Pregl* 55(9-10):371-379. ISSN: 0025-8105

Mohorovic L. (2004). First two months of pregnancy-critical time for PTD and low birth weight caused by adverse effects of coal combustion toxics. *Early Hum Dev* 80(2):115-123. ISSN: 0378-3782

Morello-Frosch R, Jesdale BM, Sadd JL, Pastor M. (2010). Ambient air pollution exposure and full-term birth weight in California. *Environ Health;*9:44. ISSN: 1476-069X

Mulholland JA, Butler AJ, Wilkinson JG, Russell AG, Tolbert PE. (1998). Temporal and spatial distributions of ozone in Atlanta: regulatory and epidemiologic implications. *J Air Waste Manag Assoc.* 48(5):418-426. ISSN: 1096-2247

Narahara H, Johnston JM. (1993). Effects of endotoxins and cytokines on the secretion of platelet-activating factor acetylhydrolase by human decidual macrophages. *Am J Obstet Gynecol* 169:531-537. ISSN: 0002-9378

Nemmar A, Hoet PH, Vanquickenborne B, Dinsdale D, Thomeer M, Hoylaerts MF,Vanbilloen H, Mortelmans L, Nemery B. (2002). Passage of inhaled particles into the blood circulation in humans. *Circulation* 105(4):411-4.

Ojima T, Uehara R, Watanabe M, Tajimi M, Oki I, Nakamura Y. (2004). Population attributable fraction of smoking to low birth weight in Japan. *Pediatr Int* 46(3):264-267. ISSN: 1328-8067.

Parker J, Rich DQ, Glinianaia SV, Leem JH, Wartenberg D, Bell ML, Bonzini M, Brauer M, Darrow L, Gehring U, Gouveia N, Grillo P, Ha E, van den Hooven EH, Jalaludin B, Jesdale BM, Lepeule J, Morello-Frosch R, Morgan GG, Slama R, Pierik FH, Pesatori AC, Sathyanarayana S, Seo J, Strickland M, Tamburic L, Woodruff TJ. (2011).The International Collaboration on Air Pollution and Pregnancy Outcomes: Initial Results. *Environ Health Perspect.* 2011 Feb 9. . ISSN: 0091-6765

Perera FP, Hemminki K, Gryzbowska E, Motykiewicz G, Michalska J, Santella RM, Young TL, Dickey C, Brandt-Rauf P, De Vivo I. (1992).Molecular and genetic damage in humans from environmental pollution in Poland. *Nature* 360:256–8. ISSN: 0028-0836

Perera FP, Whyatt RM, Jedrychowski W, Rauh V, Manchester D, Santella RM,Ottman R. (1998). Recent developments in molecular epidemiology: a study of the effects of environmental polycyclic aromatic hydrocarbons on birth outcomes in Poland. *Am J Epidemiol* 147:309–14. ISSN: 0002-9262

Perera FP, Jedrychowski W, Rauh V, Whyatt RM. (1999). Molecular epidemiologic research on the effects of environmental pollutants on the fetus. *Environ Health Perspect* 107 Suppl 3:451-60. . ISSN: 0091-6765

Peters A, Doring A, Wichmann HE, Koenig W. (1997). Increased plasma viscosity during an air pollution episode: a link To mortality? *Lancet* 349(9065):1582-1587. ISSN: 0140-6736

Petruzzelli S, Celi A, Pulera N, Baliva F, Viegi G, Carrozzi L, Ciacchini G, Bottai M, Di Pede F, Paoletti P, Giuntini C. (1998). Serum antibodies to benzo(a)pyrene diol epoxide-DNA adducts in the general population: effects of air pollution, tobacco smoking, and family history of lung diseases. *Cancer Res* 15;58(18):4122-6. ISSN: 0008-5472

Pikhart H, Bobak M, Gorynski P, Wojtyniak B, Danova J, Celko MA, Kriz B, Briggs D, Elliott P. (2001). Outdoor sulphur dioxide and respiratory symptoms in Czech and Polish school children: a small-area study (SAVIAH). Small-Area Variation in Air Pollution and Health. *Int Arch Occup Environ Health.* 74(8):574-578. ISSN: 0340-0131

Pope DP, Mishra V, Thompson L, Siddiqui AR, Rehfuess EA, Weber M, Bruce NG. (2010). Risk of low birth weight and stillbirth associated with indoor air pollution from solid fuel use in developing countries. *Epidemiol Rev.* 32(1):70- 81. ISSN: 0193-936X

Prescott GJ, Lee RJ, Cohen GR, Elton RA, Lee AJ, Fowkes FG, Agius RM. (2000). Investigation of factors which might indicate susceptibility to particulate air pollution. *Occup Environ Med.* 57(1):53-7. ISSN: 1351-0711

Reagan PB, Salsberry PJ.(2005). Race and ethnic differences in determinants of preterm birth in the USA: broadening the social context. *Soc Sci Med.* 60(10):2217-2228. ISSN: 0277-9536

Ritz B, Yu F. (1999). The effect of ambient carbon monoxide on low birth weight among children born in southern California between 1989 and 1993. *Environ Health Perspect.* 107(1):17-25. ISSN: 0091-6765

Ritz B, Yu F, Chapa G, Fruin S. (2000). Effect of air pollution on preterm birth among children born in Southern California between 1989 and 1993. *Epidemiology* 11(5):502-511. ISSN: 1044-3983

Ritz B, Wilhelm M, Hoggatt KJ, Ghosh JK. (2007). Ambient air pollution and preterm birth in the environment and pregnancy outcomes study at the University of California, Los Angeles. *Am J Epidemiol.*166(9):1045-52. ISSN: 0002-9262

Rogers JF, Thompson SJ, Addy CL, McKeown RE, Cowen DJ, Decouflé P. (2000). Association of very low birth weight with exposures to environmental sulfur dioxide and total suspended particulates. *Am J Epidemiol.*15;151(6):602- 13. ISSN: 0002-9262

Sakai M, Tanebe K, Sasaki Y, Momma K, Yoneda S, Saito S.(2001). Evaluation of the tocolytic effect of a selective cyclooxygenase-2 inhibitor in a mouse model of lipopolysaccharide-induced PTD. *Mol Hum Reprod* 7(6):595-602. ISSN: 1360-9947

Sagiv SK, Mendola P, Loomis D, Herring AH, Neas LM, Savitz DA, Poole C. (2005). A time-series analysis of air pollution and preterm birth in Pennsylvania,1997-2001. *Environ Health Perspect.* 113(5):602-6. ISSN: 0091-6765

Sangalli MR, Mclean AJ, Peek MJ, Rivory LP, Le Couteur DG. (2003). Carbon monoxide disposition and permeability-surface area product in the foetal circulation of the perfused term human placenta. *Placenta.*24(1):8-11. ISSN:0143-4004

Salam MT, Millstein J, Li YF, Lurmann FW, Margolis HG, Gilliland FD. (2005). Birth outcomes and prenatal exposure to ozone, carbon monoxide, and particulate matter: results from the Children's Health Study. *Environ Health Perspect.*113(11):1638-44. ISSN: 0091-6765

Salafia CM, Ernst LM, Pezzullo JC, Wolf EJ, Rosenkrantz TS, Vintzileos AM. (1995). The very low birthweight infant: maternal complications leading to preterm birth, placental lesions, and intrauterine growth. *Am J Perinatol.* 12(2):106-10. ISSN: 0735-1631

Schatz M. Asthma and pregnancy. *J Asthma.* 1990;27(6):335-9. ISSN:0277-0903

Seo JH, Leem JH, Ha EH, Kim OJ, Kim BM, Lee JY, Park HS, Kim HC, Hong YC, Kim YJ. (2010). Population-attributable risk of low birthweight related to PM10 pollution in seven Korean cities. *Paediatr Perinat Epidemiol.* 24(2):140-8. ISSN: 0269-5022

Sexton KG, Jeffries HE, Jang M, Kamens RM, Doyle M, Voicu I, Jaspers I. (2004). Photochemical products in urban mixtures enhance inflammatory responses in lung cells. *Inhal Toxicol.* 16 (Suppl 1):107-114. ISSN: 0895-8378

Shah PS, Balkhair T; (2011). Knowledge Synthesis Group on Determinants of Preterm/LBW births. Air pollution and birth outcomes: a systematic review. *Environ Int.* 2011 Feb;37(2):498-516. ISSN:0160-4120.

Singh J. (1989). Neonatal development altered by maternal sulfur dioxide exposure. *Neurotoxicology.* 10(3):523-7.ISSN: 0161-813X

Srám RJ, Binková B, Dejmek J, Bobak M. (2005). Ambient air pollution and pregnancy outcomes: a review of the literature. *Environ Health Perspect* 113(4):375-382. . ISSN: 0091-6765

Sram JR., BinKova B, Dejmek J, Shah PS, Balkhair T. (2011). Knowledge Synthesis Group on Determinants of Preterm/LBW births. Air pollution and birth outcomes: a systematic review. *Environ Int.* 37(2):498-516. ISSN: 0160-4120

Suzuki K, Tanaka T, Kondo N, Minai J, Sato M, Yamagata Z. (2008). Is maternal smoking during early pregnancy a risk factor for all low birth weight infants? *J Epidemiol.* 18(3):89-96. ISSN:0917-5040

Tabacova S, Nikiforov B, Balabaeva L. (1985). Postnatal effects of maternal exposure to nitrogen dioxide. *Neurobehav Toxicol Teratol.*7(6):785-9. ISSN: 0275-1380

Tsai SS, Yu HS, Liu CC, Yang CY. (2003). Increased incidence of PTD in mothers residing in an industrialized area in Taiwan. *J Toxicol Environ Health* A 66(13):987-994. ISSN: 1528-7394

Topinka J, Binkova B, Mrackova G, Stávková Z, Peterka V, Benes I, Dejmek J, Lenícek J, Pilcík T, Srám RJ. (1997). Influence of GSTM1 and NAT2 genotypes on placental DNA adducts in an environmentally exposed population. *Environ Mol Mutagen* 30:184–95. ISSN: 0893-6692

Topinka J, Milcova A, Libalova H, Novakova Z, Rossner P Jr, Balascak I, Sram RJ. (2009). Biomarkers of exposure to tobacco smoke and environmental pollutants in mothers and their transplacental transfer to the foetus. Part I: bulky DNA adducts. *Mutat Res.* 2009 Oct 2;669(1-2):13-9. ISSN: 0027-5107

United Nations Children's Fund and World Health Organization. *Low Birthweight: Country, Regional and Global Estimates.* UNICEF, 2004.

Waller LA, Gotway CA.(2004). Applied spatial statistics for public health data Wiley series in probability and statistics: John Wiley & Sons, Inc.

Walters DM, Breysse PN, Wills-Karp M. (2001). Ambient urban Baltimore particulate-induced airway hyperresponsiveness and inflammation in mice. *Am J Respir Crit Care Med.* 2001 Oct 15;164(8 Pt 1):1438-43. ISSN: 1073-449X

Wang X, Ding H, Ryan L, Xu X. (1997), Association between air pollution and low birth weight: a community-based study, *Environ Health Perspect* 105: 514–520. ISSN: 0091-6765

Werler MM, Pober BR, Holmes LB. (1985). Smoking and pregnancy. *Teratology.* 32(3):473-81. ISSN: 0040-3709

Whyatt RM, Santella RM, Jedrychowski W, Garte SJ, Bell DA, Ottman R, Gladek-Yarborough A, Cosma G, Young TL, Cooper TB, Randall MC, Manchester DK, Perera FP. (1998). Relationship between ambient air pollution and DNA damage in Polish mothers and newborns. *Environ Health Perspect.* 106 Suppl 3:821-6. ISSN: 0091-6765

Wilhelm M, Ritz B. (2005). Local variations in CO and particulate air pollution and adverse birth outcomes in Los Angeles County, California, USA. *Environ Health Perspect.* 113(9):1212-21. ISSN: 0091-6765

Windham GC, Hopkins B, Fenster L, Swan SH. (2000). Prenatal active or passive tobacco smoke exposure and the risk of preterm delivery or low birth weight. *Epidemiology.* 11(4):427-33.ISSN: 1044-3983

Woodruff TJ, Parker JD, Kyle AD, Schoendorf KC. (2003). Disparities in exposure to air pollution during pregnancy. *Environ Health Perspect* 111(7):942-946. ISSN: 0091-6765

Wu J, Ren C, Delfino RJ, Chung J, Wilhelm M, Ritz B. (2009). Association between local traffic-generated air pollution and preeclampsia and preterm delivery in the south coast air basin of California. *Environ Health Perspect.* 2009 Nov;117(11):1773-9. ISSN: 0091-6765

Wu J, Wilhelm M, Chung J, Ritz B. (2011). Comparing exposure assessment methods for traffic-related air pollution in an adverse pregnancy outcome study. *Environ Res.* Mar 29. ISSN: 0013-9351

Xu X, Ding H, Wang X. (1995). Acute effects of total suspended particles and sulfur dioxides on PTD: a community- Based cohort study. *Arch Environ Health* 50(6):407-415. ISSN: 0003-9896

Xu X, Sharma RK, Talbott EO, Zborowski JV, Rager J, Arena VC, Volz CD. (2011). PM10 air pollution exposure during pregnancy and term low birth weight in Allegheny County, PA, 1994-2000. *Int Arch Occup Environ Health.*;84(3):251-7. ISSN: 0340-0131

Yang CY, Cheng BH, Hsu TY, Chuang HY, Wu TN, Chen PC. (2002a). Association between petrochemical air pollution and adverse pregnancy outcomes in Taiwan. *Arch Environ Health* 57(5):461-465. ISSN: 0003-9896

Yang CY, Chiu HF, Tsai SS, Chang CC, Chuang HY. (2002b). Increased risk of PTD in areas with cancer mortality problems from petrochemical complexes. *Environ Res* 89(3):195-200. ISSN: 0013-9351

Yang CY, Chang CC, Chuang HY, Ho CK, Wu TN, Tsai SS. (2003) Evidence for increased risks of preterm delivery in a population residing near a freeway in Taiwan. *Arch Environ Health.* 58(10):649-54. ISSN: 0003-9896

Yang CY, Chang CC, Tsai SS, Chuang HY, Ho CK, Wu TN, Sung FC.(2003). PTD among people living around Portland cement plants. *Environ Res* 92(1):64-68. ISSN: 0013-9351

Yang CY, Chang CC, Chuang HY, Ho CK, Wu TN, Chang PY.(2004). Increased risk of PTD among people living near the three oil refineries in Taiwan. *Environ Int* 30(3):337-342. ISSN: 0160-4120

Yi O, Kim H, Ha E. (2010). Does area level socioeconomic status modify the effects of PM(10) on preterm delivery? *Environ Res.*110(1):55-61. ISSN: 0013-9351

Yorifuji T, Naruse H, Kashima S, Ohki S, Murakoshi T, Takao S, Tsuda T, Doi H.(2011). Residential proximity to major roads and preterm births. *Epidemiology.*22(1):74-80. ISSN: 1044-3983

Zondervan HA, Oosting J, Hardeman MR, Smorenberg-Schoorl ME, Treffers PE. (1987). The influence of maternal whole blood viscosity on fetal growth. *Eur J Obstet Gynecol Reprod Biol* 1987;25:187–94. ISSN: 0301-2115

Zondervan HA, Oosting J, Smorenberg-Schoorl ME, Treffers PE. (1988). Longitudinal changes in blood viscosity are correlated with fetal outcome. *Acta Obstet Gynecol Scand.* 1988;67(3):253-7. ISSN:0001-6349

3

Maternal Exposure to Particulate Air Pollution and Engineered Nanoparticles: Reproductive and Developmental Effects

Petra Jackson, Ulla Vogel, Håkan Wallin and Karin Sørig Hougaard
National Research Centre for the Working Environment, Copenhagen
Denmark

1. Introduction

Man has always been exposed to ambient airborne nanoparticles, e.g. from wildfires or volcanic eruptions. With the events of industrialization and urbanization, ambient air pollution has grown from being a localized and temporal problem to a more regional and recurring problem. The emergence of nanotechnology provides a new source of exposure to airborne nanoparticles (Oberdöster et al., 2005).

1.1 Ambient air pollution

Uncontrolled dispersion of engineered nanoparticles may affect human health, similarly to what has been found for exposure to particles in ambient air. Particulates (in contrast to gases) are suspected to be the major factor driving the adverse effects of e.g. traffic air pollution. Human exposure to ultrafine particles in the ambient air has been associated with adverse health effects e.g. lung cancer, allergy, pulmonary and cardiovascular disease.

Less is known of the effects of exposure to nanoparticles during pregnancy. Epidemiological studies indicate that exposure to environmental air pollutants (especially particulates) is associated with adverse pregnancy outcomes, such as premature birth, reduced birth weight, small size for gestational age (Shah & Balkhair, 2010) and stillbirth (Pope et al., 2010). Several mechanisms, including particle induced oxidative stress and pulmonary and placental inflammation, have been suggested (Kannan et al., 2007). DNA damage is reported to increase after maternal exposure to ambient air pollution during pregnancy, reviewed in (Kannan et al., 2007). Effects include increased total DNA adducts in placental tissue (Šrám et al., 1999), bulky DNA adducts and micronuclei in umbilical blood of newborns (Pedersen et al., 2009), and stable chromosomal aberration frequencies in cord blood (Bocskay et al., 2005). Increased systemic inflammation and levels of urinary 8-oxodeoxyguanosine (a marker of oxidative damage to DNA) have been reported in children born and raised in areas with high air pollution (Calderon-Garciduenas et al., 2008; Švecová et al., 2009). Maternal exposure during pregnancy to traffic-related or industrial air pollution has furthermore been associated with adverse effects in the children related to cognitive and perceptual performance, motor function and language skills (Freire et al., 2010; Tang et al., 2008). It is difficult to causally link exposure to air pollution to negative reproductive outcomes in man. Animal studies are important for testing this hypothesis.

Developmental toxicity of diesel exhaust has been assessed in animal studies. Prenatal exposure to diesel exhaust is not necessarily fetotoxic, but may affect other endpoints. Gestational exposure to diesel exhaust has changed immunological pattern in mouse placentas (Fujimoto et al., 2005) and altered offspring birth weight, immunological function and sexual differentiation, summarized in (Hougaard et al., 2008). More recent animal data indicate that diesel exhaust may affect male reproductive function after birth (Hemmingsen et al., 2009; Li et al., 2008; Ono et al., 2007). Several animal studies report changes in the brain and neurofunction due to diesel exhaust exposure (Masao Sugamata, 2006b; Suzuki et al., 2010; Takeda et al., 2004; Yokota et al., 2009). Most studies used whole diesel exhaust, such as exposure to exhaust gases from diesel engines. However, a numerically large fraction of the particulates in diesel exhaust is ultrafine in size, and the particulate fraction may be important for toxicity (Hougaard et al., 2008). Adverse effects of prenatal exposure to engineered nanoparticles *in vitro* and *in vivo* have been summarized by (Ema et al., 2010; Hougaard et al., 2011). The effects observed in animals include altered immune response, development of the nervous system and reproductive function (Fedulov et al., 2008; Lamoureux et al., 2010; Shimizu et al., 2009; Takahashi et al., 2010; Takeda et al., 2009; Yoshida et al., 2009; Yoshida et al., 2010).

1.2 Nanotoxicology and nanoparticles

'Nanotechnology' has allowed for creation of new materials with new, exciting properties for a vast range of applications (energy production, electronics, biomaterials, medicine, cosmetics and others). 'Nanomaterial' is defined as insoluble or bio-persistent and intentionally manufactured material with one or more external dimensions, or an internal structure, on the scale from 1 to 100 nm (Article 2.1.k of Regulation EC/1223/2009 on Cosmetic Products). Solid matter in the atomic and molecular size range has a large, specific surface reactivity and surface area to volume ratio. Therefore, nanomaterials have unique properties compared with their larger counterparts of similar particle mass (Duffin et al., 2007). Changes in material properties affect the biological interactions, and therefore 'Nanotoxicology' has emerged a new discipline in toxicology.

Based on the size, particles deposit in different regions of the pulmonary tract. Nanoparticles deposit deeper in the respiratory system than do larger particles. Because of inefficient clearance by alveolar macrophages and the mucociliary escalator, nanoparticles interact with epithelial and interstitial sites, thereby increasing biological reactivity (Oberdöster et al., 2005). Translocation of nanoparticles from the lung to the circulation is considered to be small and happen at a slow pace. It has been reported that only a fraction of a percent gets beyond the lung cavity and regional lymph nodes, depending on the particle or agglomerate size (Kreyling et al., 2009; Sadauskas et al., 2009). A portion of inhaled particles are swallowed (Jacobsen et al., 2009), however nanoparticles do not seem to pass readily over the gastrointestinal mucosa in rodents (Carr et al., 1996). Some airborne nanoparticles are known to elicit oxidative stress and inflammatory reactions (Borm et al., 2006), which can further induce oxidative damage to DNA (Møller et al., 2010a).

1.3 Questions addressing maternal pulmonary exposure to nanoparticles

Based on findings in epidemiological and animal studies, the questions addressing the effects of maternal pulmonary exposure to nanoparticles during gestation were:
Does maternal exposure to nanoparticles affect gestation and development during lactation?
Does prenatal exposure to nanoparticles induce genotoxicity in the exposed offspring?
Does exposure to nanoparticles affect development of the nervous system?

2. Particle exposure

Two types of nanoparticles and two means of exposure were used in assessment of developmental toxicity.

2.1 Titanium dioxide (UV-Titan)

Titanium dioxide (TiO_2) is a naturally occurring mineral. It is mined in large quantities world-wide, and serves as a white pigment in cosmetics, food, plastics, paints and other products, and as a UV-filter in cosmetics and sunscreens. The titanium dioxide used in the present study was UV-Titan L181 (Kemira, Pori, Finland), supplied by The Danish Association of the Paint and Lacquer industry. This material was chosen as a model nanoparticle for application by the paint industry in future paints and lacquers. TiO_2 particles are generally considered to be inert. Physico-chemical characteristics such as particle size, crystal phase, crystalline form and surface modifications determine particulate toxicity (Johnston et al., 2009); in reality, it is therefore difficult to generalise with respect to toxicity of particulate TiO_2. Nanosized TiO_2 particles have been detected in lung tissue up to four weeks after inhalation exposure (Hougaard et al., 2010), why acute as well as chronic effects of exposure must be considered. Pulmonary exposure to TiO_2 nanoparticles induces lung inflammation in mice and rats. TiO_2 nanoparticles are suspected to be both genotoxic (Johnston et al., 2009) and carcinogenic (Mohr et al., 2006). Currently, there is no evidence of TiO_2 related cancer in the occupational setting TiO_2. However, based on scientific evidence, inhaled TiO_2 has been classified by the International Agency for Research on Cancer (IARC) as possibly carcinogenic to humans (Group 2B) (Baan et al., 2006).

2.2 Carbon black (Printex 90)

Carbon black is a well-characterized, carbonaceous particle that has been used extensively as a model for diesel emission particles, but without adhered chemicals and metals. The carbon black used in this study was Printex 90 (Degüssa, Frankfurt Germany). This material is marketed as printing ink. Printex 90 has been used as a positive control in many studies of nanotoxicology. It consists of carbon with less than 1% organic and inorganic impurities. Health effects reported after exposure to carbon black are assumed to be caused by the insoluble particle core rather than by associated compounds. It is well known that pulmonary exposure to carbon black by instillation or inhalation induces lung inflammation *in vivo* in rats and mice, summarized in (Jackson et al., 2011b). Carbon black nanoparticles possess an intrinsic potential to generate reactive oxygen species, summarized in (Jacobsen et al., 2010), induce DNA strand breaks and oxidatively generated DNA damage (Jacobsen et al., 2008b; Jacobsen et al., 2009), are reported to be mutagenic (Jacobsen et al., 2010), and induce lung tumors in rats (Mohr et al., 2006). It is uncertain whether occupational exposure to carbon black is related to cancer risk, but carbon black has been classified by the IARC as possibly carcinogenic to humans (Baan et al., 2006).

2.3 Inhalation and instillation exposure

In occupational settings, exposure to nanoparticles primarily occurs via inhalation (Maynard & Kuempel E.D., 2005). The dustiness of nanoparticles is in order of magnitude larger than the dustiness of fine particles of similar chemical composition (Schneider & Jensen, 2008). Aerosolization of particles for inhalation exposure is therefore the preferred exposure regimen in toxicological studies of nanoparticles, relating to occupational settings.
However, it is not always possible to perform inhalation studies, e.g. when: 1) there is a potential health risk associated with occupational exposure during management of particles in during the exposure procedure or on animals after exposure (e.g. carbon nanotubes); 2) only small amounts of particles are available; and 3) inhalation facilities are not available. Instillation of particles presents an alternative method for pulmonary administration in particle toxicology. Particles are suspended in a liquid and are subsequently injected as a solution into the lumen of the trachea while the animals are under general anaesthesia. Particle instillation is widely a used and accepted procedure, even though inhalation and instillation methods may not be fully comparable (Driscoll et al., 2000).

3. Developmental toxicity testing of nanoparticles

Until now, relatively little attention has been given to the potential adverse effects of prenatal exposure to nanoparticles. In USEPA's nanomaterial research strategy, *in vivo* toxicity testing in animals also includes reproductive and developmental toxicity (USEPA, 2009). One of the anticipated outcomes of this strategy is identification of testing methods to predict *in vivo* toxicity of nanomaterials. The EU chemical legislature, REACH, requires reproductive and developmental studies from producers or importers of chemicals with tonnages above the ten-ton limit. It is currently discussed whether a precautionary principle should be applied in the EU regulation of nanoparticles; possibly nanoparticles should be exempt from the tonnage rule and be fully tested at much lower tonnage limits. Developmental toxicity testing would then be one of the expected requirements.

3.1 Animal model

Pregnancy is a complicated biologic process involving many developmental stages of the fetus. Chemical exposure can negatively interfere with the course of pregnancy, depending on the timing. We exposed mice on gestation days (GD) 7(8)-18. Exposure began after implantation and initial organ development had taken place. This exposure roughly corresponds to the first two trimesters of human pregnancy, where fetal organs are formed and developed. Exposure terminated two days before expected delivery to avoid stressing the animals during birth preparations and thus to prevent negative birth outcomes. Maternal and offspring weights were recorded as a classical toxicity endpoint. To confirm particle effects in the mother, we evaluated maternal lung inflammation by cellular response in bronchoalveolar lavage, gene expression in lung and liver, and genotoxicity in lung and liver. Genotoxicity was also evaluated in offspring liver. Offspring toxicity was furthermore assessed at birth, weaning, adolescents, and neurotoxicity was assessed in adulthood.
The maternal lung effects were assessed at two time points. Time-mated females that did not give birth, or only gave birth to a few offspring, were used to assess lung inflammation soon after birth (3-5 days after end of exposure). Littering dams were assessed at the end of lactation after weaning (23-27 days after end of exposure). Pregnancy is reported to alter

inflammatory response (Fedulov et al., 2008; Lamoureux et al., 2010), thus, different background levels in these sets of time-mated females were anticipated.

3.2 Mechanisms of nanoparticle toxicity during fetal development

It has been suggested that the fetus could be affected either: 1) directly by particle and/or impurities translocating through the placenta; 2) by altered placental function; or 3) indirectly by circulating cytokines or other secondary messengers from an inflammatory process in the mother (Hougaard et al., 2011).

To affect the offspring directly, nanoparticles have to translocate through the placenta and enter the embryonic/fetal tissue. Several studies has investigated placental transfer of nanoparticles *in vivo* in mice and rats (Kennison et al., 1971; Sadauskas et al., 2007; Takahashi & Matsuoka, 1981), as well as *ex vivo* in perfused human placentas collected by caesarean sections (Myllynen et al., 2008; Wick et al., 2010), and *in vitro* in a placenta trophoblast model (BeWo choriocarcinoma cells) (Myllynen et al., 2008). In addition, *in vitro* translocation and the effects on embryonic tissue have been assessed (Bosman et al., 2005; Tian et al., 2009). Nanoparticle translocation depends strongly on particle size, concentration, surface modifications and loading, as well as maternal exposure route and duration, and timing in pregnancy. Translocation of nanoparticles from the maternal lung to the circulation is considered to be slow, and probably only a fraction of a percent gets beyond the lung cavity, regional lymph nodes or over the gastrointestinal mucosa, as discussed above. After pulmonary exposure, once in circulation, the distribution across the placenta and into the fetus may also be negligible (Sadauskas et al., 2007). It is however likely that impurities and surface modifications could dissociate from nanoparticles and traverse the placenta, as reported for polycyclic aromatic hydrocarbons leached from coal fly ash and administered intratracheally to pregnant rats on gestation day 18 and 19 (Srivastava et al., 1986). Transplacental genotoxins have thus been reported to induce damage to the DNA in the offspring (Brunborg et al., 1996; Tripathi et al., 2008).

Nanoparticles can potentially affect placental function. Nanoparticles may be internalized by placental cells, when attempting to cross the placenta (Myllynen et al., 2008; Wick et al., 2010). There are currently no published reports identifying alterations in exposed placentas, though it can be expected that placenta function, and maybe even morphology, could be changed by such an intracellular particle burden.

Nanoparticles may also affect the fetus indirectly. The general effects of pulmonary exposure to nanoparticles include lung inflammation and genotoxicity. The mother is the route of exposure for offspring exposed *in utero*. Maternal inflammation has been reported to potentially affect the nervous system in developing offspring (Graciarena et al., 2010).

3.3 Maternal effects of pulmonary exposure to nanoparticles

3.3.1 Maternal inflammatory response

Maternal lung inflammation was evaluated by analysis of the cellular profile in bronchoalveolar lavage fluid (BAL) (Hunninghake et al., 1979). Lung and liver gene expression were analyzed by DNA microarrays as described (Halappanavar et al., 2011).

The response to nanoparticle exposure is characterized by activation of alveolar macrophages and recruitment of polymorphonuclear neutrophils, as a sign of ongoing lung injury. Neutrophil infiltration is usually observed during the acute phase response; however patients with lung fibrosis or asbestosis exhibit persistent accumulation of neutrophils

(Hunninghake et al., 1979), supporting the validity of this biomarker in particle toxicology screening. Cellular infiltration into the alveolar spaces is associated with the release of acute inflammatory mediators, which may ultimately lead to a state of systemic inflammation. Bronchoalveolar lavage fluid was collected from mice under Hypnorm-dormicum anaesthesia. The number of macrophages, neutrophils, lymphocytes, eosinophils and epithelial cells were determined by differential count, as described (Jackson et al., 2011b). Counts were presented relative to the total cell number in the BAL fluid determined in a NucleoCounter, following the standard kit procedure (Chemometec, Denmark).

To follow the current developments in toxicology and to obtain a deeper understanding of mechanisms of nanoparticle toxicity, the entire transcriptome was analysed by microarray to identify the changes in pathways and allow for generation of hypotheses about toxicity mechanisms and endpoints. Microarray analysis was performed in total ribonucleic acid (RNA) purified from examined tissue analyzed in Agilent mouse 4 x 44 oligonucleotide microarrays as described (Halappanavar et al., 2011). The most pronounced and statistically significant changes to genes were validated by RT-PCR. RNA isolation and cDNA synthesis were performed as described (Saber et al., 2009). Gene expression analysis is a sensitive method quantifying pathway perturbations and therefore it identifies processes induced by the chemical exposure. Changes to gene expression do not always lead to protein changes in the exposed organism. Protein analysis may therefore not identify early or marginal effects, which may have significant biological significance.

3.3.2 Maternal levels of DNA strand breaks

Maternal lung and liver levels of DNA strand breaks were analyzed by the comet assay (McNamee et al., 2000). Particularly concerning about nanoparticle exposure is that some types of nanoparticles have the ability to induce genotoxicity (Jacobsen et al., 2009; Møller et al., 2010a). The primary genotoxicity of nanoparticles is believed to relate to their ability to induce reactive oxygen species (ROS) (Jacobsen et al., 2008b), or to form DNA adducts due to surface-bound organic compounds (Jacobsen et al., 2008a), or both. Secondary genotoxicity may occur due to particle inflammation (Knaapen et al., 2004).

The comet assay is becoming a common screening strategy in nanotoxicology (Karlsson, 2010). The strand breaks measured by the assay represent a mixture of direct strand breaks, alkaline labile sites and transient breaks in the DNA due to repair processes (Collins, 2009). It is a temporary endpoint, as repair mechanisms can rejoin DNA breaks shortly after exposure (Bornholdt et al., 2002). The comet assay is based on a relatively simple protocol by which cell suspensions are embedded in agarose. The cells are subsequently lysed and subjected to alkaline electrophoresis, where DNA fragments migrate away from the nuclei and form 'comets'. The comets are microscopically analyzed and reported as mean comet tail length, %DNA in tail, or tail moment per sample. Data can be recalculated to the number of lesions per million base pair (Forchhammer et al., 2010). The protocol is being internationally validated to achieve standardized and reliable results (Forchhammer et al., 2010; Møller, 2006; Møller et al., 2010b).

3.4 Developmental endpoints

The main focus of nanotoxicology research is on effects and their mechanisms in the adult organism, whereas effects arising during pregnancy, prenatal and postnatal development are poorly investigated. We assessed maternal toxicity during gestation and lactation and a range of endpoints in the offspring, i.e. developmental toxicity.

3.4.1 Gestation and lactation

Endpoints to assess toxicity during gestation and lactation included:

1. Maternal weight gain during gestation and lactation and offspring weight at birth, during lactation and maturation. Growth is a classical toxicological parameter, and changes in growth pattern indicate toxicity in the exposed animals.
2. Gestation length. Since particles are linked to prematurity in epidemiologic studies (Shah & Balkhair, 2010), this time point is highly relevant to particle toxicology.
3. Relative number of pregnant animals (pregnancy rate), number of implantations and litter size.
4. Postnatal viability. Maternal exposure or effects of exposure extending into lactation can influence viability after birth.
5. Gender ratio. Skewed gender ratio can be an indication of effect on a specific sex of exposed animals. It can be altered by either interference during prenatal sexual development or by selective mortality in one sex.

3.4.2 Offspring levels of DNA strand breaks

Offspring liver levels of DNA strand breaks were analyzed by the comet assay by the method described earlier. During development, frequent cell divisions allow only short time for repair of DNA damage and the immune system is not fully functional. In humans, maternal exposure to air pollution during pregnancy has been associated with genotoxicity in the children. Early-life exposure might therefore predispose to cancer and other diseases later in life (Barton et al. 2005).

3.4.3 Developmental neurotoxicity

An increasing number of studies indicate that prenatal exposure to chemicals is able to influence development of the nervous system. Developmental neurotoxicity testing is therefore one of the additional endpoints included in the new 'extended one-generation reproductive toxicity study' developed under REACH. As described in this chapter, changes in the brain and neurofunction are reported from animal studies of prenatal exposure to diesel exhaust and titanium dioxide particles. On the one hand, the developing brain may not be exposed to high doses of nanoparticles directly, due to restricted transport through several membrane barriers, even if nanoparticles have been observed in the brain of offspring weeks after prenatal exposure (Gao et al., 2011; Takeda et al., 2009). On the other hand, the developing brain may be especially vulnerable to nanoparticle induced oxidative stress, because of reduced anti-oxidant capacity in immature cerebral white matter, as suggested in (Hougaard et al., 2011).

The following methods were applied to test for developmental neurotoxicity:

1. The Acoustic startle test is a test of sensorimotor processes. Animals were placed in a small chamber and exposed to a series of startle stimuli (sounds), were designed to elicit the startle reflex (response to unexpected stimuli, measured by the magnitude of movement). The startle reflex follows the basic auditory pathway and terminates as a contraction of the skeletal muscles. The used acoustic stimuli consisted of short pulses of white noise, sometimes supplemented by pre-pulse stimuli (warning stimuli).

Habituation to noise pulses, amplitude of the basal startle response, and the reaction after pre-pulse stimulation were analyzed.

2. The Open field test evaluates temperament and emotionality exhibited by locomotor activity. The animal was placed centrally in the brightly illuminated arena with a diameter of one meter and ambulation was automatically recorded for a three-minute period. Distance of locomotion and duration of stay in pre-defined zones were analyzed.
3. The Morris water maze is a test of place-learning and working memory. Animals were placed in a circular water pool surrounded by visual landmarks. An invisible escape platform was situated below the surface. The animal was put in the water and searched for the platform for one minute or until the platform was located, four trials a day for a learning period of 4 days. Memory was tested after some weeks with the same spatial setup, and then platform positions were changed for assessment of new learning. Animal memory and learning success were analyzed.

4. Experimental design

Three developmental studies were included in this work. Two studies assessed maternal inhalation exposure to titanium dioxide (UV-Titan) and carbon black (Printex 90), respectively. Printex 90 was also investigated in a dose-response study with maternal exposure by intratracheal instillation. Prior to initiation of the Printex 90 dose-response study, the intratracheal instillation protocol with short-term isoflurane anaesthesia was evaluated for effects in the mothers and their offspring.

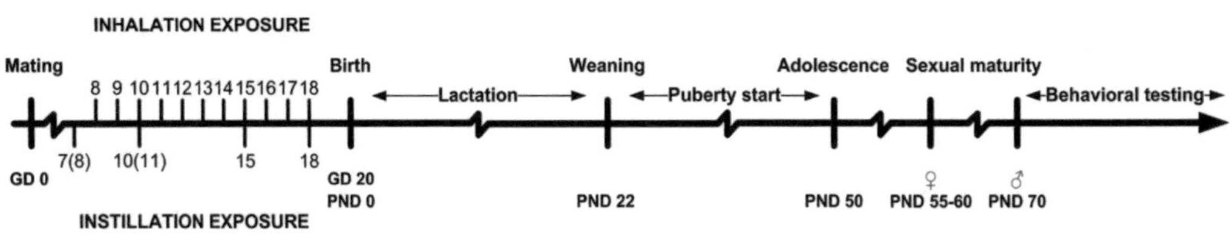

Fig. 1. Overview of the experimental design. Gestation day (GD); Postnatal day (PND)

Study	Organs collected (PND)					Neurobehavioral testing (PND)		
	Females	Newborns	Weaned Offspring	Dams	Adoles-cents	Water maze	Open field	Startle
5.1	3	2	23-24	24-25	-	♂ 71, 99 ♀ 78, 106	94	♂ 122 ♀ 129
5.2	3	2	22-23	22-23	50	-	-	-
5.3	1-2	2	23	24-25	47	-	72-73	74-75

Table 1. Timing of tissue collection and neurobehavioral testing, in postnatal days (PND). PND 0 = Gestation Day 20, which was two days after the last exposure

4.1 Titanium dioxide inhalation study

We investigated the developmental toxicity after maternal pulmonary exposure to titanium dioxide (UV-Titan). Time-mated mice (C57BL/6BomTac) were exposed by inhalation for 1 h/day to 42 mg UV-titan/m^3 aerosolized powder or filtered air on GD 8-18. Endpoints included cell composition in maternal BAL early after exposure (5 days after exposure) and after weaning (26-27 days after exposure), UV-titan deposition and distribution in maternal and offspring tissues, classical gestational and lactational parameters, offspring neurofunction (Hougaard et al., 2010), and toxicogenomics in non-littering time-mated females (5 days after exposure) (Halappanavar et al., 2011). Maternal and offspring genotoxicity was evaluated as level of DNA strand breaks in BAL and liver cells (unpublished, preliminary results).

4.2 Carbon black inhalation study

Carbon black (Printex 90) was evaluated for developmental effects of the maternal pulmonary exposure during gestation. Similar to the titanium dioxide study, time-mated mice (C57BL/6BomTac) were exposed by inhalation for 1 h/day to 42 mg Printex 90/m^3 aerosolized powder or filtered air during GD 8-18. Endpoints included cell composition in BAL early after exposure (5 days after exposure) and after weaning (24-25 days after exposure), classical gestational and lactational parameters. Genotoxicity was evaluated as the level of DNA strand breaks in BAL and liver cells (Jackson et al., 2011b).

4.3 Carbon black instillation study

After assessment of the effects of maternal inhalation exposure to carbon black (Printex 90), a study was initiated to characterize the dose-response relationship. Time-mated C57BL/6BomTac mice were exposed by four intratracheal instillations on GD 7, 10, 15 and 18, with total doses of 11, 54 and 268 μg Printex 90/animal. The highest dose was chosen to compare to the estimated deposited dose in the pulmonary (alveolar) region in mice from the inhalation study. Endpoints included cell composition in BAL early after exposure (3-4 days after exposure) and after weaning (26-27 days after exposure), and classical gestational and lactational parameters. Genotoxicity was evaluated as the level of DNA strand breaks in BAL and liver cells (Jackson et al., 2011b). In addition, offspring neurofunction was evaluated (Jackson et al., 2011c).

5. Results

Results from all studies are summarized in Table 2.

5.1 Titanium dioxide inhalation study

Inhalation exposure to surface-coated nanosized titanium dioxide (42.4 ± 2.9 mg UV-Titan/m^3 for 1 h/day on gestation days 8-18) induced persistent maternal inflammation measured by large neutrophil influx in BAL in the exposed animals. Inflammation persisted 27 days after exposure termination (non-littering females early after exposure: 19-fold increase; dams after weaning: 3-fold increase), which was the last tested time point (Hougaard et al., 2010). The analysis of total gene expression in the lungs of exposed non-littering females, revealed increased levels of lung mRNA for acute phase serum amyloid A-1 and serum amyloid A-3, and higher levels of several lung C-X-C and C-C motif chemokines and miRNAs (Halappanavar et al., 2011). Despite the inflammation and acute phase response in the lung, gene expression in the female liver was relatively unchanged (Halappanavar et al., 2011). Exposed offspring exhibited minor behavioural changes after reaching adulthood (somewhat avoided the central zone of the open field, and exposed

female offspring displayed enhanced pre-pulse inhibition in the acoustic startle test (Hougaard et al., 2010). Maternal pulmonary exposure to UV-titan did not significantly affect gestation, lactation and offspring development (endpoints: maternal weight gain during gestation and lactation, gestation length, offspring weight at birth, during lactation and maturation, litter size, gender ratio, number of implantations, and postnatal viability) (Hougaard et al., 2010). In addition, UV-Titan did not seem to induce DNA strand breaks in time-mated mice or their offspring (unpublished, preliminary results).
A fraction of nanoparticles would be assumed to translocate from the maternal lung, based on the particle size and properties. To evaluate particle translocation, titanium (Ti) content in maternal lung and liver, in offspring liver, and in milk content in newborn stomachs was analysed (Hougaard et al., 2010). Maternal lung tissue contained 34-30% of the predicted deposited UV-Titan, 5 and 26-27 days after exposure, respectively. In the remaining tissues, Ti-levels were below the detection limit. The detection limit was probably too high to measure small amounts of translocated nanoparticles.

5.2 Carbon black inhalation study

Inhalation of nanosized carbon black (42.4 ± 2.9 mg Printex 90/m^3 for 1 h/day on gestation days 8-18) induced persistent lung inflammation. Inflammation persisted 24-26 days after exposure termination (non-littering females early after exposure: 11.4-fold increase and dams after weaning: 11.6-fold increase). Maternal inhalation exposure to Printex 90 induced DNA strand breaks in the liver of time-mated mice and in the offspring even weeks after end of exposure. Despite this, we did not register gestational or developmental toxicity (Jackson et al., 2011b).

5.3 Carbon black instillation study

A follow-up study with exposure by four intratracheal instillations of Printex 90 dispersed in Nanopure filtered UV water (on gestation days 7, 10, 15 and 18, with total doses of 11, 54 and 268 μg/animal) evaluated the relationship between dose and response. The protocol described in (Jackson et al., 2011a) was used. Exposure to the highest dose induced persistent maternal lung inflammation. The lung inflammation in the instillation-exposed mice was higher compared to similar inhaled pulmonary dose (non-littering females early after exposure: 28.7-fold increase and dams after weaning: 60.9-fold increase). This confirms that instillation induces stronger lung inflammation compared to inhalation. There were no changes in the levels of DNA strand breaks after the intratracheal instillation of doses comparable to those applied in the carbon black inhalation study. We did not observe changes in gestational or lactational parameters (Jackson et al., 2011b). The female offspring prenatally exposed to Printex 90 displayed minor behavioural changes in the open field test (Jackson et al., 2011c).

6. Discussion

The purpose of the presented work was to assess the effects of pulmonary exposure to nanoparticles during pregnancy on the mouse dam and her offspring. Classical gestational, lactational and developmental parameters were evaluated, supported by assessment of maternal inflammation, maternal and offspring levels of DNA strand breaks, and offspring neurotoxicity.

Table 2. Selected result overview

Study	Maternal		Gestation & Lactation	Offspring		
	Non-littering Females (3-5 days after exposure)	**Dams (24-27 days after exposure)**		**Newborn**	**At weaning**	**Adolescents**
5.1	*Lung inflammation:* 19-fold ↑ PMN *Ti content in tissues:* 38±6 mg/kg Ti in lung <0.5 mg/kg Ti in liver *Gene expression:* *Lungs* - Inflammation, immune response, acute phase response *Liver* - Minor changes	*Lung inflammation:* 3-fold ↑ PMN *Ti content in tissues:* 33±18mg/kg Ti/lung 0.5±0.3 mg/kg Ti/liver	No effect	*Ti content in tissues:* <0.4 mg/kg Ti/liver <1 mg/kg Ti/milk	*Ti content in tissues:* <0.4 mg/kg Ti/liver	*Behavioural changes:* Activity Sensorimotor processes
5.2	*Lung inflammation:* 11-fold ↑ PMN *Genotoxicity:* BAL - none Liver – 1.3-fold ↑	*Lung inflammation:* 12-fold ↑ PMN *Genotoxicity:* BAL - none Liver - 1.6-fold ↑	No effect	*Genotoxicity:* Liver – none (↑ background)	*Genotoxicity:* Liver - 1.4-fold ↑	*Genotoxicity:* Liver - 1.5-fold ↑
5.3	*Lung inflammation:* 29-fold ↑ PMN *Genotoxicity:* BAL - none Liver – none	*Lung inflammation:* 61-fold ↑ PMN *Genotoxicity:* BAL - none Liver – none	No effect	*Genotoxicity:* Liver – none (↑ background)	*Genotoxicity:* Liver - none	*Genotoxicity:* *Liver* – none *Behavioural changes:* Habituation

Polymorphonuclear neutrophil (PMN), Gestation and lactation (Endpoints assessed: weight gain during gestation and lactation, gestation length, offspring weight at birth, during lactation and maturation, litter size, gender ratio, number of implantations, and postnatal viability), Genotoxicity (levels of DNA strand breaks by comet assay)

5.1 TiO2 inhalation study (42 mg/m^3 UV-titan 1h/day on GD 8-18, 1.7*106 n/cm^3; peak-size 97 nm)

5.2 Carbon black inhalation study (42 mg/m^3 Printex 90 1h/day on GD 8-18, 4.1*106 n/cm^3; peak-size 45 nm)

5.3 Carbon black instillation study (11, 54, 268 μg/animal on GD 7, 10, 15 & 18, zeta-size 140 nm; peak-size 55 nm)

6.1 Exposure characterization

Both UV-Titan and Printex 90 (Figure 2) were evaluated after similar whole-body inhalation exposures 42 mg/m^3 for 1 h/day on gestation days 8-18. Effects of Printex 90 were also assessed after intratracheal instillation to three dose levels of Printex 90 instilled on gestation days 7, 10, 15 and 18. The highest instilled dose level was set to compare with the expected inhaled pulmonary (alveolar) dose in the Printex 90 inhalation study. Instillation could be expected to be perceived as more stressful by the mated mice; however, no observable effects were observed in pregnant mice or their offspring (Jackson et al., 2011a). Because there are virtually no data on developmental toxicity of engineered nanoparticles, the chosen exposure levels were relatively high. Still, the daily exposures corresponded to approximately one-and-a-half day exposure that Danish workers would experience at the current 8-hours time weighed average occupational exposure limit (6 mg titanium dioxide/m^3 and 3.5 mg carbon black/m^3, The Danish Working Environment Authority 2007).

In our studies, the gravimetric doses of UV-Titan and Printex 90 were very similar and the total inhaled masses were estimated to compare between studies. The exposures did differ with respect to particle sizes. The UV-Titan dispersion contained large agglomerates, while the Printex 90 air dispersion was more in the nanoscale. This distribution naturally affected the particle number concentration, and there were half the amount of nanoparticles in the UV-Titan aerosol compared to that of Printex 90 (Figure 3, Table 3, Table 4).

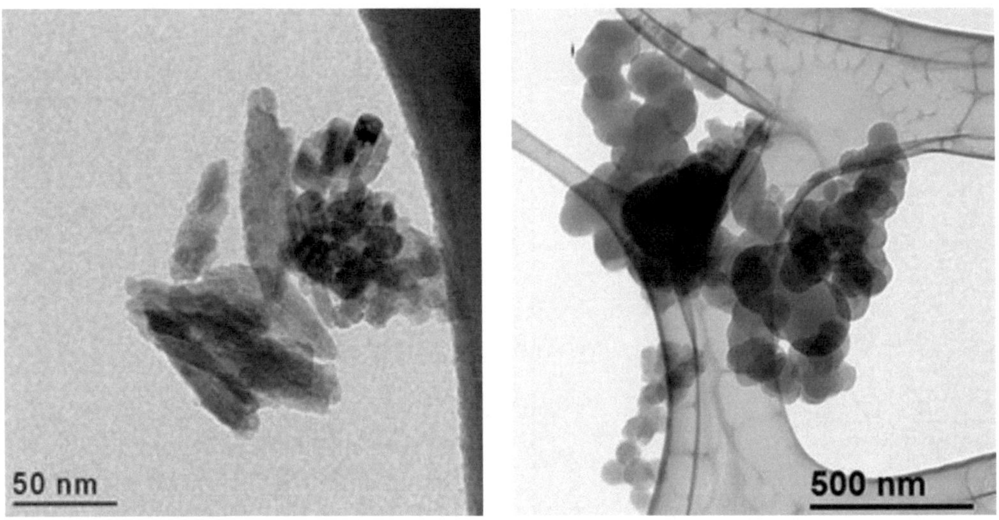

A) B)

Fig. 2. Transmission electron microscopy (TEM) pictures of A) UV-Titan B) Printex 90. Courtesy of Dr. Keld A. Jensen from National Research Centre for the Working Environment, Denmark

Particle size affects the deposition in the lung. Particles from the extra-pulmonary regions are transported up by the mucociliary escalator and swallowed. We have estimated the particle deposition based on the model revised from (Jacobsen et al., 2009). A lower pulmonary deposition was predicted for UV-Titan compared to Printex 90, whereas the estimated intragastric deposition of UV-Titan was higher compared to Printex 90. Because the time-mated mice were exposed by whole-body inhalation exposure to avoid restraint stress during pregnancy, it can be expected that the received dose in the gastrointestinal tract was even greater due to fur grooming. However, based on previously published work, we assume that the agglomerates likely passed though the digestive system without major translocation to the liver or the rest of the body and most was excreted by faecal excretion.

UV-Titan particle characteristics:	
Declared particle size	17 nm
Phase	Rutile
Crystallite size	20.6±0.3 nm (14.4-15.5 [100]; 38.4 [001])
Surface area (BET)	70 m^2/g 107.7 m^2/g
Chemical composition	Na_2O 0.60 wt% SiO_2 12.01 wt% Al_2O_3 4.58 wt% ZrO_2 1.17 wt%
Loss of ignition	5.19 wt%
Termogravimetric analysis (N_2 atmosphere, 40-800°C, 10°C/min)	6.1±0.4 wt%
Inhalation exposure:	
Particle size distribution in number	Aggregates and agglomerates were equidimensional to needle-shaped TiO2 crystallites with diameters from less than 10 nm to more than 100 nm along the shortest and longest axis, 50%<97 nm.
Particle number concentration	1.70±0.20x10^6/cm^3
Geometric mean size	97 nm
Mass-size distribution	75%>1600 nm and <1%<100 nm
Total inhaled dose	840 µg/animal
Estimated deposition	73 µg/animal in pulmonary region 315 µg/animal in extra-pulmonary region 365 µg/animal in gastro-intestinal tract

Table 3. Key physico-chemical characteristics of titanium dioxide (UV-Titan L181). Based on data from (Hougaard et al., 2010)

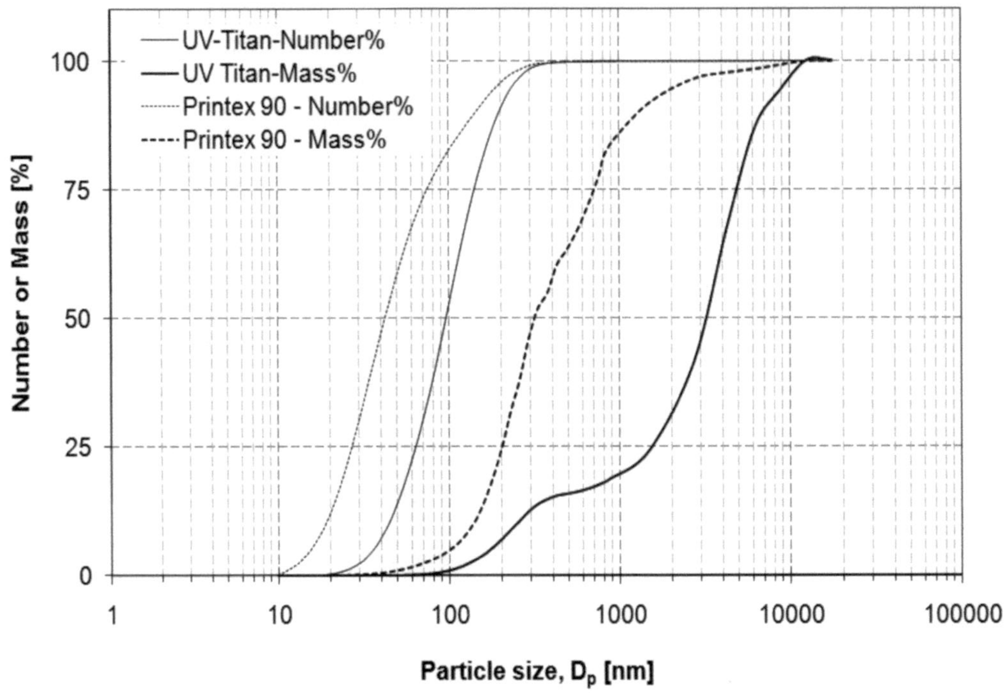

Fig. 3. Characteristics of the exposure atmosphere for UV-Titan and Printex 90 inhalation exposures. Accumulated number and mass concentration of particle concentrations in the exposure chamber. Courtesy of Dr. Keld A. Jensen from NRCWE, Denmark

To make inhalation and instillation exposure to Printex 90 comparable (Table 4), we estimated the dose deposited in the pulmonary (alveolar) region by inhalation (287 µg/animal), and intratracheally instilled a similar dose (268 µg/animal) as the highest of the three instilled dose levels. Therefore, the total received maternal dose was lower in the instillation study, compared to the total inhaled mass (826 µg/animal). When comparing the particle size during the two exposures, the instilled Printex 90 nanoparticles dispersed in Millipore water resembled the aerosolized Printex 90. The hydrodynamic number size-distribution peaked between 50-60 nm in the instilled solution, which compared to the particle size distribution number of 45 nm measured in the aerosol exposure atmosphere. Thus, the particle effects in maternal lung could be expected to be similar. The Printex 90 instillation exposure probably also resulted in small intragastric deposition, however smaller compared to the inhalation exposure.

6.2 Maternal effects

We assessed the particle effects in the exposed time-mated females, i.e. maternal lung inflammation and genotoxicity, to establish a broader understanding of the offspring exposure.

Printex 90 particle characteristics	
Declared particle size	14 nm
Geometric mean size	65 nm (carbon spheres)
Surface area (BET)	295-338 m^2/g
Pycnometric particle density	2.1 g/cm^3
Chemical composition	99% C, 0.8% N and 0.01% H_2
The total PAH content (Carbon black extract - Soxhlet)	0.0742 μg/g
The total PAH content (DEP extract - NIST SRM 1650)	216 μg/g
Inhalation exposure:	
Particle size distribution in number	Average distribution was multimodal and highly dominated by sub-100 nm particles, aggregates were most commonly 41 nm and that was also the average size, 50%<45 nm.
Particle number concentration	$4.09 \pm 0.03 \times 10^6/cm^3$
Mass-size distribution	310 nm (bimodal; 290 and 1500 nm) 75%>200nm and 5%<100 nm
Total inhaled dose	826 μg/animal
Estimated deposition	287 μg/animal in pulmonary region 166 μg/animal in extra-pulmonary region 137 μg/animal in gastro-intestinal tract
Instillation exposure:	
Morphology	The agglomerates consisted of spherical to sub-spherical carbonaceous particles as well as minor amounts of free single primary spheres.
Average zeta-size	140 nm
Hydrodynamic number	50-60 nm
Volume distributions	Peaks 50-60 nm and 200-400 nm

Table 4. Key physico-chemical characteristics of carbon black (Printex 90). Data from Degussa-Hüls, (Saber et al., 2005; Jacobsen et al., 2007; Jacobsen et al., 2008b; Jackson et al., 2011b)

6.2.1 Maternal inflammatory response

Inhalation exposure to nanoparticles (UV-Titan and Printex 90) induced potent and persistent lung inflammation (evident by polymorphonuclear neutrophil infiltration), observed as late as 24-27 days after termination of exposure. Thus, extensive inflammation persisted throughout lactation.

When comparing the inhalation and instillation exposures (the Printex 90 studies), both induced a massive influx of neutrophils in the BAL fluid, which persisted for 24-27 days after the end of exposure. The pulmonary inflammation observed after inhalation of Printex 90 compared in magnitude to that in mice instilled with the medium dose of Printex 90. The instilled medium dose was one fifth of the instilled highest dose, which was originally estimated to equal the pulmonary deposition in the inhalation study. Generally, instillation of particles tends to induce stronger lung inflammation than do inhalation of particles (Jacobsen et al., 2009). Instilled particles are forced into the alveoli, resulting in a relatively smaller deposition in the bronchia and bronchioles, which slows down clearance.
Global gene expression was analyzed in UV-Titan exposed lungs of non-littering time-mated females five days after exposure using DNA microarrays. Gene profiling revealed increased levels of several genes involved in inflammation, acute phase response and immune response along with significant changes in expression of several miRNAs (Halappanavar et al., 2011). Significantly upregulated genes included several chemokines involved in chemotaxis, infiltration of neutrophils, and epithelial proliferation. Collectively, these molecules act in host defence by promoting phagocytosis and inflammation. In addition, the expressed genes are involved in activation of the pulmonary immune response. The expressed genes are indicative of persistent pulmonary inflammation and initiation of a secondary lung response. In a genomic study with pregnant mice intranasally exposed to micro-sized 50 µg/mouse TiO_2 (2 mg/kg) pregnant mice demonstrated enhanced lung inflammation compared to non-pregnant mice (Lamoureux et al., 2010). This suggests that the observed response in the pregnant mice might have been even stronger, than that observed in the non-littering mice. Interestingly, the gene expression in liver from UV-Titan exposed non-pregnant females was relatively unaffected five days after exposure (Halappanavar et al., 2011). These findings agree with a previous report, where few changes in global hepatic transcriptome were observed after inhalation exposure to carbon black or diesel exhaust particles (20 mg/m^3 for 90 min/day for 4 consecutive days), although pulmonary inflammation was present (Saber et al., 2009).

6.2.2 Maternal levels of DNA strand breaks

The primary genotoxicity of nanoparticles is related to their ability to induce reactive oxygen species (ROS) (Jacobsen et al., 2008b), and nanoparticles can induce secondary genotoxicity mediated by inflammation (Knaapen et al., 2004).
In our studies, we did not observed DNA strand breaks in BAL cells after inhalation exposure to Printex 90 at the time of analysis, 5 and 26-27 days after the last exposure. We assume that it is possible that induced DNA damage was repaired at the time (Jackson et al., 2011b). Similarly, it was observed (Bornholdt et al., 2002) that levels of DNA strand breaks peaked before induction of proinflammatory cytokines and were repaired subsequently.
In the liver cells, DNA strand breaks were increased after inhalation exposure to Printex 90 even weeks after exposure, while no genotoxicity was found instillation exposure to Printex 90. Both exposures induced a strong and persistent inflammatory response in the exposed lung, so it is unlikely that the DNA strand breaks were affected by circulating cytokines (Jackson et al., 2011b). Similarly to our studies, pulmonary exposure to Printex 90 (nose-only inhalation exposure to 20 mg/m^3 for 90 min/day on 4 consecutive days) resulted in pulmonary production of cytokines (Saber et al., 2005), but no liver inflammation or acute phase response was found in liver after to Printex 90 (Saber et al., 2009). Thus, it is unlikely

that the observed DNA strand breaks were caused by liver inflammation induced by pulmonary exposure.

The exposure procedure is a key determinant for particle size-distribution and consequently for deposition and uptake, which probably determined genotoxicity. The DNA damage observed in the liver of Printex 90 inhalation exposed mice may be a result of the inhalation-associated gastrointestinal exposure rather than from exposure in the lungs. It has been reported that an intra-gastric exposure to 0.64 mg/kg Printex 90 induced DNA damage in the liver of rats 24 hours after exposure, whereas the same dose administered by intratracheal instillation caused no DNA damage in the liver or lung (Danielsen et al., 2010). Similarly, intra-gastric administration of other carbonaceous nanoparticles (such as single wall carbon nanotubes, C_{60} fullerenes and diesel exhaust particles) at the same or even lower doses, caused DNA base oxidation damage in the liver and lung of rats (Danielsen et al., 2008; Folkmann et al., 2009).

Printex 90 instillation exposure would result in a small intragastric deposition; however, a smaller instilled dose would end in less intragastric dose. It is possible that most Printex 90 particles that would reach the circulation were expected to accumulate in the liver. Nanoparticles would persist in the Kupffer cells of the liver for months (Sadauskas et al., 2007; Sadauskas et al., 2009). Consequently, few liver cells would be directly exposed to ROS generated from Printex 90. ROS production and oxidative stress are linked to damage to the DNA (Borm et al., 2006; Møller et al., 2010a). Thus, it is likely that observed DNA damage in Printex 90 exposed liver cells is caused by ROS-induced primary genotoxicity.

6.3 Developmental toxicity of nanoparticles

As discussed previously, we expect that only a small fraction of nanoparticles translocate outside the lung cavity or the gastrointestinal mucosa. Therefore, only few particles would reach the placental barrier, and even fewer would reach the fetus and affect the fetus directly. Analysis of UV-Titan in the maternal lung, liver and offspring liver confirmed our expectation, and significant amounts of UV-Titan were restricted to the maternal lung (Hougaard et al., 2010). However, small amounts of nanoparticles would not be detected due to a high analytical detection limit for titanium. The UV-Titan was coated with Zr, Si, Al, Na as well complex polyalcohols. It is possible that impurities leached from the particle coating reached the blood stream, crossed the placenta, and affected the offspring. Printex 90 contained minute amounts of polycyclic aromatic hydrocarbons (PAH) that are also reported to leach from the particle and cause effects in the prenatally exposed offspring (Srivastava et al., 1986). However, we assumed that the concentration of PAHs in Printex 90 was negligible and effects of Printex 90 were not caused by leached impurities.

Pulmonary exposure to nanoparticles induced inflammation and acute phase response in the lungs of exposed time-mated mice (Halappanavar et al., 2011; Hougaard et al., 2010; Jackson et al., 2011b). The inflammatory cytokines and acute phase proteins induced in the maternal lungs may have crossed the placenta and induced effects in the offspring. The inflammatory cytokines interleukin-1b (*IL-1b*), interleukin-6 (*IL-6*) and tumor necrosis factor (*TNF-a*) are important for fetal development. IL-1b regulates embryogenesis and fetal development, and inflammatory cytokines increase as a signal for the onset of labour (similarly to 'transplant rejection'). However, excessive inflammation during pregnancy negatively affects the offspring. Maternal inflammation during pregnancy, induced by exposure to lipopolysaccharide (LPS), resulted in reduced activity of the hypothalamic pituitary adrenal axis and altered the immune response in offspring later in life (Lasala & Zhou, 2007). Pro-inflammatory insult during gestation may also affect neuronal development and memory (Graciarena et al., 2010; Lasala & Zhou, 2007). Based on these results, it can be hypothesised that nanoparticle induced inflammatory signals from the mother could interfere with development of the offspring.

6.3.1 Gestation and lactation

Despite the persistent inflammation and acute phase response in the maternal lung, inhalation exposure to UV-Titan and Printex 90, and instillation exposure to Printex 90, did not affect gestational and developmental parameters (Hougaard et al., 2010; Jackson et al., 2011b). To our knowledge, only one other study reported gestational parameters after airway exposure. Time-mated mice were intratracheally instilled with carbon black dispersed in saline solution with 0.05% Tween 80 (0.2 mg/animal total dose) on GD 7 and 14. Similar to our studies, no effects on gestation were observed (Yoshida et al., 2010). In conclusion, pulmonary exposure to nanoparticles repeatedly did not induce maternal toxicity or fetotoxicity and developmental toxicity.

6.3.2 Offspring levels of DNA strand breaks

We evaluated liver DNA damage in the offspring prenatally exposed to Printex 90 at birth, weaning and during adolescence. DNA damage was quantified by the alkaline comet assay, an analysis of the level of DNA strand breaks.

The background level of DNA strand breaks was higher in liver cells from newborns compared to older siblings in all three studies. These DNA strand breaks might be related to a high proliferation rate during tissue maturation and/or the naturally occurring high level of oxidative stress at birth (McArt et al., 2010; Randerath et al., 1996). This may have reduced the sensitivity of the comet assay to detect differences between the exposure groups at this time point.

Prenatal exposure to Printex 90 after maternal inhalation exposure increased levels of DNA strand breaks even weeks after birth, while no effects were found after exposure to Printex 90 after maternal instillation exposure. A few molecular genotoxins have been demonstrated to pass from the mother to the fetus and generate DNA damage in fetal tissues (Brunborg et al., 1996; Tripathi et al., 2008). However, we expect that only a small fraction of Printex 90 particles translocated from the lungs of the mothers to the fetuses because the particles would have to pass two compartmental barriers, i.e. in the lung and placenta. The observed effects of prenatal exposure were therefore more likely due to changes in signalling cascades. It is possible that inflammatory molecules were transferred from the maternal to the fetal compartment and affect the fetus. Thus, the increased levels of DNA strand breaks in liver tissue of the offspring may be caused by maternally induced inflammatory mediators after Printex 90 inhalation exposure.

DNA strand breaks in offspring liver of the inhalation-exposed dams were still evident in fifty-day old offspring. At this time, the offspring were independently fed and had no contact with the dams. Therefore, it is unlikely that secondary genotoxicity caused by inflammatory signalling from the dams caused the observed DNA strand breaks in the older offspring. Further work is needed to establish the basis of this extended damage to the DNA in the Printex 90 prenatally exposed offspring.

6.3.3 Developmental neurotoxicity

Offspring prenatally exposed to UV-Titan (after maternal inhalation exposure) exhibited changes in activity and in sensori-motor processes (tested in the Open field and the Startle test, respectively); no changes in learning and memory tested in the Morris water maze were observed (Hougaard et al., 2010). Prenatal exposure to Printex 90 (after maternal intratracheal instillation exposure) induced small behavioural changes in the open field (Jackson et al., 2011c).

A few other studies have assessed the effects of prenatal exposure to nanoparticles on the nervous system. Titanium dioxide nanoparticles (100 µL of 1 mg/ml) were injected subcutaneously into pregnant mice on GD 3, 7, 10, and 14. Nanoparticles were found in the brain, and they induced apoptotic changes in the olfactory bulb in the prenatally exposed male offspring six weeks after the prenatal exposure (Takeda et al., 2009). A comparable exposure increased levels of dopamine in the offspring brain (Takahashi et al., 2010). In addition, gene expression was altered in the offspring brain related to development and function of the central nervous system. However, only one offspring per group was analysed, which hampers the interpretation of these results (Shimizu et al., 2009). After an oral dose (100 mg/kg BW per day) administered to rats during gestation (GD 2-21) or lactation (PND 2-21), titanium was increased in the hippocampus. Lactational exposure attenuated synaptic plasticity in the hippocampus (associated with learning and memory), while a lesser effect was observed in the prenatally exposed offspring (Gao et al., 2011). We have reported previously that female offspring prenatally exposed to diesel exhaust particles (dams inhaled 19 mg DEP/m^3 1 h/day on GD 8-18) exhibited increased activity in the Open field (Hougaard et al., 2009). In conclusion, prenatal exposure to nanoparticles have been repeatedly associated with changes in behaviour related to activity level and pathological changes in the brain, suggesting that nanoparticles may impact development of the nervous system.

7. Conclusions

There is probably a limited uptake of insoluble nanoparticles over epithelial barriers in lung and intestines. However, particles that reach the blood stream may reach the placenta and the fetus. Furthermore, the well established effects of nanoparticles, inflammation and genotoxicity, may be especially damaging to the developing fetus. There is a lack of information of particle effects during pregnancy, and the consequences of prenatal exposure later in the life.

Prenatal exposure to titanium dioxide (UV-Titan) and carbon black (Printex 90) nanoparticles was evaluated in similar whole-body inhalation exposures. Printex 90 was also evaluated by intratracheal instillation. The chosen exposures were relatively high, where the daily exposures corresponded approximately to the 8-hr time weighed average occupational exposure limit (6 mg titanium dioxide/m^3 and 3.5 mg carbon black/m^3, The Danish Working Environment Authority 2007). The UV-Titan exposure atmosphere contained more agglomerates, compared to the Printex 90 exposures. This affected the particle number in the inhaled air, as well as the estimated pulmonary and intragastric exposure. All exposures induced persistent maternal pulmonary inflammation, lasting for weeks after the end of exposure. Gene expression analysis also indicated that maternal lungs also exhibited signs of acute phase response.

DNA damage due to chemical exposure is associated with an increased rate of mutations and ultimately increased risk of cancer. Especially the fast developing fetus may be sensitive to DNA damage, because frequent cell divisions may not allow sufficient DNA repair and thus exposure may be predisposition to cancer. In our study, the nanoparticle-induced genotoxicity was particle and exposure specific. Inhalation exposure to Printex 90 induced DNA strand breaks in the liver of exposed time-mated females and their offspring lasting weeks after the end of exposure. Genotoxicity was not observed in time-mated mice or their offspring exposed to Printex 90 by intratracheal instillation. Differences in genotoxicity were likely caused by different particle deposition, uptake and reactivity.

The developing brain may be much more sensitive to chemical exposure compared to the adult brain. Despite this, developmental neurotoxicity is not a commonly tested endpoint included in chemical assessment. We report that prenatal exposure to nanoparticles induced developmental neurotoxicity in the exposed offspring after reaching adulthood. Exposure to UV-Titan induced changes in activity and sensorimotor processes and exposure to Printex 90 affected offspring activity. Our results combined with results reported by others suggest that prenatal exposure to nanoparticles may affect the development of the nervous system.

Despite the observed effects, maternal exposure to nanoparticles did not affect gestation, lactation and prenatally exposed offspring survived and developed normally. The results suggest that the traditional endpoints in the existing guidelines may not fully evaluate nanoparticle effects, because nanoparticles do not seem to be fetotoxic, as such. Nanoparticle effect seem be more subtle, such as altering functional domains of the offspring that require testing beyond the traditional scope.

8. Further research

Many scientific questions regarding prenatal exposure to nanoparticles have been answered in the presented work, and more questions are waiting to be answered.

The nanoparticles used in the current work were of relatively low toxicity. The question remains whether more toxic nanoparticles, e.g. carbon nanotubes or different metal oxides, would affect gestational and developmental parameters. Moreover, the exposure regime used in the presented work excluded early pregnancy and thus did not assess effects on fertilization, early embryonic development and implantation. It remains to be answered whether parental (maternal or paternal) exposure to nanoparticles before fertilization would affect parental reproductive function, gestation, lactation and offspring development. Furthermore, what types of nanoparticles cross the placenta and the effects on placenta function also need to be further addressed. Our results and that of others suggest that prenatal exposure to nanoparticles may affect neurodevelopment of the offspring. More research addressing the mechanisms of developmental neurotoxicity is needed.

We have begun gene expression profiling of newborn offspring liver. However, analysis on exposed offspring later in life could envision possible consequences of the exposure in the adult organism.

It has been suggested that the fetus makes physiological adaptations in response to changes in its environment to prepare itself for postnatal life: 'developmental origins of disease – Baker hypothesis'. These changes may include epigenetic modification of gene expression. Several environmental chemicals are reported to induce epigenetic changes, reviewed by (Baccarelli & Bollati, 2009). The question remains, does prenatal exposure to nanoparticles lead to epigenetic alterations and are the epigenetic alterations transmitted across generations?

9. References

Baan R, Straif K, Grosse Y, Secretan B, El GF, Cogliano V. 2006. Carcinogenicity of carbon black, titanium dioxide, and talc. *Lancet Oncology* 7:295-296.

Baccarelli A and Bollati V. 2009. Epigenetics and environmental chemicals. *Current Opinion in Pediatrics* 21:243-251.

Barton HA, Cogliano VJ, Flowers L, Valcovic L, Setzer RW, Woodruff TJ. 2005. Assessing susceptibility from early-life exposure to carcinogens. *Environmental Health Perspectives* 113:1125-1133.

Bocskay KA, Tang D, Orjuela MA, Liu X, Warburton DP, Perera FP. 2005. Chromosomal aberrations in cord blood are associated with prenatal exposure to carcinogenic polycyclic aromatic hydrocarbons. *Cancer Epidemiology, Biomarkers & Prevention* 14:506-511.

Borm PJ, Robbins D, Haubold S, Kuhlbusch T, Fissan H, Donaldson K, Schins R, Stone V, Kreyling W, Lademann J. 2006. The potential risks of nanomaterials: a review carried out for ECETOC. *Particle and Fibre Toxicology* 3:11.

Bornholdt J, Dybdahl M, Vogel U, Hansen M, Loft S, Wallin H. 2002. Inhalation of ozone induces DNA strand breaks and inflammation in mice. *Mutation Research* 520:63-71.

Bosman SJ, Nieto SP, Patton WC, Jacobson JD, Corselli JU, Chan PJ. 2005. Development of mammalian embryos exposed to mixed-size nanoparticles. *Clinical & Experimental Obstetrics & Gynecology* 32:222-224.

Brunborg G, Soderlund EJ, Holme JA, Dybing E. 1996. Organ-specific and transplacental DNA damage and its repair in rats treated with 1,2-dibromo-3-chloropropane. *Chemico-Biological Interactions* 101:33-48.

Calderon-Garciduenas L, Villarreal-Calderon R, Valencia-Salazar G, Henriquez-Roldan C, Gutierrez-Castrellon P, Torres-Jardon R, Osnaya-Brizuela N, Romero L, Torres-Jardon R, Solt A. 2008. Systemic inflammation, endothelial dysfunction, and activation in clinically healthy children exposed to air pollutants. *Inhalation toxicology* 20:499-506.

Carr KE, Hazzard RA, Reid S, Hodges GM. 1996. The effect of size on uptake of orally administered latex microparticles in the small intestine and transport to mesenteric lymph nodes. *Pharmaceutical Research* 13:1205-1209.

Collins AR. 2009. Investigating oxidative DNA damage and its repair using the comet assay. *Mutation Research* 681:24-32.

Danielsen PH, Loft S, Jacobsen NR, Jensen KA, Autrup H, Ravanat JL, Wallin H, Møller P. 2010. Oxidative stress, inflammation and DNA damage in rats after intratracheal instillation or oral exposure to ambient air and wood smoke particulate matter. *Toxicological Sciences* 118:574-585.

Danielsen PH, Risom L, Wallin H, Autrup H, Vogel U, Loft S, Møller P. 2008. DNA damage in rats after a single oral exposure to diesel exhaust particles. Mutat Res 637:49-55.

Driscoll KE, Costa DL, Hatch G, Henderson R, Oberdöster G, Salem H, Schlesinger RB. 2000. Intratracheal instillation as an exposure technique for the evaluation of respiratory tract toxicity: uses and limitations. *Toxicological Sciences* 55:24-35.

Ema M, Kobayashi N, Naya M, Hanai S, Nakanishi J. 2010. Reproductive and developmental toxicity studies of manufactured nanomaterials. *Reproductive Toxicology* 30:343-352.

Fedulov AV, Leme A, Yang Z, Dahl M, Lim R, Mariani TJ, Kobzik L. 2008. Pulmonary exposure to particles during pregnancy causes increased neonatal asthma susceptibility. *The American Journal of Respiratory Cell and Molecular Biology* 38:57-67.

Folkmann JK, Risom L, Jacobsen NR, Wallin H, Loft S, Møller P. 2009. Oxidatively damaged DNA in rats exposed by oral gavage to C60 fullerenes and single-walled carbon nanotubes. *Environmental Health Perspectives* 117:703-708.

Forchhammer L, Johansson C, Loft S, Møller L, Godschalk RW, Langie SA, Jones GD, Kwok RW, Collins AR, Azqueta A. 2010. Variation in the measurement of DNA damage by comet assay measured by the ECVAG inter-laboratory validation trial. *Mutagenesis* 25:113-123.

Freire C, Ramos R, Puertas R, Lopez-Espinosa MJ, Julvez J, Aguilera I, Cruz F, Fernandez MF, Sunyer J, Olea N. 2010. Association of traffic-related air pollution with cognitive development in children. *J Epidemiology Community Health* 64:223-228.

Fujimoto A, Tsukue N, Watanabe M, Sugawara I, Yanagisawa R, Takano H, Yoshida S, Takeda K. 2005. Diesel exhaust affects immunological action in the placentas of mice. *Environmental Toxicology* 20:431-440.

Gao X, Yin S, Tang M, Chen J, Yang Z, Zhang W, Chen L, Yang B, Li Z, Zha Y. 2011. Effects of Developmental Exposure to TiO(2) Nanoparticles on Synaptic Plasticity in Hippocampal Dentate Gyrus Area: an *In Vivo* Study in Anesthetized Rats. *Biological Trace Element Research in press.*

Graciarena M, Depino AM, Pitossi FJ. 2010. Prenatal inflammation impairs adult neurogenesis and memory related behavior through persistent hippocampal TGFbeta1 downregulation. *Brain, Behavior, and Immunity* 24:1301-1309.

Halappanavar S, Jackson P, Williams A, Jensen KA, Hougaard KS, Vogel U, Yauk CL, Wallin H. 2011. Pulmonary response to surface-coated nanotitanium dioxide particles includes induction of acute phase response genes, inflammatory cascades, and changes in microRNAs: A toxicogenomic study. *Environmental and Molecular Mutagenesis* doi: 10.1002/em.20639.

Hemmingsen JG, Hougaard KS, Talsness C, Wellejus A, Loft S, Wallin H, Møller P. 2009. Prenatal exposure to diesel exhaust particles and effect on the male reproductive system in mice. *Toxicology* 264:61-68.

Hougaard KS, Fadeel B, Gulumian M, Kagan VE, Savolainen K. 2011. Developmental toxicity of engineered nanoparticles. In: Gupta RC, editor. *Reproductive and Developmental Toxicology.* Academic Press, Amsterdam. pp 269-290

Hougaard KS, Jackson P, Jensen KA, Sloth JJ, Loschner K, Larsen EH, Birkedal RK, Vibenholt A, Boisen AM, Wallin H, Vogel U. 2010. Effects of prenatal exposure to surface-coated nanosized titanium dioxide (UV-Titan). A study in mice. *Particle and Fibre Toxicology* 7:16.

Hougaard KS, Jensen KA, Nordly P, Taxvig C, Vogel U, Saber AT, Wallin H. 2008. Effects of prenatal exposure to diesel exhaust particles on postnatal development, behavior, genotoxicity, and inflammation in mice. *Particle and Fibre Toxicology* 5:3.

Hunninghake GW, Gadek JE, Kawanami O, Ferrans VJ, Crystal RG. 1979. Inflammatory and immune processes in the human lung in health and disease: evaluation by bronchoalveolar lavage. *American Journal of Pathology* 97:149-206.

Jackson P, Lund SP, Kristiansen G, Andersen O, Vogel U, Wallin H, Hougaard KS. 2011a. An Experimental Protocol for Maternal Pulmonary Exposure in Developmental Toxicology. *Basic and Clinical Pharmacology Toxicology* 108:202-207.

Jackson P, Hougaard KS, Boisen AMZ, Jacobsen NR, Jensen KA, Møller P, Brunborg G, Gutzkow KB, Andersen O, Loft S, Vogel U and Wallin H. 2011b. Pulmonary exposure to carbon black by inhalation or instillation in pregnant mice: Effects on liver DNA strand breaks in dams and offspring. *Nanotoxicology in press.*

Jackson P, Vogel U, Wallin H, Hougaard KS. 2011c. Prenatal exposure to carbon black (Printex 90): Effects on sexual development and neurofunction. *Basic and Clinical Pharmacology Toxicology in press.*

Jacobsen NR, Møller P, Cohn CA, Loft S, Vogel U, Wallin H. 2008a. Diesel exhaust particles are mutagenic in FE1-MutaMouse lung epithelial cells. *Mutation Research* 641:54-57.

Jacobsen NR, Møller P, Jensen KA, Vogel U, Ladefoged O, Loft S, Wallin H. 2009. Lung inflammation and genotoxicity following pulmonary exposure to nanoparticles in ApoE-/- mice. *Particle and Fibre Toxicology* 6:2.

Jacobsen NR, Pojana G, White P, Møller P, Cohn CA, Korsholm KS, Vogel U, Marcomini A, Loft S, Wallin H. 2008b. Genotoxicity, cytotoxicity, and reactive oxygen species induced by single-walled carbon nanotubes and C(60) fullerenes in the FE1-Mutatrade markMouse lung epithelial cells. *Environmental and Molecular Mutagenesis* 49:476-487.

Jacobsen NR, White PA, Gingerich J, Møller P, Saber AT, Douglas GR, Vogel U, Wallin H. 2010. Mutation spectrum in FE1-MUTA(TM)Mouse lung epithelial cells exposed to nanoparticulate carbon black. *Environmental and Molecular Mutagenesis* 52:331-337.

Johnston HJ, Hutchison GR, Christensen FM, Peters S, Hankin S, Stone V. 2009. Identification of the mechanisms that drive the toxicity of TiO(2)particulates: the contribution of physicochemical characteristics. *Particle and Fibre Toxicology* 6:33.

Kannan S, Misra DP, Dvonch JT, Krishnakumar A. 2007. Exposures to airborne particulate matter and adverse perinatal outcomes: a biologically plausible mechanistic framework for exploring potential. *Cien Saude Colet* 12:1591-1602.

Karlsson HL. 2010. The comet assay in nanotoxicology research. *Analytical and Bioanalytical Chemistry* 398:651-666.

Kennison RD, Bardawil WA, Mitchell GW, Jr. 1971. Passage of particles across the mouse placenta. *Surgery Forum* 22:392-394.

Knaapen AM, Borm PJ, Albrecht C, Schins RP. 2004. Inhaled particles and lung cancer. Part A: Mechanisms. *International Journal of Cancer* 109:799-809.

Kreyling WG, Semmler-Behnke M, Seitz J, Scymczak W, Wenk A, Mayer P, Takenaka S, Oberdöster G. 2009. Size dependence of the translocation of inhaled iridium and carbon nanoparticle aggregates from the lung of rats to the blood and secondary target organs. *Inhalation Toxicology* 21 Suppl 1:55-60.

Lamoureux DP, Kobzik L, Fedulov AV. 2010. Customized PCR-array analysis informed by gene-chip microarray and biological hypothesis reveals pathways involved in lung inflammatory response to titanium dioxide in pregnancy. *Journal of Toxicology and Environmental Health* A 73:596-606.

Lasala N and Zhou H. 2007. Effects of maternal exposure to LPS on the inflammatory response in the offspring. *Journal of Neuroimmunology* 189:95-101.

Li C, Taneda S, Taya K, Watanabe G, Li X, Fujitani Y, Nakajima T, Suzuki AK. 2009. Effects of *in utero* exposure to nanoparticle-rich diesel exhaust on testicular function in immature male rats. *Toxicology Letters* 185(1): 1-8.

Masao Sugamata. Maternal Exposure to Diesel Exhaust Leads to Pathological Similarity to Autism in Newborns. Journal of Health Science 52[4], 486-488. 2006.

Maynard A and Kuempel E.D. Airborne nanostructured particles and occupational health. *Journal of Nanoparticle Research* 7, 587-614. 2005.

McArt, D.G., McKerr, G., Saetzler, K., Howard, C.V., Downes, C.S. & Wasson, G.R. (2010). Comet sensitivity in assessing DNA damage and repair in different cell cycle stages. *Mutagenesis*, Vol. 25, No. 3, pp. 299-303.

McNamee JP, McLean JR, Ferrarotto CL, Bellier PV. 2000. Comet assay: rapid processing of multiple samples. *Mutation Research* 466:63-69.

Mohr U, Ernst H, Roller M, Pott F. 2006. Pulmonary tumor types induced in Wistar rats of the so-called "19-dust study". *Experimental and Toxicologic Pathology* 58:13-20.

Møller P. 2006. The alkaline comet assay: towards validation in biomonitoring of DNA damaging exposures. *Basic and Clinical Pharmacology Toxicology* 98:336-345.

Møller P, Jacobsen NR, Folkmann JK, Danielsen PH, Mikkelsen L, Hemmingsen JG, Vesterdal LK, Forchhammer L, Wallin H, Loft S. 2010a. Role of oxidative damage in toxicity of particulates. *Free Radical Research* 44:1-46.

Møller P, Møller L, Godschalk RW, Jones GD. 2010b. Assessment and reduction of comet assay variation in relation to DNA damage: studies from the European Comet Assay Validation Group. *Mutagenesis* 25:109-111.

Myllynen PK, Loughran MJ, Howard CV, Sormunen R, Walsh AA, Vahakangas KH. 2008. Kinetics of gold nanoparticles in the human placenta. Reprod Toxicol 26:130-137.

Oberdöster G, Oberdöster E, Oberdöster J. 2005. Nanotoxicology: an emerging discipline evolving from studies of ultrafine particles. *Environmental Health Perspectives* 113:823-839.

Oberdöster G, Sharp Z, Atudorei V, Elder A, Gelein R, Lunts A, Kreyling W, Cox C. 2002. Extrapulmonary translocation of ultrafine carbon particles following whole-body inhalation exposure of rats. *Journal of Toxicology and Environmental Health* A 65:1531-1543.

Ono N, Oshio S, Niwata Y, Yoshida S, Tsukue N, Sugawara I, Takano H, Takeda K. 2007. Prenatal exposure to diesel exhaust impairs mouse spermatogenesis. *Inhalation Toxicology* 19:275-281.

Pedersen M, Wichmann J, Autrup H, Dang DA, Decordier I, Hvidberg M, Bossi R, Jakobsen J, Loft S, Knudsen LE. 2009. Increased micronuclei and bulky DNA adducts in cord blood after maternal exposures to traffic-related air pollution. *Environmental Research* 109:1012-1020.

Pope DP, Mishra V, Thompson L, Siddiqui AR, Rehfuess EA, Weber M, Bruce NG. 2010. Risk of low birth weight and stillbirth associated with indoor air pollution from solid fuel use in developing countries. *Epidemiologic Reviews* 32:70-81.

Randerath, E., Zhou, G.D. & Randerath, K. (1996). Organ-specific oxidative DNA damage associated with normal birth in rats. *Carcinogenesis*, Vol. 17, No. 12, pp. 2563-2570.

Saber AT, Bornholdt J, Dybdahl M, Sharma AK, Loft S, Vogel U, Wallin H. 2005. Tumor necrosis factor is not required for particle-induced genotoxicity and pulmonary inflammation. *Archives of Toxicology* 79:177-182.

Saber AT, Halappanavar S, Folkmann JK, Bornholdt J, Boisen AM, Møller P, Williams A, Yauk C, Vogel U, Loft S. 2009. Lack of acute phase response in the livers of mice exposed to diesel exhaust particles or carbon black by inhalation. *Particle and Fibre Toxicology* 6:12.

Sadauskas E, Jacobsen NR, Danscher G, Stoltenberg M, Vogel U, Larsen A, Kreyling W, Wallin H. 2009. Biodistribution of gold nanoparticles in mouse lung following intratracheal instillation. *Chemistry Central Journal* 3:16.

Sadauskas E, Wallin H, Stoltenberg M, Vogel U, Doering P, Larsen A, Danscher G. 2007. Kupffer cells are central in the removal of nanoparticles from the organism. *Particle and Fibre Toxicology* 4:10.

Schneider T and Jensen KA. 2008. Combined single-drop and rotating drum dustiness test of fine to nanosize powders using a small drum. *Annals of Occupation Hygiene* 52:23-34.

Shah PS and Balkhair T. 2010. Air pollution and birth outcomes: A systematic review. *Environment International* 37(2):498-516.

Shimizu M, Tainaka H, Oba T, Mizuo K, Umezawa M, Takeda K. 2009. Maternal exposure to nanoparticulate titanium dioxide during the prenatal period alters gene expression related to brain development in the mouse. *Particle and Fibre Toxicology* 6:20.

Šrám RJ, Binková B, Rossner P, Rubes J, Topinka J, Dejmek J. 1999. Adverse reproductive outcomes from exposure to environmental mutagens. *Mutation Res* 428:203-215.

Srivastava VK, Chauhan SS, Srivastava PK, Kumar V, Misra UK. 1986. Fetal translocation and metabolism of PAH obtained from coal fly ash given intratracheally to pregnant rats. *Journal of Toxicology and Environmental Health* 18:459-469.

Suzuki T, Oshio S, Iwata M, Saburi H, Odagiri T, Udagawa T, Sugawara I, Umezawa M, Takeda K. 2010. *In utero* exposure to a low concentration of diesel exhaust affects spontaneous locomotor activity and monoaminergic system in male mice. *Particle and Fibre Toxicology* 7:7.

Švecová V, Rossner P, Jr., Dostál M, Topinka J, Solansky I, Šrám RJ. 2009. Urinary 8-oxodeoxyguanosine levels in children exposed to air pollutants. *Mutation Research* 662:37-43.

Takahashi S and Matsuoka O. 1981. Cross placental transfer of 198Au-colloid in near term rats. *Journal of Radiation Research* (Tokyo) 22:242-249.

Takahashi Y, Mizuo K, Shinkai Y, Oshio S, Takeda K. 2010. Prenatal exposure to titanium dioxide nanoparticles increases dopamine levels in the prefrontal cortex and neostriatum of mice. *Journal of Toxicological Sciences* 35:749-756.

Takeda K, Suzuki K-I, Ishihara A, Kubo-Irie M, Fujimoto R, Tabata M, Oshio S, Nihei Y, Ihara T, Sugamata M. 2009. Nanoparticles transferred from pregnant mice to their offspring can damage the genital and cranial nerve systems. *Journal of Health Science* 55:95-102.

Takeda K, Tsukue N, Yoshida S. 2004. Endocrine-disrupting activity of chemicals in diesel exhaust and diesel exhaust particles. Environ Sci 11:33-45.

Tang D, Li TY, Liu JJ, Zhou ZJ, Yuan T, Chen YH, Rauh VA, Xie J, Perera F. 2008. Effects of prenatal exposure to coal-burning pollutants on children's development in China. *Environmental Health Perspectives* 116:674-679.

Tian F, Razansky D, Estrada GG, Semmler-Behnke M, Beyerle A, Kreyling W, Ntziachristos V, Stoeger T. 2009. Surface modification and size dependence in particle translocation during early embryonic development. *Inhalation Toxicology* 21:1:92-96.

Tripathi DN, Pawar AA, Vikram A, Ramarao P, Jena GB. 2008. Use of the alkaline comet assay for the detection of transplacental genotoxins in newborn mice. *Mutation Research* 653:134-139.

USEPA (2009). Nanomaterials Research Strategy (EPA 620/K-09/011), USEPA

Wick P, Malek A, Manser P, Meili D, Maeder-Althaus X, Diener L, Diener PA, Zisch A, Krug HF, von MU. 2010. Barrier capacity of human placenta for nanosized materials. *Environmental Health Perspectives* 118:432-436.

Yokota S, Mizuo K, Moriya N, Oshio S, Sugawara I, Takeda K. 2009. Effect of prenatal exposure to diesel exhaust on dopaminergic system in mice. *Neuroscience Letters* 449:38-41.

Yoshida S, Hiyoshi K, Ichinose T, Takano H, Oshio S, Sugawara I, Takeda K, Shibamoto T. 2009. Effect of nanoparticles on the male reproductive system of mice. *International Journal of Andrology* 32:337-342.

Yoshida S, Hiyoshi K, Oshio S, Takano H, Takeda K, Ichinose T. 2010. Effects of fetal exposure to carbon nanoparticles on reproductive function in male offspring. *Fertility and Sterility* 93:1695-1699.

Part 2

Air Pollution and Impact on Human Health

4

Molecular Markers Associated with the Biological Response to Aromatic Hydrocarbons from Urban Air in Humans

Francisco Arenas-Huertero[1,2], Elisa Apátiga-Vega[2], Gabriela Miguel-Pérez[2], David Villeda-Cuevas[2] and Jimena Trillo-Tinoco[3]

[1]*Hospital Infantil de México Federico Gómez,*
[2]*Universidad Simón Bolívar,*
[3]*Hospital General de México*
México

1. Introduction

Morbidity and mortality attributable to air pollution continue to be a growing problem in several parts of the world. Both epidemiologic and clinical studies have demonstrated a strong link between exposure to particulate matter (PM) and adverse effects on health. From the PM generated in the atmospheres of several countries, the respirable fraction of $PM_{2.5}$ ($PM_{2.5}$) and diesel exhaust particles (DEP) represent some of the largest products of vehicular- and industrial-emitted airborne PM that can persist in the air, where they are readily inhaled and deposited throughout the respiratory tract. $PM_{2.5}$ and DEP have been associated with cardiac and pulmonary alterations. Also, exposure to DEP has been associated with lung cancer, pulmonary inflammation, an increased susceptibility to respiratory infections and the exacerbation of asthma and chronic obstructive pulmonary diseases. Furthermore, the effect of tobacco and its smoke, a complex mixture, represents another source of polycyclic aromatic hydrocarbons (PAH). The three, $PM_{2.5}$, DEP and tobacco smoke, are the main foci of several studies that evaluate the principal effects of PAH, and represent the via to/source of PAH exposure. Normally, black carbon particles, also products of incomplete fuel combustion, act as condensation nuclei for organic chemicals, such as aromatic aliphatic compounds, including PAH, but they are not considered in this chapter. Rather, this chapter will centre on a molecular description of the cellular responses (AHR pathway) after exposure to PAH from urban air, including relatively new markers, microRNAs and their utility as new biomarkers of exposure to PAH. We propose the use of lung tissue embedded in paraffin as a source of biological material to perform any kind of study: retrospective and prospectives. First, however, an important general description of PAH, its main sources and its concentrations in urban air and their metabolism will be presented in order to contextualize the main objective of this study of molecular markers associated with PAH exposure.

2. Polycyclic aromatic hydrocarbons (PAH)

Polycyclic aromatic hydrocarbons (PAH) are a complex class of condensed multi-cyclic of benzoic ring compounds, fused in linear, angular, or cluster arrangements that sometimes include a five-member ring, containing only carbon and hydrogen atoms. They have a basic structure and substituent moieties such as the alkyl, amino, chloro, cyano, hydroxyl, or thiol groups, and/or contain atoms such as nitrogen, oxygen, or sulphur in their aromatic structure. The most well known PAH is benzo[a]pyrene (BaP) (Gachanja, 2009; Wang et al., 2010).

The physico-chemical properties of PAH largely determine their environmental behaviour. They are semi-volatile substances under atmospheric conditions and frequently occur both in the vapour phase and attached to particles, depending on the vapour pressure of each PAH. Low molecular weight PAH, containing two or three fused rings, are more water soluble and volatile, and are found predominantly in the vapour phase. High molecular weight PAH containing more than three fused rings, which are primarily associated with particles, are found mainly absorbed in PM (Rajput et al., 2010). Generally, between 80% and almost 100% of PAH with 5 rings or more (which are predominantly particle-bound in the atmosphere) can be found associated with particles with an aerodynamic diameter of less than 2.5 µm, $PM_{2.5}$.The physico-chemical properties of PAH vary considerably, their semi-volatility makes them highly mobile throughout the environment; their deposition and re-volatilisation distribute them among air, soil and water bodies and most can be photo-oxidized and degraded to simpler substances (Office for Official Publications of the European Communities, 2001).

PAH are ubiquitous in the environment and are suspected or known mutagenic and carcinogenic agents. The United States Environment Protection Agency has listed 16 PAH as priority pollutants. These are: naphthalene, acenaphthylene, acenaphthene, fluorene, phenanthrene, anthracene, fluoranthene, pyrene, benzo[*a*]anthracene, chrysene, benzo[*b*]fluoranthene, benzo[*k*]fluoranthene, benzo[*a*]pyrene, indeno[1,2,3-cd]pyrene, dibenzo-[*a,h*]anthracene, and benzo[*ghi*]perylene (United States Environment Protection Agency [USEPA], 1993).

2.1 Main sources of PAH

PAH are produced in all processes of incomplete combustion of organic substances. Their production is favoured by an oxygen-deficient flame, temperatures in the range of 650-900°C and fuels that are not highly oxidized (Maliszewska-Kordybach, 1999). At the temperature of pyrolisis (about 700°C), aromatic ring systems are the most stable among the structural types present, while aliphatic C-C bond and C-H bonds readily break down to yield molecular fragments of a free radical character which then undergo recombination (Rajput & Lakhani, 2010).

Natural sources of PAH such as volcanic activity and forest fires do not significantly contribute to present-to-overall PAH emissions. Anthropogenic sources can be divided into two categories: the combustion of materials for energy supplies (*e.g.* coal, oil, gas, wood, biodiesel, *etc.)*; and combustion for waste elimination (*e.g.* incineration). The first category includes stationary sources that are considered the main producers of PAH in outdoor air, such as industry (mainly coke and carbon production, petroleum processing, aluminium sintering, etc.), refineries, residential heating (furnaces, fireplaces and stoves, gas and oil burners), power and heat generation (coal, oil, wood and peat power plants), and mobile sources like cars, lorries, trains, airplanes and sea traffic (gasoline and diesel engines). The second category includes the incineration of municipal and industrial wastes. Other miscellaneous sources are unregulated fires, such as agricultural burning, recreational fires, crematoria, cigarette smoking, as well as volatilisation from soils, vegetation and other surfaces (Maliszewska-Kordybach, 1999).

2.2 Formation of polycyclic aromatic hydrocarbons from tobacco

There are two generally accepted mechanisms in the literature; the first involves the pyrogenesis of PAH by the thermal degradation of organic tobacco components into small reactive molecules and/or free radicals during the high temperature pyrolysis processes, followed by recombination reactions of these reactive species to yield PAH. The second mechanism involves the unimolecular cyclization, dehydration, aromatization and ring growth of high molecular weight tobacco components, such as phytosterols, long-chained saturated and unsaturated hydrocarbons, alcohols and esters. It has been suggested that both mechanisms exist during the high temperature pyrolysis/combustion of tobacco. Of the many individual constituent classes of tobacco (alkaloids, reductor sugars, polysaccharides, long-chain hydrocarbon waxes, amino acids, proteins, *etc.*), lipophilic components, such as phytosterols, saturated aliphatic hydrocarbons, and terpenoid compounds were believed to be the major precursors of PAH formed from a burning cigarette. There is no doubt that the pyrolysis of tobacco, tobacco extracts, and individual components of tobacco at temperatures greater than 700°C leads to the formation of PAH (McGrath et al., 2006).

Cigarette smoking and environmental tobacco smoke are other sources of air exposure. Smoking one cigarette can yield an intake of 20-40 ng of BaP. Smoking one pack of unfiltered cigarettes per day yields 0.7 μg/day of BaP, while smoking a pack of filtered cigarettes per day yields 0.4 μg/day (Agency for Toxic Substances and Disease Registry´s [ASTDR], 1995).

2.3 PAH in urban air

PAH concentrations in air can vary from less than 5 to 200,000 nanograms/cubic meter (ng/m^3). Although environmental air levels are lower than those associated with specific occupational exposures, they are a public health concern when spread over large urban populations. The background levels of 17 of the Agency for Toxic Substances and Disease Registry's (ATSDR) toxicological profile priority PAH in ambient air are reported to be 0.02-1.2 ng/m^3 in rural areas and 0.15-19.3 ng/m^3 in urban zones [ATSDR, 1995].

The results for 37 countries were compared with other PAH emission inventories. It was estimated that the total global atmospheric emission of these 16 PAH in 2004 was 520 gigagrams per year (Gg y_1) with biofuel (56.7%), wildfires (17.0%) and consumer product usage (6.9%) being the major sources; China (114 Gg y_1), India (90 Gg y_1) and United States (32 Gg y_1) were the three countries with the highest PAH emissions (Zhang & Tao, 2008).

Levels of individual PAH are monitored in outdoor air, but are always present with other PAH as part of a complex mixture. Levels of PAH in outdoor air are much lower than those encountered in occupational settings and there is no convincing evidence, as yet, to suggest that PAH in outdoor air are a significant cause of lung cancer in the general population.

Estimating the health effects due to PAH in outdoor air is made difficult by cigarette smoking, which is a source of PAH in itself, and which can affect outdoor sources in population studies. Therefore, people are always exposed to a number of different PAH, rather than single PAH on their own (Comittee on the Medical Effects of Air Pollutants [COMEAP], 2011). Furthermore, a specific PAH can be characteristic of the air in each country. For example, in Mexico City, benzo[*ghi*]pyrelene and benzo[*123cd*]pyrelene are the most representative in the atmosphere with annual medians from 1.119 ng/m^3 in the southwest zone, to 1.84 ng/m^3 in the central zone, where an important number of automobiles circulate because it is the downtown area. Otherwise, BaP was present from 0.265 ng/m^3 in the southwest zone to 0.455 ng/m^3 in the downtown area of Mexico City. In these same zones, total heavy PAH were 6.089 ng/m^3 in the downtown zone and 3.402 in the southwest sector (Amador-Muñoz et al., 2011). These measurements were obtained from the total PAH contained in $PM_{2.5}$. Again, this example reveals how important it is to take into account each geographical region with its characteristic PAH, which are significant in evaluating health risks.

3. Metabolism of PAH: background

Despite the fact that the human body has evolved inducible enzymatic detoxification and DNA repair systems over millions of years for efficient protection against natural toxic non-polar exogenous chemicals, given the tremendous amount and diversity of chemical pollutants that have recently permeated the environment, these systems may be saturated by excess toxicants without being fully adapted for a complete detoxification of all man-made molecules. Because the organism can not fully metabolize and inactivate all non-polar exogenous chemicals, this would explain why lipophilic carcinogenic environmental pollutants such as PAH can bio-accumulate in the adipose tissue and be toxic (Irigaray & Belpomme, 2010). In this way, mammals and many lower organisms metabolize PAH primarily by enzymatic oxygenation into epoxides, phenols, dihydrodiols, quinones, and water-soluble conjugates, in an attempt to make them more soluble and thus facilitate their excretion from the organism. This converts the pre-carcinogen into the ultimate carcinogen that covalently binds mainly to DNA, forming the DNA-adducts, and leads to carcinogenesis, v.g. lung cancer. The biological activity of PAH is dependent upon the molecular structure and their isomers show diverse carcinogenic activity (Gerhard, 2005).

In most cases, oxidation of PAH by cytochrome P450 (CYP) enzymes is an initial step of the activation process to produce the polar biochemically reactive electrophilic species (ultimate carcinogenic metabolites), capable of interacting with cellular macromolecules, particularly nucleic acids, as mentioned above, and proteins (Nebert et al., 2004; Weiling & Warshawsky, 2005).

Since the 2-4-ring PAH is poor enzyme inducer, it appears that the PAH-metabolizing pathways are mainly induced by BaP-type minor constituents. Gene-environmental interactions that magnify the polymorphic variability in the pulmonary bioactivation/detoxification capacity probably play a key role in individual susceptibility to or protection against, chemically-induced lung cancer. Hence, human exposure to PAH mixtures with a high content of BaP-type hydrocarbons confers a potentially higher health risk than PAH mixtures with a low content of pro-carcinogens (Elovaara et al., 2006).

3.1 Metabolism of PAH in liver

PAH are metabolized into various products by xenobiotic (drug)-metabolizing enzymes such as CYPs, epoxide hydrolase (EH), glutation transferase (GST), UDP-glucuronosyltransferase (UGT), sulfotransferase (SULT), NAD(P)H-quinone oxidoreductase 1 (NQO1), and aldoketoredustases (AKR). As was mentioned above, CYP play key roles in the initial step of oxidation of PAH. First, BaP is converted into radical cations through a one-electron oxidation mechanism by peroxidase activity and the resultant products can react with DNA to form unstable depurinating adducts. Second, AKR has been shown to convert B[a]p-7,8-diol to reactive B[a]P o-quinone via B[a]P-catechols. The reactive B[a]P o-quinone is able to interact with DNA, forming stable and depurinating DNA adducts. As mentioned: PAH can be activated by two enzymatic actions; one forms PAH diol-epoxides by CYPs, while the other produces PAH-*o*-quinones by AKR. Finally, it is suggested that the quinines formed are carcinogenic metabolites. These quinine derivatives are detoxified by NQO1. These reactive metabolites of PAH are further converted by so-called phase II enzymes such as EH, UGT, GST, SULT, NQO1 and AKR into more polar and detoxified metabolites. The EH catalyzes the hydrolysis of various oxides of numerous endobiotic and xenobiotic chemicals into less reactive and more polar dihyrodiols. GST is constitutively expressed in various human tissues. Cytosolic GST encoded by polymorphic members of the alpha, mu, pi and theta gene families and play important roles in the metabolism of a variety of toxic and carcinogenic compounds. Most of the glutathione conjugates are less toxic and more polar and can be excreted from the body. Glucuronidation is a major pathway for the detoxification of numerous carcinogens such as PAH and aryl and heterocyclic amines. The UGT superfamily consists of two families, UGT1 and UGT2. These enzymes convert the PAH intermediates into molecular forms that are more likely to be excreted from the organism. Sulfonation performed by SULT is generally thought to be a detoxification process. In humans, SULT consists of three families. However, sometimes SULT can activate certain promutagens to highly reactive sulphate esters that bind covalently to DNA. NQO1 catalyzes two-electron reduction of a wide variety of substrates, including the PAH o-quinone, into inactive products, such as PAH-hydroquinones. NQO1 has been shown to convert benzene-derived quinines into more inactive hydroquinones, thus protecting benzene-induced hematotoxicity. Finally, AKR converts PAH trans-dihydrodiols into reactive PAH o-quinones that form stable and depurinating DNA adducts. Thus, PAH-diols are shown to be activated by two enzymatic actions (Shimada, 2006; Shimada & Fujii-Kuriyama, 2004).

Human CYP1A2 is notable among family 1 enzymes for its capacity to *N*-oxidize arylamines, the major metabolic process involved in the bioactivation of arylamines to potent mutagenic or carcinogenic compounds. CYP1A2 is the principal family 1 enzyme expressed in the human liver, and CYP1A2 contributes significantly to the hepatic metabolism of drugs. Among liver CYP drug-metabolizing enzymes, CYP1A2 plays a predominant role in the metabolic elimination of caffeine and melatonin, as well as commercial drugs such as flutamide, lidocaine, olanzapine, tacrine, theophylline, triamterene, and zolmitriptan. This enzyme is also of great importance in the bioactivation of mutagens, including the *N*-hydroxylation of arylamines. These enzymes are generally distinguished from CYPs in other families by their capacity to oxidize a variety of polynuclear aromatic hydrocarbons. The induction is mediated by a ligand-activated transcription factor, the aryl hydrocarbon receptor, (AHR), which binds to enhancer elements flanking the *CYP1A1*, *CYP1A2*, and *CYP1B1* genes and stimulates transcription (Shimada & Fujii-Kuriyama, 2004). This mechanism will be described below.

CYP1A1 and 1B1 are expressed in a wide range of extra-hepatic tissues and catalyze both the activation and detoxification reactions of PAH metabolism (Shimada & Fujii-Kuriyama, 2004). In addition, both enzymes are inducible by the PAH in cigarette smoke (Nebert et al., 2004). Induction of these two enzymes is generally mediated by the AHR, but differences may exist in the mode of induction for each enzyme.
Research on human hepatoma cells gene expression changes caused by BaP at 12 time points after exposure, in relation to DNA adduct and cell cycle, the temporal profiles for functional gene sets demonstrate both early and late effects in the up- and down-regulation of the relevant gene sets involved in cell cycle, apoptosis, DNA repair, and the metabolism of amino acids and lipids. Many significant transcription regulation networks appeared to be performed by transcription factors, TF, that are proto-oncogenes or tumour suppressor genes. Most correlations are with DNA adduct levels, which is an early response, and less with the later responses on G1 and S phase cells. The majority of the modulated genes are regulated by several of these TF, e.g., 73% by nuclear factor-kappa B and 34–42% by c-MYC, SRF, AP1, and E2F1. All these TF can also regulate one or more of the others. The data indicate that a complex network of a few TFs is responsible for the majority of the transcriptional changes induced by BaP. This network hardly changes over time, despite the fact that the transcriptional profiles clearly alter, suggesting that other regulatory mechanisms are also involved (Joost et al., 2010). These include the AHR pathway, for example.

3.2 The metabolism of tobacco smoke

Tobacco from both active and passive smoking has the main point of entry into the body via the airways; some constituents are dissolved in saliva and absorbed or swallowed. Virtually all organs and tissues are reached by the active products of smoking. Data from epidemiological studies confirm the widespread presence of active products of tobacco smoke on tissues and organs (Taioli, 2008). After binding the AHR, they induce Phase I and Phase II mRNAs and their corresponding proteins both *in vitro* and in experimental animals. The CYP superfamily members CYP1B1 and CYP1A1 have been reported to be expressed in human lungs. The highly reactive bioactivated intermediates of inhaled PAH carcinogens, such as benzo(*a*)pyrene-7,8-dihydrodiol-9,10-epoxide, confer "hot spots" for benzo(*a*)pyrene-7,8-dihydrodiol-9,10-epoxide-induced mutations in the tumour suppressor p53 gene *in vitro* and closely match the overall p53 gene mutation spectra found in a wide array of epithelial cancers *in vivo*. There is coordinate metabolism of estradiol and inhaled PAH by CYP1B1, suggesting a need to assay gender-associated factors impacting carcinogen metabolism expression in the human lung (Spivack et al., 2003).
The balance between the metabolic activation and detoxification of carcinogens varies among individuals and likely affects cancer susceptibility. Persons with a higher activation and lower detoxification capacity are at the highest risk for smoking-related cancers (Taioli, 2008).

4. The aryl hydrocarbon receptor in lung and the polycyclic aromatic hydrocarbons

The lung serves as a primary site for the xenobiotic metabolism of environmental toxicants and airborne pollutants. The lung is composed of more than 40 different cell types and is known to activate pro-carcinogens (e.g. polycyclic aromatic hydrocarbons or N-nitrosamines) into more reactive intermediates that easily form DNA adducts (Pavek & Dvorak, 2008). Epidemiological studies suggest that exposure to PAH has been associated with an increased risk of lung cancer (Lin et al., 2003). The body has numerous molecular mechanisms that induce and restrict the activity of cell receptors in order to respond to these environmental compounds. These studies recognized the AhR (protein, AhR; gene, *AHR*) as a receptor that binds a wide variety of endogenous and exogenous compounds; it was identified as a mediator of the induction of certain xenobiotic drug metabolizing enzymes and the toxicity elicited by halogenated aromatic hydrocarbons (Gasiewicz & Collins, 2008). AHR is both a player in chemical toxicity and an important component of normal development (Nguyen & Bradfield, 2008). It was discovered due to its stimulation by a variety of planar aromatic hydrocarbons with BAP as prototype. Currently, more than 400 exogenous ligands have been identified. The best studied and one of the most potent ligands knows so far is 2,3,7,8-tetrachlorodibenzo-p-dioxin (TCDD) (Dietrich & Kaina, 2010). Human lung bronchial-epithelial cells (NL-20) expresses mRNA of AHR (Fig 1A) with important levels of the protein delocalized in the cell (Fig 1B) and important translocation of it into the nucleus after exposition of solvent extractable organic matter (SEOM) 0.1 µg/µl (Fig. 1C). Interaction of the antagonist of the AhR induces both accumulation of the mRNA (Fig. 1D) and the protein (Figs 1E-1F).

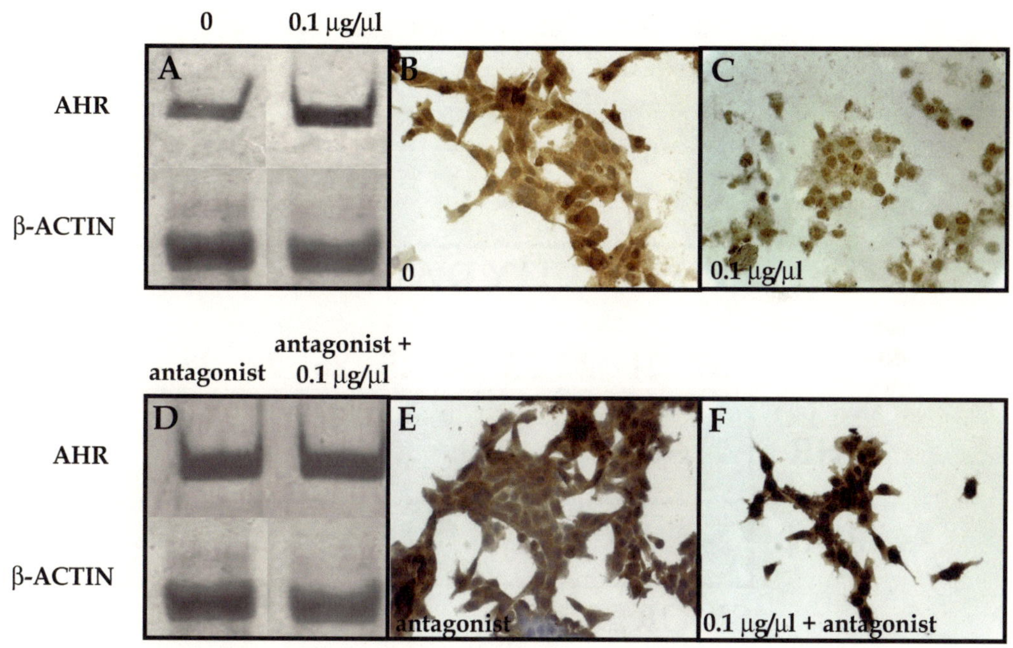

Fig. 1. Effect of PAH contained in SEOM on NL-20 human bronchial epithelial cells in the expression of the mRNA of *AHR* (A and D) and the protein AhR in cells exposed to 0.1 µg/µl of SEOM alone (from A to C), and in presence of SEOM plus an antagonist of the AhR (from D to F). PCR products resolved in polyacrylamide-gel electrophoresis and silver staining. Immunohistochemistry of AhR. Micrographies at 200X

AHR can be a molecular marker of exposition to different levels of air pollutants and it can reveal different populations exposed. Thus, residents in Mexico out of big cities as Mexico City (Fig. 2A) express high levels of the mRNA of the *AHR*. It trends to lowering if the lung is of Mexico City dwellers (Fig. 2B) and it decrease importantly in lung of active smokers (Fig. 2C). These results correlate with the cycle of the AhR, which is degraded if is interacting with its ligand (PAH) and it is maintained when there is no ligand present in the air. The main issue in these results is that mRNA of the *AHR* can be a molecular marker of air pollution, mainly to PAH in lung tissue aged for several years, 5 years as is illustrated in the Fig. 2.

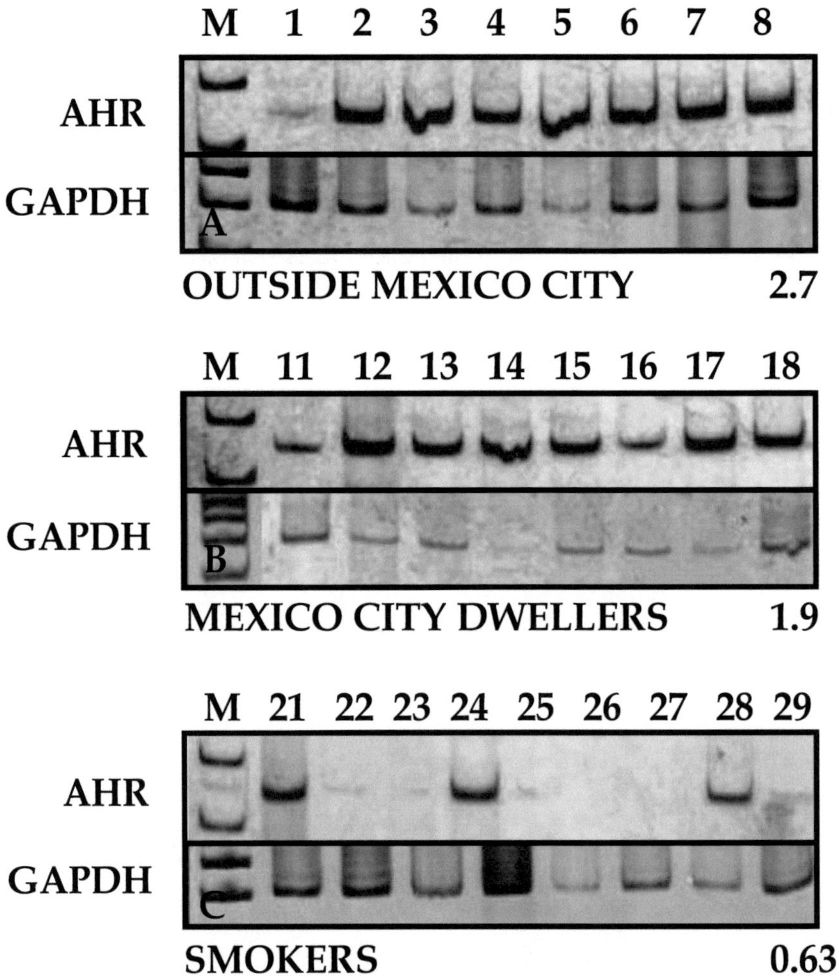

Fig. 2. RT-PCR of *AHR* in human lung tissue from outside residents (A), Mexico City residents (B) and in smokers (C). mRNA was obtained from paraffin embedded tissue. The number below of each PCR-gel indicates the relation of expression of the gene. GAPDH is gliceraldehyde phosphate dehydrogenase as gene control. Each lane is a different case. M, molecular markers. PCR products resolved in polyacrylamide-gel electrophoresis and silver staining

Finally, AHR is an important gene whose expression is observed along of the ages of the human been: from 40 hours post-natal (Fig. 3) to adults more than 60 years-old (some cases in Fig. 2).

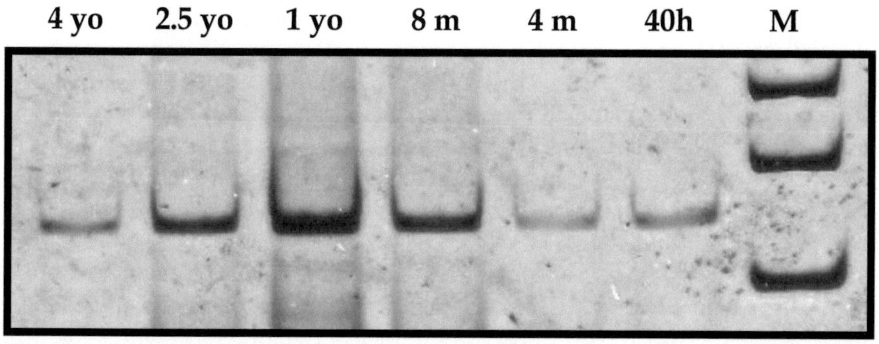

Fig. 3. RT-PCR of *AHR* in human lung tissue of children from 40 hours (h), 4 and 8 months (m) to 1, 2.5 and 4 years old (yo). M, molecular markers. RT of constitutive gene, glyceraldehyde-3-phosphate dehydrogenase was equal in each sample (not shown). PCR products resolved in polyacrylamide-gel electrophoresis and silver staining

4.1 The components of the AHR pathway

The AhR has been shown to bind, as have similar highly lipophilic halogenated and non-halogenated hydrocarbons, leading to cardiovascular, carcinogenic, and endocrine effects. At the cellular level, activated AhR interacts with various signal transduction pathways, induces biotransformation enzymes, alters the cell cycle, cell adhesion, and migration, and causes apoptosis or aberrant cell growth. *In vivo*, AhR plays significant roles in connection with development, immunological and reproductive functions, and adaptive responses to light and xenobiotics (Wincent et al., 2009). For these reasons, AhR regulates several important functions in the cell and is recognized by all participants as an AHR pathway.

AhR is a ligand-dependent transcription factor belonging to the basic helix-loop-helix/Per/ARNT/Sim (PAS) family, that regulates the expression of a battery of genes in a wide range of species and tissues (Dietrich, 2010; Gasiewicz & Collins, 2008).

The inducing chemical enters the responsive cell and binds with high affinity to the cytosolic AhR, which exists as a multiprotein complex containing two molecules of the chaperone protein hsp90 (a heat shock protein of 90 kDa) (Denison & Nagy, 2003), which is required to maintain the receptor in a conformation that facilitates ligand binding. The X-associated protein 2 [XAP2], also called AIP and ARA9, appears to function by stabilizing the interaction between hsp90 and the receptor, has some role in regulating the rate of AhR turnover, intracellular localization and interaction with other proteins in the nucleus (Gasiewicz & Collins, 2008), and an identified 23-kDa co-chaperone protein referred to as p23 (Denison, 2003) to stabilize an intermediary complex that contains the ligand-occupied hsp90-associated AhR (Gasiewicz & Collins, 2008). Binding of the ligand results in nuclear translocation of the AhR, dissociation from the chaperone proteins, heterodimerization with Arnt (Aryl hydrocarbon Receptor Nuclear Translocator) and the subsequent binding of the Ahr-Arnt heterodimer to dioxin-responsive elements (DREs) with the consensus core recognition sequence 5´- TNGCGTG-3´ (Fig. 4), also known as Xenobiotic-Responsive Elements (XREs)(Dietrich, 2010).

Arnt is a basic helix-loop-helix (bHLH) protein that also contains a PAS domain. In addition to forming heterodimers with many other bHLH-PAS proteins, including the AhR and hypoxia-inducible factors 1alpha, 2alpha and 3alpha, Arnt can also form homodimers when expressed from its cDNA *in vitro* or *in vivo* (Wang et al., 2006). AhR/Arnt binds with the TATA-binding protein (TBP) and several TBP-associated factors (TAFs), leading to general transcriptional machinery with RNA polymerase II (RNA pol II) and the transcriptional activation of target genes. The AhR is then exported to the cytosol and degraded by the 26S proteasome pathway. A mechanism of negative feedback regulation of the AhR function is performed by the aryl hydrocarbon receptor repressor (AhRR). Ligand-activated AhR/Arnt heterodimer transactivates the expression of target genes including the *AHRR* gene. AhRR suppresses AhR transcriptional activity by competing with AhR for dimerizing with Arnt and binding to the DRE sentence of target genes (Pavek, 2008) (Fig. 4). The interaction of the AhR with Arnt increases their capacity to bind specific enhancer sequences adjacent to target promoters called DREs. An assembly of coactivators and general transcription factors, including p300, SCR-1, p/CIP and transcription for IIB, then interacts with gene promoters and potentiates the expression of target loci (Nguyen, 2008).

4.2 The target genes of the AHR pathway

The interactions of PAH ligands with the AhR may explain the pathway of CYP induction (Lee, 2008) and several genes (Marley et al., 2005), thus leading to their detoxication and excretion and, at the same time, to their metabolic activation to genotoxic compounds (Dietrich & Kaina, 2010). The nuclear AhR complex, ligand-AhR-Arnt, interacts with consensus dioxin or XREs in the CYP 1 promoter and in promoters of other Ah-responsive genes, and the subsequent recruitment of coactivators and the general transcription of coactivator and general transcriptions factors results in the expression of target genes, such as genes phase I xenobiotic metabolizing enzymes (CYP1A1,1A2 and 1B1) and phase II enzymes (NQO1, GSTA2, UGT1A1 and UGT1A6) (Pavek & Dvorak, 2008) (Fig. 4).
Their action in the metabolism of PAH was explained in the previous section. Cyclo-oxigenase-2 (COX-2) has a XRE site in its promoter region, and PAH and TCDD induce COX-2 and prostaglandin synthesis (Marley, 2005). However, the mechanism by which AhR activation may result either in carcinogenic or protective effects is not clear. Moreover, many intracellular interactions of AhR and Arnt with various regulatory transcription factors, such as retinoblastoma protein-1, NF-kB, estrogen receptors and SP1, as well as with different coactivators and repressors, may modify the transcriptional activity of the AhR–Arnt heterodimeric complex, and this might explain why chemicals that bind to AhR may elicit detoxification agonist or antagonist responses (Irigaray P, 2010).
The roles of AhR in the carcinogenic process may be positive and negative. The ultimate response may be dependent on the level and length of exposure, as well as on the cellular context, stage of differentiation and the presence of other conditions in the tissue environment (Gasiewicz, 2008).

5. New molecular markers, microRNAs: definition and biogenesis

MicroRNAs or miRNAs are small non-coding RNAs almost 22 nucleotides long, involved in negative post-transcriptional gene regulation via the RNA interference mechanism. The sequences of miRNAs are highly conserved among plant-microorganisms-animals, suggesting that miRNAs represent a relatively old and important regulatory pathway. Up to a third of the human genes are regulated by miRNAs. They are important regulators of several genes in many and broad biological processes, from developmental timing to cellular proliferation and apoptosis (Tomankova et al., 2010).

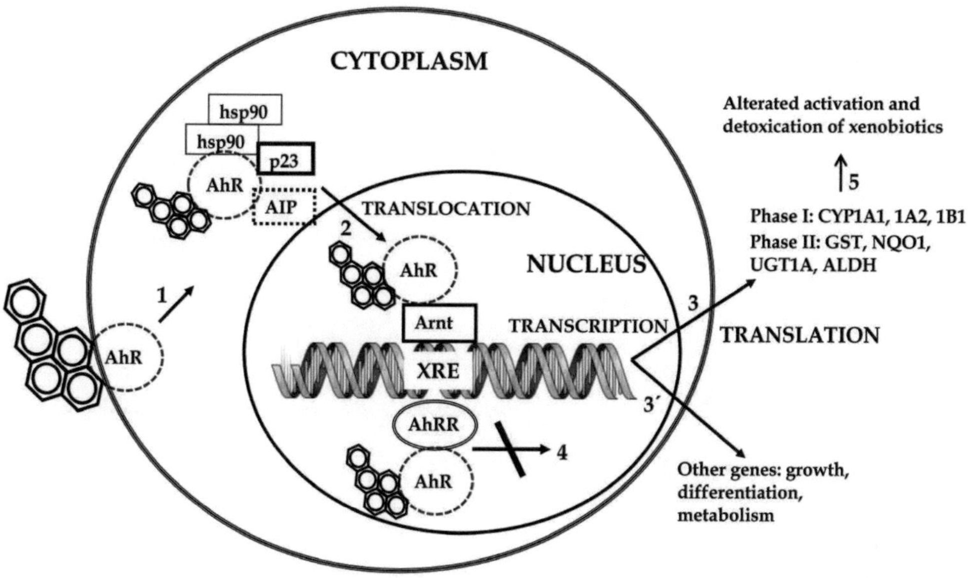

Fig. 4. Activation of the AhR and the AHR pathway. PAH interact with the receptor AhR (1) an it is assisted with other proteins as hsp90, p23 and AIP. AhR-PAH complex is translocated to the nucleus (2) and it makes complex with Arnt in order to activate transcription of several genes as Phase I and Phase II (3), and other genes participating in other cellular responses (3´). If AhRR interacts with the AhR-PAH complex, the transcription is inhibited (4). The transcription is performed in order to response the detoxification of xenobiotics and other molecules (5)

MiRNAs genes are localized in the non-coding regions, 3´, or in the introns or exons of protein-coding genes in the genomic DNA. They can be codified singly or in clusters. The miRNA genes are much longer than biologically active, mature miRNAs that originate through a multistep process (Fig. 5). They are transcribed by RNA polymerase II and lead to hundred- or thousand- nucleotides-long primary miRNAs transcripts (pri-miRNA) (Kim, 2005).

A local stem-loop structure of pri-miRNA is then cleaved in the nucleus by the dsRNA-specific ribonuclease Drosa/Pasha to the 70 nucleotide-long precursor miRNA (premiRNA). PremiRNA is actively transported from the nucleus to the cytoplasm. In the cytoplasm, pre-miRNA is subsequently cleaved by the RNase III Dicer into an almost 22-nt miRNA duplex. One strand of the miRNA duplex is degraded ("passenger, miR"), whereas the other is incorporated into the RNA-induced silencing complex (RISC) and serves as a functional, mature miRNA. Depending on the complementarity between miRNA and the 3´untranslated region (UTR), of the target mRNA there are two known mechanisms of miRNA action: 1) target mRNA degradation; and, 2) translational inhibition with little or no influence on mRNA levels (Fig. 5). The deadenylation and subsequent degradation of the target mRNA occurs when miRNA is near-perfectly complementary with target mRNA. This represents the major mechanism of miRNA regulation. About 84% of all protein-coding mRNA targets undergo degradation when recognized by their cognate miRNA. Otherwise, the translational inhibition occurs when miRNA is only partially complementary to its target mRNA. This mechanism does not represent a predominant reason for reduced protein output. Several other factors may influence the action of miRNA, such as impaired processing, methylation, gene polymorphisms, gene amplification, deletion of the Dicer, translocations and others (Kim, 2005; Tomankova et al., 2010).

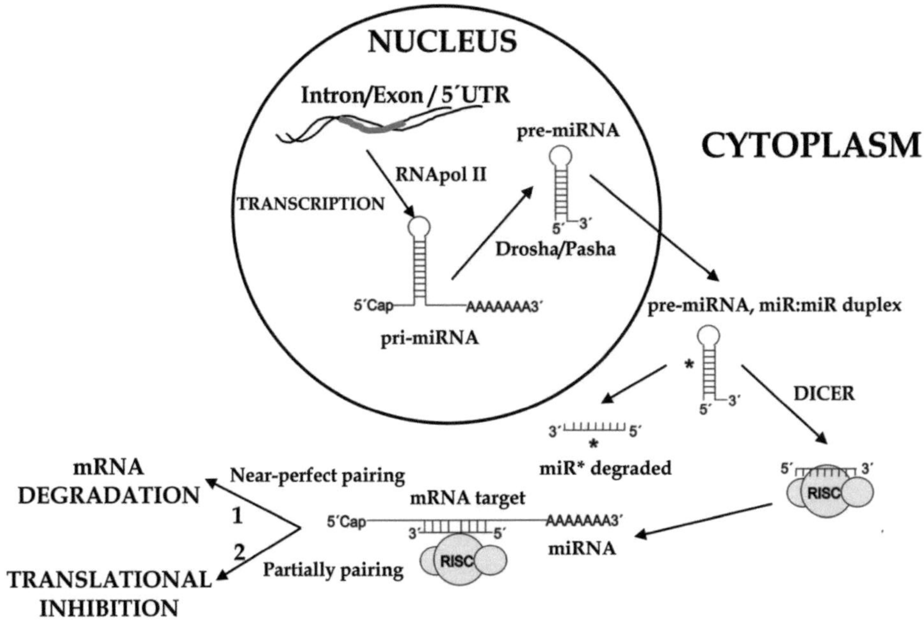

Fig. 5. MiRNAs are transcribed by RNA polII from genomic DNA as long primary miRNA transcripts (pri-miRNAs). Pre-miRNAs are then transported from the nucleus to the cytoplasm where is subsequently cleaved by RNase III Dicer into ~22 nt miRNA duplexes, consisting of the "guide" strand (miR) and"passanger" strand (miR*) which is degraded. MiR is incorporated into RISC and serves as a functional mature miRNA acting by two different mechanisms according to the complementarity with the target mRNA. According to Kim, 2005

It is evident that a single miRNA may regulate the translation of numerous downstream mRNA, and mRNA is likely to be regulated by several miRNAs simultaneously. The identification of miRNA target genes is important and due to high similarities in miRNA sequences, computational algorithms may predict a large number of putative miRNA binding sites on mRNA targets. Thus, experimental validation in biological systems is fundamental to completing the target prediction (Kuhn et al., 2008; Tomankova et al., 2010). Modulation of miRNA repression has been shown to be a highly dynamic process, as evidenced by studies that demonstrate the rapid alleviation of miRNA-mediated translational repression in response to specific needs.

5.1 The role of microRNAs in lung development

The lung has a very specific miRNA expression profile, which is highly conserved across mammalian species. The roles of miRNAs on physiological and pathological conditions in the lung compartment are still limited and based mainly on studies of animal models. The increased expression of 8 miRNAs in neonatal and fetal lungs, miR-134, -154, -214, -296, -299, -323, -337 and -370; and the up-regulation of 5 miRNAs in the adult mouse and human lung, miR-26b, -29a, -29b,-142-3p and -187; were found to be conserved during development in both species. The examination of their genomic localization shows that 6 of the miRNAs are highly expressed in the developing lungs of mice and humans: miR-134, -154, -299, -323, -337 and -370, and map to the *Gtl2-Dio3* domain at human chromosome 14q32.21, a region that is highly conserved between the two species. Importantly, imprinting means that these miRNAs are only expressed from the maternally inherited chromosome and their expression is regulated by an intergenic germline-derived differentially methylated region located ~200 kp upstream from the miRNA cluster. Imprinting also results in the *Gtl-2* gene being expressed from the maternal chromosome (Williams et al., 2007).

MiRNA deregulation may contribute to various pulmonary diseases. Several miRNAs, such as miR-155, miR-26a, let-7, miR-29, miR-15/miR-16, miR-223, miR-146a/b and the cluster miR-17-92, have been shown to be involved in homeostasis and in lung development. For example, miR-26a has been shown to be selectively expressed in the bronchial and alveolar epithelial cells in the murine lung. MiR-29a and miR-29b are up-regulated in adult tissue and expressed at a lower level in developing lungs in mice and humans. These miRNAs genes map to human chromosome 7q32.3, a region that maps to a fragile site that has been associated with a number of cancers (Nana-Sinkam et al., 2009).

The miR-17-92 cluster is believed to regulate lung development because its expression is high in embryonic development and steadily declines during development into adulthood. Overexpression of the miR-17-92 cluster in murine models resulted in an abnormal phenotype manifested by the absence of terminal air sacs, which were replaced by highly proliferative, undifferentiated pulmonary epithelia. Finally, miR-223 has been shown to be crucial for normal granulocyte development and long function lung (Lu et al., 2007).

There is evidence that the up-regulation or down-regulation of miRNAs is critical for lung development/homeostasis and thus may contribute to the development of pathological pulmonary conditions, smoking-related diseases including lung carcinogenesis, fibrosis and other disorders, such as allergies. The effect of such environmental factors as organic pollutants and PAHs emitted from burning food or produced by the incomplete combustion of fossil fuels in automotive engines can be studied. Some can be highly specific, such as the reduced miR-146a expression that results in a prolonged mRNA half-life of cyclo-oxygenase-2 that increases prostaglandin E2 in the fibroblast of COPD (chronic obstructive pulmonary disease) subjects. However, the majority of miRNA studies in smoking-related diseases focus on the role of miRNAs in lung cancer (Tomankova et al., 2010).

5.2 Responses of microRNAs to smoking

The three PM can induce a response by the lung in inhabitants of large cities that can activate several molecular responses. The first lung response evaluated has been that due to smoking. Smoking induces the expression of airway genes involved in the regulation of oxidant stress, xenobiotic metabolism and oncogenesis, while suppressing those involved in the regulation of inflammation and tumour suppression. This pattern can be described in cytologically normal airway epithelial cells, thus these airway genes can serve as biomarkers for lung cancer and even in the early stages of neoplasic transformation and, of course, in other diseases (Spira et al., 2007). These studies suggest that gene expression changes in the airway epithelium reflect host responses to, and damage from, cigarette smoke.

In the case of microRNAs responses to smoking, Schembri et al. (2009) examined whole-genome microRNA and mRNA expressions in bronchial airway epithelia of current and never smokers, finding 28 miRNAs to be differentially expressed with the majority being down-regulated in smokers. Previously, 26% of the mRNAs that are differentially expressed in smokers were predicted to be targets of only these 28 miRNAs/miRNAs families. This suggests that a relatively small number of miRNAs in response to smoking could potentially contribute to several of the smoking-associated changes in mRNA expression, and that modulation of specific miRNAs might therefore represent a mechanism that contributes to the overall host response to tobacco smoke exposure. Gene ontology analysis of the anti-correlated targets of these miRNAs reveals that these genes are involved in cell-to-cell adhesion, cellular signalling and the cytoskeletal structure. These findings must be confirmed by other studies. Using two strategies, one from the mRNA microarrays expression dataset of normal human bronchial epithelial cells exposed to cigarette smoke, and the other by the exposure to cigarette smoke condensates of normal human bronchial epithelial cells, both studies indicated that the targets of mir-128 are induced upon acute exposure to cigarette smoke and suggest that mir-128 expression likely decreases upon such exposure (Fig. 6). Manipulating the levels of mir-128 by several strategies that increase or decrease its levels indicates that they modulate the airway epithelial gene expression response to cigarette smoke and supports a role for miRNAs in regulating host responses to environmental pollutants. The finding that most miRNAs were differentially expressed, with most (82%) being down-regulated in smokers, is similar to studies of miRNA in cancer. In both cases, most miRNAs were down-regulated when compared to normal tissues or cells from never smokers. This may be because miRNA levels are closely linked to the degree of cellular differentiation and because the reduced expression of miRNAs in cancer is associated with more dedifferentiated tumours, and the same alteration in the bronchial cells of smokers. These results suggest that the down-regulation of miRNAs in smokers could be associated with the development of tobacco-related cancers. Airway miRNA expression could potentially serve as an indicator of smoking-induced disease processes. Using experimental methods, Schembri *et al.* (2009) performed an experimental validation in biological systems of the role of mir-128 in regulating the expression of transcriptional regulators such as MAFG. These findings suggest that miRNAs may play a role in regulating the gene expression response to tobacco exposure in airway epithelial cells. Finally, the analysis of miRNAs in lung cancer in never smokers, is important to show the molecular characteristics of lung cancer in never-smokers: changes in the expression of a relatively small number of miRNAs are involved in lung carcinogenesis, miR-138 is down-regulated preferentially and miR-21 is one of the most aberrantly increased miRNAs in both never smokers and cases of smokers.

5.3 Responses of microRNAs against 3, 4-(methylnitrosamino)-1-(3-pyridyl)-1-butanone (NNK)

Lung cancer is the most common cause of cancer deaths in the world. Approximately 90% of all lung cancers are directly attributable to smoking. Animal models are invaluable tools for studying the initiation and progression of human disease. To mimic tobacco carcinogenesis, Kalscheuer et al. (2008) used a male model with F344 rats that were chronically treated with 4-(methylnitrosamino)-1-(3-pyridyl)-1-butanone (NNK), a carcinogen present in tobacco products, for up to 20 weeks. NNK along with polycyclic aromatic hydrocarbons are the most prevalent and potent carcinogens in tobacco products and smoke. Microarray analyses of miRNA expression revealed that a small number of miRNAs, such as miR-101, miR-126*, miR-199a, miR-199b and miR-34b, were down-regulated in NNK-treated rats when compared with controls. These results were also confirmed by Northern blot and real-time PCR, at least for miR-34b, miR-101, miR-126 and miR-199a.

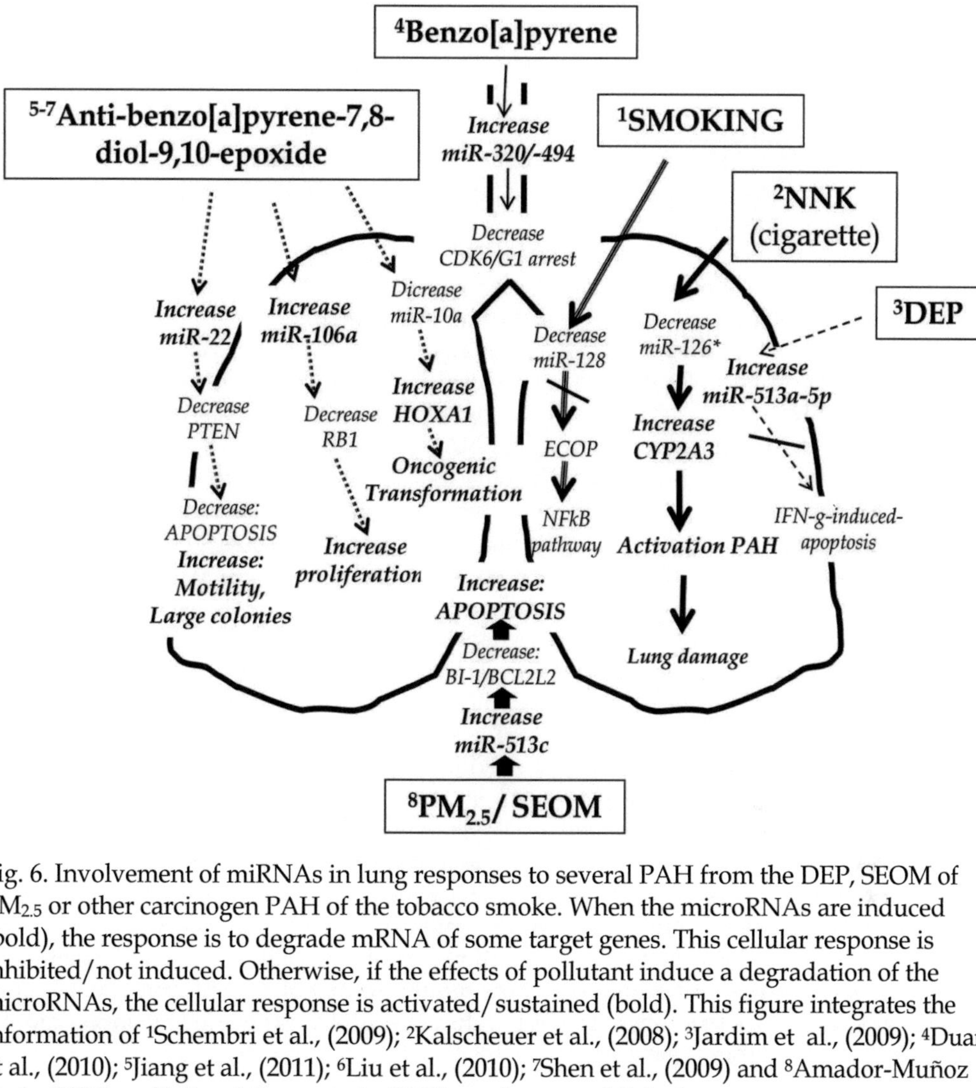

Fig. 6. Involvement of miRNAs in lung responses to several PAH from the DEP, SEOM of $PM_{2.5}$ or other carcinogen PAH of the tobacco smoke. When the microRNAs are induced (bold), the response is to degrade mRNA of some target genes. This cellular response is inhibited/not induced. Otherwise, if the effects of pollutant induce a degradation of the microRNAs, the cellular response is activated/sustained (bold). This figure integrates the information of [1]Schembri et al., (2009); [2]Kalscheuer et al., (2008); [3]Jardim et al., (2009); [4]Duan et al., (2010); [5]Jiang et al., (2011); [6]Liu et al., (2010); [7]Shen et al., (2009) and [8]Amador-Muñoz et al., (2011) and Miguel-Perez et al., (2011) results not published

In order to validate one of these findings, human and other species of *CYP2A13* genes were shown to be prominently predicted as targets by for miR-126*. Deleting much of the predicted miR-126*-binding site in the 3´UTR abrogated the inhibitory effect of miR-126 on reporter expression, demonstrating the specificity of the interaction between the *CYP2A3* mRNA of the rat and miR-126*. This model mimicked the early stages of lung cancer development. These miRNAs could indicate early biomarkers for lung cancer. Many miRNAs might not change their expression levels until later stages of pulmonary

tumorigenesis. The finding that miR-126* may regulate the expression of *CYP2A3* suggests that the early changes in miRNA expression indeed have important biological consequences (Fig. 6). Finally, miR-34 is another miRNA with clearly interesting implications due to the fact that it inhibits the expression of a large number of genes involved in DNA damage response, cell cycle progression and apoptosis.
The results of this study open the possibility that some of these miRNAs and the target genes or pathways regulated by the miRNAs might constitute early diagnostic markers for human lung cancers.

5.4 Responses of microRNAs to Diesel Exhaust Particles (DEP)

Jarim et al. (2009) revealed a disruption of the expression of microRNAs in human airway cells exposed to 10 mg/cm^2 of DEP for 24 h. They analysed total RNA for miRNA expressions using microarray profile analysis and quantitative real-time polymerase chain reactions. DEP treatment causes differential expression patterns of miRNAs in differentiated human bronchial epithelial cells. A total of 130 miRNAs showed an increase of ≥-1.5-fold in response to DEP, whereas in the case of smoking, 67 showed a decrease ≥ 1.5-fold in expression.
After an analysis of the miRNAs that were more altered in their expression, those authors found that a family of miRs, such as miR-513c, miR-513b and miR-513a-5p, are over-expressed 16-, 13- and 11.3-fold than controls. Other miRNAs, miR-923, miR-494 and miR-338-5p, had 10.6-, 9.2- and 4.6-fold expressions. On the other hand, the lowered miRNAs were miR-31, miR-26b, miR-96, miR-27a, miR-135b and miR-374a. Several functions related to lung physiology/response are involved, such as miR-513a-5p, in the repression of IFN–γ–induced apoptosis, and miR-26b was down-regulated in the lungs of rats exposed to cigarette smoke (Fig. 6). MiR-96 has been shown to regulate levels of the protein arginine methyltransferase (PRMT5), which interacts with the chromatin remodelling complex and may play a crucial role in silencing tumour suppressors. Identifying the network analysis of miRNAs highly modulated by DEP, using TargetScan and miRDB for highly modulated miRNAs, reveals molecular networks that were enriched by inflammatory responses that correlated well with canonical signalling pathways, such as interleukin (IL-8), nuclear factor kappa B (NF-kB), and the chemokine (C-X-C motif) receptor 4, where these targets could be acting. To further analyse the putative effects of differential miRNA expression profiles due to DEP exposure, Jarim *et al.* used a TargetScan and miRDB analyses to identify possible mRNA targets.
There are different types of DEP, all of which have been found to possess mutagenic potential. Although DEP are classified as potential causal carcinogenic agents, the mechanism by which they alter cellular function remains largely unknown. Furthermore, the carbonaceous core of DEP alone can induce tumour formation (Wichmann, 2007). Future studies will show whether aberrant DEP-induced miRNA expression changes can be functionally linked to tumorigenesis

5.5 Responses of microRNAs to Benzo[a]pyrene, BaP

There is an important work group at the Institute for Chemical Carcinogenesis of the Guangzhou Medical University in China that has described several miRNAs in *in vitro* models of exposure to PAH. When Duan et al (2010) exposed with 1 mM BaP for 24h, they revealed an important increase in miR-320 and miR-494 expression. Since carcinogenesis is considered to be the result of the deregulation of the cell cycle machinery, the transitions between cell-cycle phases are mediated by cyclin-dependent kinases and their modulators; and little is known about the functions of miRNAs in the process of BaP-exposed cell cycle alteration. In this sense, when primary murine bronchial epithelial cells were exposed to BaP, the analysis of the cell cycle progression revealed that G1 arrest occurred in response to BaP. This depends, at least partially, on reduced expression levels of CDK6. These results were confirmed using antago-miRs corresponding to miR-320 and miR-494. Some important additional data are that miR-494 is highly expressed in retinoblastoma and several solid tumours. Central to these events is the G1 arrest, which allows cells more time to repair DNA and/or bypass the machinery time to remove or resolve the damaged sites before the DNA synthetic phase of the mammalian cell cycle, and protects cells from mutagenesis. In this sense, G1 arrest may be related to the cellular defence mechanism, which can arrest the progress of the cells with damaged DNA through the cell cycle, prior to their entry into a critical phase, such as DNA replication. In conclusion, CDK6 may be a target of miR-320 and miR-494 affecting G1/S transition through this cycle (Fig. 6).

5.6 Responses of microRNAs to anti-BPDE and miR-106a

The polycyclic aromatic hydrocarbon BaP, which is a toxic element in the environment in general and in tobacco smoke in particular, has atherogenic and carcinogenic properties. BaP is activated by microsomal enzymes to yield anti-benzo[a]pyrene-*trans*-7,8-diol-9,10-epoxide (anti-BPDE), which binds covalently to nuclear DNA to form adducts that can initiate carcinogenesis. A malignant transformation model of the human bronchial epithelium cell line 16HBE induced by anti-BPDE was established by Jiang *et al.* (2011). The immortalized 16HBE cell line retains the specific morphology and function of normal human bronchial epithelial cells and provides a suitable resource for studying the molecular pathogenesis of lung cancer. Previous studies from this laboratory revealed 55 significantly differentially expressed miRNAs, identified by microarray in transformed 16HBE cells. This description suggests that some of these miRNAs may play important roles in the process of transformation due to anti-BPDE.

The levels of miR-106a were 2.9-fold higher in 16HBE transformed cells than in non-transformed ones. This effect produces a decrease in the proportion of cells in G1/S (40.11%) that is greater than that observed in normal cells (70.62%). Also, the apoptosis rate was different between the two kinds of cells, as the transformed cells had 4% of apoptosis, in contrast to more than 8% observed in non-transformed cells. The next step, the capacity for colony formation, was evaluated and the results showed important changes in the number of colonies from 8.39 for transformed cells to 5.79 for non-transformed ones. All these results support the suggestion that miR-106a affects cell proliferation, cell apoptosis, and colony formation in 16HBT transformed cells. Finally, the formation of xenografts was induced subcutaneously in BALB/c nude mice and they were measured after 7 days post-inoculation. The tumours derived from anti-miR-106a-transfected cells grew substantially more slowly compared to the transformed group. The tumours were harvested 42 days post-injection and the weight of the tumours from the anti-miR-106a-transfected cells was significantly less than that of those derived from transformed cells (Fig. 6). The assays validated the role of miR-106a in the tumorigenesis induced in the *in vivo* model.

Bioinformatic analysis showed that the tumour suppressor RB1 is one of the predictive targets of miR-106a. The levels of RB1 mRNA and protein in the transformed cells were below those of the cells transformed by anti-miR-106a, and vice versa. In a dual-luciferase assay, the relative luciferase activity of the reporter that contained the RB1 3′-UTR was markedly decreased compared to that seen with the parent containing the mutant sequence. These results showed that the effects of miR-106a were mediated by sequences in the 3′-UTR of RB1 mRNA, which is in concordance with the results of previous studies. Over-expression of miR-106a provides a proliferative advantage to the malignant transformation of cells. The retinoblastoma tumour suppressor gene product RB1 regulates differentiation, apoptosis, and cell cycle control by coordinating the cell cycle at G1/S with transcriptional machinery. Therefore, miR-106a might function as an oncogene and might serve as a potential target for cancer therapy.

5.7 Responses of microRNAs to anti-BPDE and miR-22

Smoke contains many carcinogens and of course PAH is one of them and is wide-spread environmental pollutant found also in DEP and charbroiled food. Anti-BPDE is one of the many metabolites of BaP and is capable of electrophilic attacks on guanine residues, forming DNA adducts identical to those produced by BaP. Based on previous information, Liu et al. (2010) used bioinformatic tools to predict miR-22 target sites on PTEN (Phospatase and ensin homologue deleted on chromosome 10), and experimental analysis was used to validate the importance of these molecules in malignancy. MiR-22 was chosen because it had the highest predictive scores. The PTEN mRNA contains one binding site that is partially complementary to miR-22 and carries sequences identical to those of humans, mice, rats and other mammals. QRT-PCR analysis showed that miR-22 was significantly up-regulated as compared to normal cells. A similar pattern was observed by miRNA microarrays. The PTEN protein level in 16HBE-transformed cells was reduced significantly, but the PTEN mRNA level between normal and transformed cells was unchanged. Using anti-miR-22 over-expression in transformed cells revealed an increase of PTEN protein in those cells and the opposite effect was observed as the over-expression of miR-22 decreased the expression of the PTEN protein by almost half. Dual-luciferase assays proved that PTEN is a target of miR-22. These results indicated that miR-22 modulates the post-transcription expression of PTEN. MiR-22 also regulates other cellular processes, such as apoptosis, cell colony formation and motility. Increased caspase-3/7 activity was found for anti-miR-22 in 16HBE cells transformed by anti-BPDE. Under fluorescence staining, the inhibition of miR-22 in cells transformed increased the number of apoptotic cells in comparison to control cells. Cells transformed by anti-BPDE, with transfection with anti-miR-22, induced both markedly smaller colonies and a lower motility in the wound-scratch healing assay. In summary, miR-22 is increased in 16HBE cells transformed by anti-BPDE; this fact resulted in a decreased expression of PTEN, leading to resistance to apoptosis but increasing colony formation and cell motility. Cell motility reflects a cell's invasive capacity (Fig. 6).

5.8 Responses of microRNAs to anti-BPDE and miR-10a

In the same line of research, anti-BPDE is the most important metabolite of BaP activation. The study of the molecular mechanism of cell transformation induced by anti-BPDE was performed in 16HBE transformed-cells. In their study, Shen et al. (2009) quantified by QRT-PCR and found that the expression of miR-10a in 16HBE-transformed cells was 0.01 times that observed in non-transformed cells: it decreased. Previously, the RNA target of miR-10a was predicted by bioinformatic analysis and located within the HOXB cluster on 17q21, and was associated with the risk of human megakaryocytopoiesis and adult acute myeloid leukaemia. It also revealed HOXA1 as a target for miR-10a. QRT-PCR results demonstrated that HOXA1 mRNA expressions were 8.75-fold greater than those of normal 16HBE cells. Those findings do not validate the role of miR-10a in high expressions of the HOXA1 gene, but it has been found that the forced expression of HOXA1 dramatically increases the anchorage-independent proliferation of immortalized human mammary epithelial cells. The down-regulated miR-10a and its up-regulated predicted target HOXA1 were expressed reciprocally in 16HBE-transformed cells, suggesting that miR-10a is potentially involved in cellular transformation, and might act as a candidate tumour suppressor by intervening in HOXA1 (Fig. 6).

5.9 Responses of microRNAs to soluble extractable organic matter (SEOM) of the $PM_{2.5}$

The organic chemicals associated with airborne particles can be evaluated through extraction, which generates solvent extractable organic matter (SEOM) that contains hundreds of compounds, among which the PAHs have been the most widely investigated in studies exploring the mutagenic and potentially carcinogenic activity of ambient particulate matter, mainly in Mexico by the Villalobos-Pietrini group (Villalobos-Pietrini et al., 2006). $PM_{2.5}$ contains PAH and Mexico City has important emissions of these particles due to mobile sources represented by four million vehicles and 35,000 industries in the Metropolitan Zone of the Valley of Mexico (MZVM). The concentrations of BaP and other PAHs and nitro-PAH are different depending of the zone of the MZVM, as the northeast, central and southeast zones have higher concentrations than the others: 0.404, 0.455 and 0.452 ng/m^3, respectively. However, benzo[*ghi*]perylene is the most abundant PAH in Mexico City's atmosphere with high levels of 1.84 ng/m^3 in the central zone and similar concentrations in the others. The C_{24}-C_{26} were the most abundant n-alkanes and 2-nitrofluoranthene and 9-nitroanthracene the most abundant nitro-PAHs (Amador-Muñoz et al., 2011). Our group evaluated the genetic responses of miRNAs due to exposure *in vitro* of human bronchial cells, NL-20, to SEOM in concentrations of 13 and 17 µg/ml for 24h. These concentrations of SEOM were proved to be mutagenic in a *Salmonella typhimurium* assay previously obtained of PM10. From several microRNAs evaluated, we demonstrated that there was an increase in the expression of miR-513c (Figure 7), due to SEOM obtained from a filter sample from the northeast station of the MZVM (San Agustín). This was similar to the results obtained by Jardim et al. (2009), which reported a 16-fold increase of miR-513c in primary human bronchial epithelial cells cultured and exposed to 10 mg/cm^2 of diesel exhaust particles (DEP). MiR-26b did not change after exposure. The evaluations of several miRNAs related to PAH response of the lung is currently under study in our laboratory using this *in vitro* model. An important response of NL-20 cells is that they suffered apoptosis, as revealed by the DNA-ladder after exposure for 24 h. After analysis, possible mRNA targets of miR-513c revealed by Targetscan are BCL2L2 (BCL2-like2) and BI-1 (Bax inhibitor-1) that have anti-apoptotic functions. One possible mechanism is miR-513c, which induces a degradation of one of these proteins (or both) and turns on apoptosis induction verified after SEOM exposure in NL-20 cells (Fig. 6). Currently, we are evaluating the responses *in vitro* in human bronchial cells after exposure to SEOM obtained from several monitoring stations, in order to view the differences in the molecular responses: miRNAs and AHR pathway activation, since $PM_{2.5}$ in the northwest and the southeast originates mainly from primary emissions of primary organic compounds. $PM_{2.5}$ in the northeast, central and southwest contains a greater proportion of secondary organic compounds, with the less oxidized organic aerosols being found in the northeast and the most aged organic aerosol in the southwest.

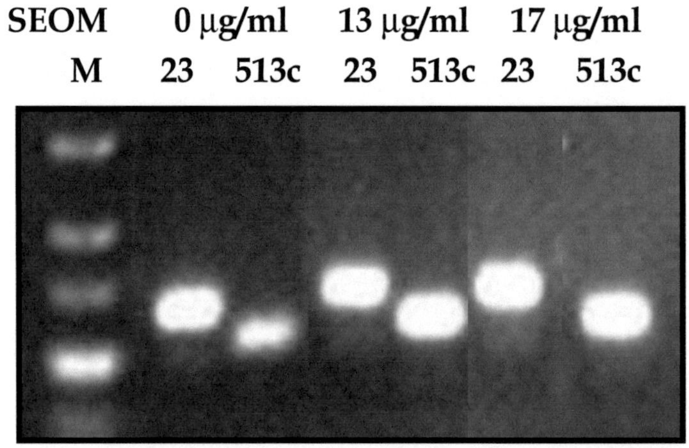

Fig. 7. Expression of miR-23 and miR-513c in human bronchial epithelial cells NL-20 exposed to 13 and 17 µg/ml of SEOM extracted of $PM_{2.5}$ from San Agustín, Northeast of México City. An important induction of miR-513c is evident. PCR products resolved in agarose 4% stained with ethidium bromide

6. Biomarkers of exposure to polycyclic aromatic hydrocarbons from environmental air pollution

6.1 Background and levels of exposure in several scenarios

The main sources of human exposure to PAH are occupational, passive and active smoking, food and water, and air pollution. The total intake of carcinogenic PAH in the general population has been estimated at 3µg/day. In smokers, BaP levels range from 0.5 to 7.8 µg/100 cigarettes when exposure is from mainstream smoke. Levels from passive smoking are lower, from 0.0028 to 0.76 µg/m^3. There is a high variation in atmospheric PAH levels across geographical areas, with BaP concentrations from 0.01 to 100 ng/m^3. In Mexico City, the mean BAP determined in 2006 was 0.3912 µg/m^3. Pollution of air by PAH is mainly due to the incomplete combustion of wood or fuel used for industrial or motor vehicle exhaust. The level of exposure to PAH through these sources is low compared to other sources, such as diet, occupation, or tobacco smoke. The half-life of airborne PAH is on the order of days, but can be longer when they are bound to small particles. Inhaled PAH are absorbed mainly thorough the bronchial epithelium. After absorption, PAH are distributed to tissues where they are biotransformed by the enzymes of phase I to chemically reactive intermediates that may bind covalently to DNA-producing DNA adducts. These give rise to mutation and, eventually, tumour initiation. PAH metabolites are mostly conjugated with glucuronic acid by phase II enzymes and excreted as hydroxylated metabolites. PAH are excreted mainly through the faeces; only about 10% are excreted in the urine (Castaño-Vinyals et al., 2004).

The difficulty in finding an index substance arises from the fact that the composition of PAH mixtures depends on the source of combustion. The most common compound used as a reference substance for carcinogenic PAH is BaP. Certain PAH and nitro-PAH are more important and characteristic of the environment of each city. For example, benzo[*ghi*]perylene was the most abundant PAH contained in the SEMO extracted from the $PM_{2.5}$ in Mexico City (Amador-Muñoz et al., 2011).

6.2 Biomarkers of PAH exposure

The most commonly used biomarkers of PAH exposure are metabolites of PAHs, particularly 1-hydroxypyrene (1-OHP) and DNA-adducts. 1-OHP is the principal product of pyrene metabolism, representing 90% of its metabolites (Brzeznicki et al., 1997). Following inhalation, the half-life of 1-OHP is on average about 18-20 hours (Buckley & Lioy, 1992). PAH adducts have been mainly employed as a measure of PAH linked to DNA in target tissues and cells. The half-life of DNA adducts in lymphocytes is on the order of months. There are protein adducts of PAH, such as albumin adducts whose half-life is around 20 days, while for haemoglobin adducts it is around 120 days (Castaño-Vinyals et al., 2004).

Occupational studies and research on active and passive smoking have reported positive correlations between BaP and some markers, such as 1-OHP and DNA-adducts, but in exposure situations where BaP concentrations were 10-to-100-fold higher than those normally found in ambient air concentrations (Dor et al., 1999). Castaño-Vinyals et al. (2004) demonstrated in their work, which integrated the highest number of studies quantifying biomarkers of PAH exposure from air pollution, that both 1-OHP and DNA-adducts can be usefully applied to assess environmental exposure to PAH. The correlation between environmental levels and 1-OHP was high; in particular when personal monitoring of BaP was done, indicating that this biomarker can distinguish between fairly small exposure gradients. For DNA-adducts, the studies analysed showed heterogeneity in several variables that probably resulted in an underestimation of the correlation between atmospheric PAH and DNA adducts.

6.3 Main markers of biomarkers of PAH exposure

PAH in the air are present at concentrations from tens (or less) of ng/m^3 in non-polluted or low-polluted areas to hundreds of ng/m^3 in polluted areas. Most studies have used BaP as a reference substance for their carcinogenic potential. However, the presence of other PAH in the environment, some of which are carcinogenic, does not allow an accurate estimation of the risk linked to a PAH mixture on the basis of BaP concentrations alone (Castaño-Vinyals et al., 2004). The major contributors to air PAH in the urban and suburban atmosphere are mobile sources, such as diesel and gasoline engines. Emissions from these sources contain mainly benzo[g,h,i]perylene, fluoranthene, and phenanthrene, so that measuring only BaP as an index substance may result in underestimating exposures. Such is the case of Mexico City, whose atmosphere contains concentrations of benzo[*ghi*]perylene (Amador-Muñoz et al., 2011) as high as those of other large cities like Rome (Cecinato et al, 2008) and Nanjing, China (Wang et al., 2006).

The correlation among different studies indicates that differences as low as 5 ng/m^3 of BaP measured through personal monitoring can be identified using these biomarkers. In this kind of study it is important to register several data, such as smoking status, dietary intake mainly in non-smokers and differences in the season of blood-drawing, all of which may be factors that affect results. The time of year is important because, in general, BaP levels in winter are higher than those measured during summer (Castaño-Vinyals et al., 2004).

The use of biomarkers to assess exposure to PAH at high levels is well-studied, but more work is needed to validate these biomarkers when exposure occurs at low environmental levels. At the group level, DNA adducts and, particularly, 1-OHP seems to reflect exposure well, even at low levels of air pollution. In contrast, protein adducts do not yet appear to be valid markers for assessing environmental exposure to BaP. The use of these biomarkers should be more widely implemented in combination with more traditional techniques to evaluate the effects of ambient air pollution (Castaño-Vinyals et al., 2004).

6.4 MicroRNAs as "new" or "complementary" biomarkers of PAH exposure

Due to the rapid and sustained response of the pulmonary cells to several different stimuli, such as PM exposure and also to PAH contained, molecular expression can be considered also as a way to response to PAH exposure and other contaminants of the air, and due to sometimes they are sustained in their expression for the PAH exposure, they could have a sensitivity as DNA-adducts has. MicroRNAs are molecules with a rapid and sustained response. They can be considered as biomarkers because:

1. they are induced in a specific way, according to the type of agent or contaminant
2. they can reveal cellular responses altered due to exposure
3. they can define a group of genes altered by the action of the microRNAs
4. they can belong to a specific group of microRNAs codified in fragile sites related to neoplasic diseases
5. the altered gene and cellular pathways allow us to predict the possible implications of several diseases, including cancer.

MicroRNAs are specific and, due to this property, we can confirm whether tissue or cells were exposed to some kind of contaminants in the air. For example, according to the studies performed by several authors, we can define one of them as a marker; e.g. miR-101, that was reported as a microRNA specific to exposure to isolated carcinogenic compounds of cigarettes (Kalscheuer et al., 2008). This miR normally decreases its expression in the presence of the tobacco carcinogen and increases its expression in cells not so exposed. We demonstrated that this miR-101 is expressed in lung tissue from non-smokers who lived in the 70s in Mexico, and its expression is increased (Fig. 8), thus confirming their lack of exposure to tobacco. We are using this molecular marker in order to confirm whether people were not really exposed. In that time, the passive smokers were frequent before the establishment of laws more strict of the tobacco in Mexico, and some cases can reveal low levels of miR-101 (Fig. 8, lanes 1 to 2).

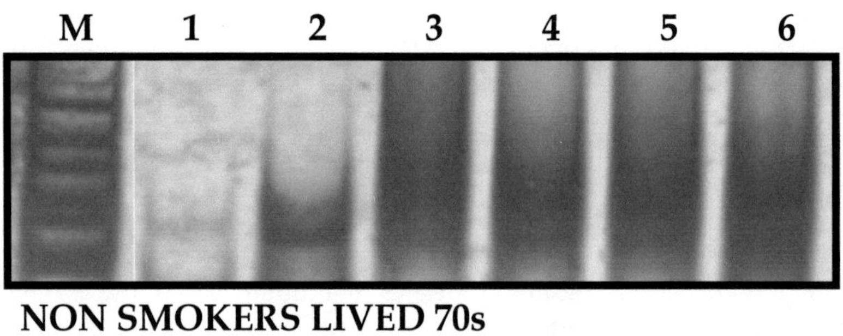

Fig. 8. Expression of miR-101 in lung tissue from people non-smoker lived at 70s in Mexico (From 1975 to 1978). Total RNA was extracted and the RT-PCR was performed to mir-101. Lines 3 to 6 have a lot of miR-101 amplification which explain the no definition of the band. Cases 1 to 2 showed less amplification of miR-101, maybe they were passive smokers. Each lane is a different case. M, molecular markers. PCR products resolved in polyacrylamide-gel electrophoresis and silver staining

7. Future research

The search for DNA adducts is problematic in terms of the quality of the DNA and its preservation, the kind of fixative, long time periods of preservation and the integrity of the

DNA in the cells. The use of paraffin-embedded tissue represents an easy, economical and practical form of preserving biological samples, and performing studies of the impact of climatic change and several other analyses. PriRNAs are RNA duplex that give stability. They can be stored in P-bodies and favour their preservation. They can be present in extracellular vesicles and maybe intracellular ones. These properties preserve their structure and permit their isolation in optimal conditions. They do not demand sophisticated, extra requirements or procedures, other than those that are standard in analyzing expressions following the general methodology of Molecular Biology. To consolidate the measurement of the levels of microRNAs, quantitive RT-PCR improve the results and analysis of its study. Some of the microRNAs described in Fig. 6 can be used as probable new molecular markers associated to PAH exposure in urban air. Also the information described by Schembri et al., (2009); Kalscheuer et al., (2008); Jardim et al., (2009); Duan et al., (2010); Jiang et al., (2011); Liu et al., (2010) and Shen et al., (2009) is important to be considered in future analysis of this kind of markers.

We can not leave the expression of *AHR* gene and AhR due that both have demonstrated be a sensitive and specific molecular sensors of the PAH exposure in urban air. Future work has to be started in other genes of the AHR pathway and it could explain the response of the target genes widely studied as CYPs genes due to PAH contained in urban air.

Finally, the information obtained from studies analyzing microarrays of microRNAs in response to particular air contaminants is important, because we can elaborate a guide of microRNAs with specific responses in the pulmonary cells against PAH, tobacco metabolites, tobacco carcinogens and SEOM obtained from $PM_{2.5}$ and PM_{10}.

8. Acknowledgment

We would like to thank to Dr. Omar Amador-Muñoz and his staff for the SEOM of the San Agustin station and all support. The data of this chapter were obtained of the experimental work supported by Fondos Federales of the Dirección de Investigación of the Hospital Infantil de México Federico Gómez HIM/2010/013 and HIM/2011/008. Also the support of the Coordinación de Investigación of Universidad Simón Bolívar.

9. References

Agency for Toxic Substances and Disease Registry (ATSDR). (1995). Toxicological profile for polycyclic aromatic hydrocarbons. In: *Toxicity of Polycyclic Aromatic Hydrocarbons (PAHs)*. 02-01-11. http://www.atsdr.cdc.gov/toxprofiles/index.asp

Amador-Muñoz, O., Villalobos-Pietrini, R., Miranda, J., & Vera-Avila, L E. (2011). Organic compounds of $PM_{2.5}$ in Mexico Valley: spatial and temporal patterns, behavior and sources. *Science Total Environ*, 409, 8, 1453-1465.

Brzeznicki, S., Jakubowski, M., & Czerski, B. (1997). Elimination of 1-hydroxypyrene after human volunteer exposure to polycyclic aromatic hydrocarbons. *Int Arch Occup Environ Health*, 70, 4, 257-260.

Buckley, T J., & Lioy, P J. (1992). An examination of the time course from human dietary exposure to polycyclic aromatic hydrocarbons to urinary elimination of 1-hydroxypyrene. *Br J Ind Med*, 49, 2, 113-124.

Castaño-Vinyals, G., D'Errico, A., Malats, N., & Kogevinas, M. (2004). Biomarkers of exposure to polycyclic aromatic hydrocarbons from environmental air pollution. *Occup Environ Med*, 61, 4, e12.

Cecinato, A., Balducci, C., Nervegna, G., Pareti, S., Tagliacozzo, G., Brachetti, A. (2008). Year time modulation of n-alkanes, PAH, nitro-PAH and organic acids at Montelibretti Rome, Italy. *Polycyclic Aromat Compd*, 28, 500-517.

Committee on the Medical Effects of Air Pollutants, COMEAP. (2011). Polycyclic aromatic hydrocarbons (PAHs). In: *Air pollutants of relevance to health.* 02-01-11, <http://www.comeap.org.uk/component/content/article/28/107-polycyclic-aromatic-hydrocarbons-pahs-and-other-toxic-organic-compounds.html>

Denison, M., & Nagy, S. (2003). Activation of the aryl hydrocarbon receptor by structurally diverse exogenous and endogenous chemicals. *Annual Rev Pharmacol Toxicol., 43,* 309-334.

Dietrich, C., & Kaina, B. (2010). The aryl hydrocarbon receptor (AhR) in the regulation of cell-cell contact and tumor growth. *Carcinogenesis,* 31, 8, 1319-1328.

Dor, F., Dap, W., & Empereur-Bissonnet, P. (1999). Validity of biomarkers in environmental health studies: the case of PAHs and benzene. *Crit Rev Toxicol,* 29, 2, 129-168.

Duan, H., Jiang, Y., Zhang, H., & Wu, Y. (2010). MiR-320 and miR-494 affect cell cycles of primary murine bronchial epithelial cells exposed to benzo[a]pyrene. *Toxicol In Vitro,* 24, 3, 928-935.

Elovaara, E M J., Stockmann-Juvala, H., Luukkanen, L., Keski-Hynnila, H., Kostiainen, R., Pasanen, M., Pelkonen, O., & Vainio H. (2006). Polycyclic aromatic hydrocabon (PAH) metabolizing enzyme activities in human lung, and their inducibility by exposure to naphtalene, phenanthrenen, pyrene, chrysene, and benzo (a)pyrene as shown in the rat lung and liver. *Arch Toxicol, 81,* 3, 169-182.

Gachanja, A N. (2009). PAHs determination. Environmental applications, In: *Regional Synthesis Report on the Status of Pollution in the Western Indian Ocean Region,* 02-01-11, www.unep.org/NairobiConvention/docs

Gasiewicz, T H., & Collins, L. (2008). Expression and activity of aryl hydrocarbon receptor in development and cancer. *Crit Rev Euk Gene Exp,* 18, 4, 279-321.

Gerhard, S. (2005). Biomonitoring of inhaled complex mixtures-ambient air, diesel exhaust and cigarette smoke. *Exp Toxicol Path, 57,* Suppl 1, 75-110.

Irigaray, P., & Belpomme, D. (2010). Basic properties and molecular mechanisms of exogenous chemical carcinogens. *Carcinogenesis, 2, 31,* 135-148.

Jardim, M J., Fry, R C., Jaspers, I., Dailey, L., & Diaz-Sanchez, D. (2009). Disruption of microRNAs expression in human airway cells by diesel exhaust particles is linked to tumorigenesis-associated pathways. *Env Health Persp,* 117, 11, 1745-1751.

Jiang, Y., Wu, Y., Greenlee, A R., Wu, J., Han, Z., Li, X., & Zhao, Y. (2011). MiR-106a-mediated malignant transformation of cells induced by anti-benzo[a]pyrene-*trans*-7,8-diol-9,10-epoxide. *Toxicol Sci* 119, 1, 1, 50-60.

Joost, H M., van Delft, K M., Yvonne, C M., Marcel, H M., van Herwijnen, K., Brauers, J J., & Kleinjans C S. (2010). Time series analysis of benzo (a) pyrene-induced transcriptome changes suggest that a network of transcription factors regulates the effects on functional genes sets. *Toxicol Sci, 117,* 2, 381-392.

Kalscheuer, S., Zhang, X., Zeng, Y., Upadhyaya, P. (2008). Differential expression of microRNAs in early-stage neoplastic transformation in the lungs of F344 rats chronically treated with the tobacco carcinogen 4-(methylnitrosamino)-1-(3-pyridyl)-1-butanone. *Carcinogenesis,* 29, 12, 2394-2399.

Kim, V N. (2005). MicroRNA biogenesis: coordinated cropping and dicing. *Nature Rev Mol Cell Biol,* 6, 5, 376-385.

Kuhn, D E., Martin, M M., Feldman, D S., Terry, A V Jr., Nuovo, G J., Elton, T S. (2008). Experimental validation of miRNA targets. *Methods*, 44, 1, 47-54.

Lee, H Y., & Yang, M. (2008). Applications of CYP-450 expression for biomonitoring in environmental health. *Env Health Prev Med*, 13, 2, 84-93.

Lin, P., Tsai, W., Wu, M., Liao, Y., Chen, J., & Su, J. (2003). Overexpression of aryl hydrocarbon receptor in human lung carcinomas. *Toxicol Path*, 31, 1, 22-30.

Liu, L., Jiang, Y., Zhang, H., Greenlee, A R., Yu, R., & Yang, Q. (2010). MiR-22 functions as a micro-oncogene in transformed human bronchial epithelial cells induced by anti-benzo[a]pyrene-*trans*-7,8-diol-9,10-epoxide. *Toxicol In vitro*, 24, 4, 1168-1175.

Lu, Y., Thomson, A G., Wong, H Y., Hammond, S M., & Hogan, B L. (2007). Transgenic over-expression of the microRNAs miR-17-92 cluster promotes proliferation and inhibits differentiation of lung epithelial progenitor cells. *Dev Biol*, 10, 2, 442-453.

Maliszewska-Kordybach, B. (1999). Sources, concentrations, fate and effects of Polycyclic Aromatic Hydrocarbons (PAHs) in the environment. Part A: PAHs in air. *Polish J Environ Studies*, 8, 3, 131-136.

Marley, C A., Gasiewicz, T A., Sime, P J., & Phipps, R P. (2005). The aryl hydrocarbon receptor is a regulator of cigarette smoke induction of the cyclooxygenase and prostaglandin pathways in human lung fibroblasts. *American J Physiol Lung Cell Mol Physiol*, 289, 3, 391-399.

McGrath, T., Wooten, J B., Chan, G., W., & Hajaligol, M R. (2007). Formation of polycyclic aromatic hydrocarbons from tobacco: The link between low temperature residual solid (char) and PAH formation. *Food Chem Toxicol*, 45, 6, 1039-1050.

Nana-Sinkam, S P., Hunter, M G., Nuovo, G J., Schmittgen, T D., Gelinas, R., Galas, D., & Marsh, C B. (2009). Integrating the microRNome into the study of lung disease. *Am J Respir Crit Care Med*, 179, 1, 4-10.

Nebert, D T., Okeys, A., & Gonzalez, F. (2004). Role of aryl hydrocarbon receptor-mediated induction of the CYP 1 enzymes in environmental toxicity and cancer. *J Biol Chem, 279*, 23, 23847-23850.

Nguyen, L., & Bradfield, C. (2008). The search of endogenous activators of the aryl hydrocarbon receptor. *Chem Res Toxicol*, 21, 1, 102-116.

Pavek, P., & Dvorak, Z. (2008). Xenobiotic-induced transcriptional regulation of xenobiotic metabolizing enzymes of the cytochromeP450 super family in human extrahepatic tissues. *Current Drug Metabolism*, 9, 2, 129-143.

Rajput, N., & Lakhani, A. (2010). Measurement of polycyclic aromatic hydrocarbons in an urban atmosphere of Agra, India. *Atmósfera*, 23, 2, 165-183.

Schembri, F., Sridhay, S., Perdomo, C., Gustafson, A M., Zhang, X., Ergun, A., Lu, J., Liu, G., Zhang, X., Bowers, J., Vaziri, C., Ott, K., Sensinger, K., Collins, J J., Brody, J S., Gatts, R., Lenburg M E., & Spira A. (2009). MicroRNAs as modulators of smoking-induced gene expression changes in human airway epithelium. *Proc Natl Acad Sci USA*, 106, 7, 2319-2324.

Shen, Y-L., Jiang, Y-G., Greenlee, A R., Zhou, L-L., Liu, L-H. (2009). MicroRNA expression profiles and miR-10[a] target in anti-benzo[a]pyrene-*trans*-7,8-diol-9,10-epoxide-transformed human 16HBE cells. *Biomed Environ Sci*, 22, 1, 4-21.

Shimada, T. (2006). Xenobiotic-metabolizing enzymes involved in activation and detoxification of carcinogenic polycyclic aromatic hydrocarbons. *Drug Metabolism Pharm, 21*, 4, 257-276.

Shimada, T., Fujii-Kuriyama, Y. (2004). Metabolic activation of polycyclic aromatic hydrocarbons to carcinogens by cytochromes P450 1A1 and 1B1. *Cancer Science, 95*, 1, 1-6.

Spira, A., Beane, J E., Shah, V., Steiling, K., Liu, G., Schembri, F., Gilman, S., Dumas, Y M., Calner, P., Sebastiani, P., Sridhar, S., Beamis, J., Lamb, C., Anderson, T., Gerry, N., Keane, J., Lenburg, M E., & Brody, J S. (2007). Airway epithelial gene expression in the diagnostic evaluation of smokers with suspect lung cancer. *Nat Med*, 13, 3, 361-366.

Spivack, S., Fasco, H G., & Kaminsky, M L. (2003). Phase I and II Carcinogen Metabolism Gene Expression in Human Lung Tissue and Tumors. *Clinical Cancer Res, 9*, 16 Pt 1, 6002-6011.

Taioli, E. (2008). Gene-Environment interaction in tobacco-related cancers. *Carcinogenesis*, 29, 8, 1467-1474.

Tomankova, T., Petrek, M., & Kriegova, E. (2010). Involvement of microRNAs in physiological and pathological processes in the lung. *Resp Res*, 11, 159, 1-10.

United States for Environmental Protection Agency, USEPA. (1993). Provisional guidance for quantitative risk assessment of polycyclic aromatic hydrocarbons. 02-01-11. EPA/600/R-93/089

Villalobos-Pietrini, R., Amador-Muñoz, O., Waliszewski, S., Hernandez-Mena, L, Munive-Colin, Z., Gómez-Arroyo, S, Bravo-Carrera J L., & Frías-Villegas, A. (2006). Mutagenicity and polycyclic aromatic hydrocarbons associated with extractable organic matter from airborne particles ≤ 10 µm in southwest Mexico City. *Atmospheric Env*, 40, 5845-5857.

Wang, F., Zhang, R., & Hankinson, O. (2006). Identifying target genes of the aryl hydrocarbon receptor nuclear translocator (Arnt) using DNA microarray analysis. *Biol Chem*, 387, 9, 1215-1218.

Wang, G., Huang, L., Zhao, X., Niu, H., & Dai, Z. (2006). Aliphatic and polycyclic aromatic of atmospheric aerosols in five locations of Nanjing urban area, China. *Atmos Res*, 81, 1, 54-66.

Wang, X Y., Li, Q B., Luo, Y M., Ding, Q., Xi, L M., Ma, J M., Li, Y., Liu, Y P., & Cheng, C L. (2010). Characteristics and sources of atmospheric polycyclic aromatic hydrocarbons (PAHs) in Shangai, China. *Environ Monit Assess*, 165, 1-4, 295-305.

Weiling, X., & Warshawsky, D. (2005). Metabolic activation of polycyclic and heterocyclic aromatic hydrocarbons and DNA damage: A review. *Toxicol Appl Pharmacol, 206*, 1, 73-93.

Wichmann, H E. (2007). Diesel exhaust particles. *Inhal Tox*, 19, suppl1, 241-244.

Williams, A E., Moschos, S A., Perry, M M., Barnes, P J., Lindsay, M A. (2007). Maternally imprinted microRNAs are differentially expressed during mouse and human lung development. *Dev Dyn*, 236, 2, 572-580.

Wincent, E A., Lueche, S., Glatt, H., Bergman, J., Cresenzi, C., Rannug, A., & Rannug, U. (2009). The suggested physiologic aryl hydrocarbon receptor activator and cytochrome P4501 substrate 6- Formylindolo[3,2-b]carbazole is present in humans. *J Biol Chem*, 284, 5, 2690-2696.

Working Group of Polycyclic Aromatic Hydrocarbons. (2001). Ambient Air Pollution by PAH´s Position Paper, In: *Office for Official Publications of the European Communities*, 02-01-11, <http://ec.europa.eu/environment/air/pdf>

Zhang, Y., & Tao, S. (2009). Global atmospheric emission inventory of polycyclic aromatic hydrocarbons. *Atmosph Environ*, 43, 4, 812-819.

5

Air Pollution and Primordial Prevention of Chronic Non-Communicable Diseases

Parinaz Poursafa[1] and Roya Kelishadi[2]
[1]*Environment Research Center, Isfahan University of Medical Sciences, Isfahan,*
[2]*Child Health Promotion Research Center, Isfahan University of Medical Sciences, Isfahan*
Iran

1. Introduction

Air pollution is a global health issue with serious public health implications particularly for children. Studying the effects of environmental factors on the early stages of atherosclerosis can serve for future studies and offer strategies for primary prevention of chronic disease.

Usually respiratory effects of air pollutants are being considered, this chapter highlights the importance of non-respiratory health hazards from early life. In addition to short-term effects, exposure to criteria air pollutants from early life might be associated with low birth weight, increase in stress oxidative and endothelial dysfunction which in turn might have long-term effects on chronic non-communicable diseases.

The independent association of air pollutants with surrogate markers of endothelial dysfunction and a possible pro-coagulant state is underscored. Similar independent associations are documented for air pollutants and hematologic parameters as well as a possible pro-inflammatory state. The presence of these associations with PM_{10} (larger than $PM_{2.5}$ usually considered as harmful) and in a moderate air quality (which is commonly considered with few or no health effect for the general population) highlights the need to re-examine environmental health policies and standards for the pediatric age group.

Atherosclerosis begins in early life, and the role of platelets is well-documented from its early stages. The concern of medical literature on atherosclerotic cardiovascular diseases is mostly about some specific inflammatory diseases, and the role of environmental factors, as air pollution is overlooked in many studies. Many studies have documented that disturbance of the inflammatory and the coagulation systems after exposure to air pollution might be a factor in endothelial dysfunction and the progression of cardiovascular diseases. The increase in platelets number and aggregation may be a surrogate marker of early hematologic and hemostatic changes due to air pollutants. The systemic pro-inflammatory and pro-coagulant response to inhalation of fine and ultrafine particulate matters suggest a role for platelet activation in this process.

Facilities should be provided for families to become aware of the quality of the air year-round and to check daily air-quality levels and air-pollution forecasts by mass media, local weather reports and other available public information sources. This is especially important for smog levels during hot weather. Protective measures should be taken into account for children and pregnant women to reduce their exposure to air pollutants, e.g. children and pregnant women should avoid congested streets and rush hour traffic, moreover families

should try to limit the amount of time their child spends outdoors in vigorous activity if the air quality is unhealthy.
In view of the emerging epidemic of chronic disease in low- and middle- income countries, the vicious cycle of rapid urbanization in such communities resulting in increasing levels of air pollution and its consequent effects on chronic diseases, as well as the limited financial resources of these countries for planning effective air pollution control programs, public health and regulatory policies for air quality protection should be integrated into the main priorities of primary health care system and into the educational curriculum of health professionals.

2. Health hazards of air pollution

Air pollutants have many adverse effects on various body organs with short- and long-term health consequences. A summary of the health hazards of air pollution is presented in Table 1.

3. Susceptibility of children health to air pollutants

Infants and children are among the most susceptible age groups for air pollutants, because they may have greater exposure than adults to air pollutants, this is especially important during summer time with highest smog levels; they have higher respiratory rates than adults, and consequently higher exposure to air pollutants. The mouth breathing of infants and children bypass the filtering effect of the nose, and they would inhale higher levels of pollutants than adults. Children generally spend significantly more time outdoors than adults, In addition, the children's immune systems and developing organs are still immature (Kim, 2004).

4. Long-term effects of air pollutants on children's health

Air pollutants have various adverse effects from early life, some of the most important harmful effects are perinatal disorders, infant mortality, respiratory disorders, allergy, malignancies, cardiovascular disorders, increase in stress oxidative, endothelial dysfunction, mental disorders and vitamin D deficiency. However, till now most focus has been on the short-term respiratory effects of air pollution on children's health. In this chapter, we highlight the wide range of hazards of air pollution from early life, and their possible implication on chronic non-communicable diseases of adulthood.
The late-onset effects of air pollution in early life may be related to many chronic diseases later in life. Most chronic non-communicable diseases originate from early life, however studies about the relationship of environmental factors, notably air pollution, with risk factors of chronic diseases are scarce in children and adolescents.

5. Exposure to air pollutants in early life and chronic diseases in adulthood

Many studies have documented the effects of criteria air pollutants on low birth weight and or prematurity .There is a growing body of evidence about the association of intrauterine growth retardation and low birth weight with increased risk of chronic non-communicable diseases such as obesity, hypertension and cardiovascular disease later in life (Sinclair et al., 2007). Furthermore, prematurity can be associated with higher risk of chronic diseases (Evensen et al., 2008).

Reference	Location	Population studied	Aims	Findings
Mengersen et al. (2011)	Lao PDR (one of the least developed countries in Southeast Asia)	the first study that investigated indoor air quality and its impact within residential dwellings in Lao PDR	study on the association between measured air pollutants and the respiratory health of resident women and children	There was a strong and consistent positive association between NO(2) and CO for almost all questionnaire-based health outcomes for both women and children. Women in dwellings with higher measured NO(2) had more than triple of the odds of almost all of the health outcomes, and higher concentrations of NO(2) and CO were significantly associated with lower PEFR
Kaplan (2010)	Review	-	This review focuses on the contribution of solid fuels to indoor air pollution .	The incomplete combustion of such materials releases byproducts with well-known adverse health effects, hence increasing the risk of many diseases and death. Among these conditions are acute respiratory infections, chronic obstructive pulmonary disease, heart disease, stroke, lung cancer, cataracts and blindness, tuberculosis, asthma, and adverse pregnancy outcomes.
Dennekamp &Carey (2010)	USA			Health effects : lung (↓function, ↑ respiratory symptoms and airway reactivity , Exacerbation of asthma and other respiratory disease) , Exacerbation of cardiovascular disease, Premature mortality.
Cao et al., (2010)	China	70,947 middle-aged men and women in the China National Hypertension Survey and its follow-up study. Baseline data were obtained in 1991 and follow-up evaluation was conducted in 1999 and 2000.	association of air pollution with mortality using proportional hazards regression model.	We found significant associations between air pollution levels and mortality from cardiopulmonary diseases and from lung cancer. Each 10μg/m(3) elevation of TSP, SO(2) and NO(x) was associated with a 0.9% (95%CI: 0.3%, 1.5%), 3.2% (95%CI: 2.3%, 4.0%), and 2.3% (95%CI: 0.6%, 4.1%) increased risk of cardiovascular mortality, respectively.
Nandi &Gorain (2010)	India	population of Durgapur town	detect the effect of pollution on human health. Two parameters, i.e., modes of transport and travelling time were chosen for this analysis.	There is pollution effect on human health at 1% and 5% level of significance. Only headache problem is considered in this paper.

Reference	Location	Population studied	Aims	Findings
Nandasena et al., (2010)	Sri Lanka	PUBMED and Medline databases, local journals and conference.	PUBMED and Medline databases, local journals and conference proceedings were searched for epidemiologic studies pertaining to air pollution and health effects in Sri Lanka	Sixteen studies investigated the association between exposure to ambient or indoor air pollution (IAP) and various health outcomes ranging from respiratory symptoms, low birth weight and lung cancers. Of the sixteen, three used a case control design. Half of the studies collected exposure data only through questionnaires. There were positive associations between air pollution and adverse health effects in all studies. Methodological limitations in most of the studies resulted in poor quantification of risk estimates.
Yoshioka et al., (2010)		investigated cytokine production and nuclear factor-kappaB (NF-kappaB) activation after stimulation of macrophage cells by exposure of urban aerosols.	evaluated the induction of airway inflammation in vitro and in vivo due to exposure of urban aerosols	urban aerosols induce respiratory inflammation and onset of inflammatory disease due to an activation of the immune system.
Layshock et al., (2010)	China	this is the first report of dibenzopyrenes in the Beijing atmosphere and among the few studies that report these highly potent PAHs in ambient particulate matter .	Size fractionated particulate matter (PM) was collected in summer and winter from Beijing, China for the characterization of an expanded list of PAHs and evaluation of air pollution metrics.	Lifetime risk calculations indicated that 1 out of 10,000 to over 6 out of 100 Beijing residents may have an increased risk of lung cancer due to PAH concentration. Over half of the lifetime risk was attributed to Σdibenzopyrenes.
Longo et al., (2010)	on the island of Hawai'i	Kilauea Volcano population . Using a within-clinic retrospective cohort design, comparisons were made for visits of acute illnesses .	assess for a relative increase in cases of medically diagnosed acute illnesses in an exposed Hawaiian community.	There were statistically significant positive associations between high vog exposure and visits for medically diagnosed cough, headache, acute pharyngitis, and acute airway problems.

Reference	Location	Population studied	Aims	Findings
Adar et al., (2010)		participants (46 to 87 years of age) were without clinical cardiovascular disease at the baseline examination (2000-2002). Subcohort of MESA cohort study .	investigate cross-sectional associations between long- and short-term air pollution concentrations and microvascular characteristics using arteriolar vessel diameter as measured by retinal photography in MESA .	greater air pollution concentrations would be associated with widened retinal venular diameters. Among the 4,607 participants with complete data, CRAE were found to be narrower among persons residing in regions with increased long- and short-term levels of PM2.5.
Balakrishnan et al., (2010)	Editorial , India		For Integrated Urban–Rural Frameworks for Air Pollution and Health-Related Research in India	In an effort to close existing data gaps, the Indian Council of Medical Research (ICMR) has established a Centre for Advanced Research in Environmental Health that will focus on air pollution and examine a range of exposures and outcomes in a rural–urban pregnant mother–child cohort and an adult endovascular disease cohort. Land-use regression modeling and select gene–environment related end points are also being examined in a nested subsample. In addition, the center will engage in capacity building to address human resource needs by developing training modules for different categories of professionals.
Siddique et al. , (2010)	Delhi , India	cross-sectional study 969 school-going children (9-17 years) and 850 age- and sex-matched children from rural areas were assessed,	The prevalence of attention-deficit hyperactivity disorder (ADHD) was assessed in two childhood populations.	ADHD was found in 11.0% of urban children in contrast to 2.7% of the control group ($p<0.001$). Major risk factors were male gender, lower socioeconomic status, 12–14 year age group, and PM10 level in breathing air. ADHD was more prevalent among boys both in urban and rural areas. It was prevalent among 18.0% of the boys enrolled in Delhi against 4.0% of the girls, giving a male/female ratio

Reference	Location	Population studied	Aims	Findings
Tung et al . (2010)	Taiwan	Total of 3741 children was enrolled in the Taiwan Children Health Study from 14 communities.	investigate the associations of EPHX1 Tyr113His, His139Arg and GSTP1 Ile105Val polymorphisms with asthma and wheezing outcomes, and focused on the functional genetic change in different ambient NO(2) levels, GSTP1 and GSTM1 genotypes.	Children with high EPHX1 activity may increase the risk of asthma and wheezing outcomes, and might be mediated through airway oxidative stress generation.
Zhou et al., (2010)	China	Meta-analysis method was used to polysynthetically analyze 16 quantitative studies about the	associations between particulate air pollution and stroke daily attack or mortality.	There are positive associations between PM(10) and stroke daily attack and mortality, increase of PM(2.5) was not associated with stroke attack and mortality. A 10 μg/m(3) increase in PM(10) was associated with a 1.09% increase in stroke daily attack and 0.70%increase in stroke daily mortality . As for PM(2.5) OR appeared to be 1.001 with a 10 μ g/m(3) increase in stroke daily attack and 1.052 for daily mortality.
Power et al., (2010)	USA	In a Cohort of Older Men 680 older men (mean ± age 71± 7 y) between 1996 and 2007.	To assess the association between black carbon, a marker of traffic-related air pollution, and cognition in older men.	The association between black carbon and cognition was non-linear and black carbon estimates were logtransformed for all analyses . Ambient traffic-related air pollution was associated with decreased cognitive function in older men .
Novaes et al. , (2010)	São Paulo, Brazil	A panel study involving 55 volunteers was carried out in São Paulo, Brazil.	To explore the clinical relevance of chronic exposure to ambient levels of traffic derived air pollution on the ocular surface.	Subjects exposed to higher levels of traffic derived air pollution reported more ocular discomfort symptoms and presented greater tear film instability, suggesting that the ocular discomfort symptoms and tear breakup time could be used as convenient bioindicators of the adverse health effects of traffic derived air pollution exposure.

Reference	Location	Population studied	Aims	Findings
Phalen et al., (2010)	USA	?	the doses delivered to subjects inhaling air-pollutant particles, the concept of a dose metric (also called an indicator) has emerged. An ideal dose metric has the following properties : it is measurable; it is expressible in physical and temporal scientific units; and it has a causal relationship to one or more biological responses	Recent advances include a better understanding related to aerosol dosimetry of the influence of susceptible populations on dose, including various diseases such as chronic obstructive pulmonary disease and bronchitis, as well as various human physiological characteristics (size, age, gender); UF PM, including transport of UF particles to the brain via the olfactory nerves and through the lung to other organs; hot spots in the lung resulting in localized high doses; and advancements in dosimetric modeling.
Puett et al., (2010)		Using two prospective cohorts, the Nurses' Health Study (NHS) and the Health Professionals Follow-Up Study (HPFS),	investigated the relationship of incident type 2 DM with PM2.5, PM10, and PM10-2.5 exposures in the prior 12 months and distance to roadways.	results did not provide strong evidence of an association between exposure to PM in the prior 12 months and incident DM, however an association with distance to road (a proxy marker of exposure to traffic-related pollution) was shown among women.
Zhuang et al., (2010)	Beijing , China	The monitoring data of daily air pollution, along with the daily numbers of outpatients visits at the Allergy Department of Beijing Shijitan Hospital from April to September in 2004 were collected.	assess the effects of ambient air pollutants on hospital outpatient visits for allergic disease and pollinosis.	significant positive associations were found between levels of airborne pollen and doctor visits, with an excess risk (ER) of 2.44% (95%CI: 0.75% - 4.13%)for allergic disease and 6.58% (95%CI: 3.82% - 9.34%) for pollinosis per 100 grains/1000 mm(2) increase in pollen, in single-pollutant models. results suggest that level of airborne pollen may have a stronger effect than ambient air pollutants on allergic disease and pollinosis.

Reference	Location	Population studied	Aims	Findings
Dadvand et al . , (2010)	northeast of England	used registry-based data on congenital heart disease for the population of the northeast of England in 1985-1996.	Investigate the association between maternal exposure to ambient air pollution and congenital heart disease	The authors found a weak association between maternal exposure to black smoke and congenital malformations of cardiac chambers and connections only when using exposure as a continuous variable. When the authors used quartiles of exposure, odds ratios did not show a dose-response relation for consecutive quartiles. For sulfur dioxide, the results were not indicative of any association.
Poursafa,&Kelishadi , (2010)			The effect of air pollution on inflammatory and pro-thrombotic factors implicated in the progression of cardiovascular diseases.	The systemic pro-inflammatory and pro-thrombotic response to the inhalation of fine and ultrafine particulate matters is seemingly associated with platelet activation.This association may have a clinical significance, particularly in the presence of cardiometabolic risk factors, and may indicate the need for anti-platelet treatment.
Szyszkowicz et al. (2010)	Canada	Emergency visit data were collected in a hospital in Vancouver, Canada	Therefore the effects of ambient air pollution on emergency department (ED) visits for suicide attempts were investigated.	The results indicate a potential association between air pollution and emergency department visits for suicide attempts. Suicide attempts and ideations as the result of depression may be linked to air pollution exposure.
Brunekreef et al.,(2009)	Netherlands	a randomly selected subcohort of 5000 older adults participating in the ongoing Netherlands Cohort Study (NLCS) on diet and cancer.	the effects of traffic-related air pollution by analyzing associations with cause-specific mortality, as well as lung cancer incidence	traffic-related air pollution, especially at the local scale, was related to cardiopulmonary mortality
Carmichael et al., (2009)	Asia	Asia calculated over a 4-year period	Aerosol distributions in Asia calculated over a 4-year period and constrained by satellite observations of aerosol optical depth (AOD) are presented.	Black carbon (BC) concentrations are high throughout Asia, representing 5-10% of the total AOD, and contributing significantly to atmospheric warming (its warming potential is approximately 55% of that due to CO_2).

Reference	Location	Population studied	Aims	Findings
Nurkiewicz et al., (2008)		closer examination by toxicologists of vascular responses following PM exposure	impairment of endothelium-dependent dilation	Increased systemic inflammation and/or vascular oxidative stress
Simpson R et al., (2005)	Austria	Brisbane, Melbourne, Perth and Sydney population	investigating the health effects of air pollution on daily mortality	Strongest associations with particulate matter, nitrogen dioxide and ozone. For example, a 10 mg/m3 elevation in PM2.5 concentration was associated with approximately a 1% increase in the daily total number of deaths.
Pope et al., (2002)	USA	The risk factor data for approximately 500 000 adults were linked with air pollution data for metropolitan areas	To assess the relationship between long-term exposure to fine particulate air pollution and all-cause, lung cancer, and cardiopulmonary mortality	each 10 mg/m3 elevation in ambient fine particulate air pollution was associated with approximately a 4%, 6% and 8% increased risk of allcause, cardiopulmonary and lung cancer mortality, respectively
Lewis PR et al ., (1998)		La Trobe Valley		Respiratory morbidity has been reported in association with outdoor air pollution .

Table 1. Summary of harmful health effects of air pollutants

The relationship of long-term traffic exposure (NO2 level by residence) and diabetes mellitus is documented (Brook et al., 2008). The first biological support for this finding comes from our study that demonstrated an independent association of exposure to air pollutants, notably PM_{10}, with markers of insulin resistance among children and adolescents (Kelishadi et al.,2009), as cited in the statement of the American Heart Association (Brook et al.,2010).

These findings suggest that the systemic responses to long-term exposure to air pollutants could potentially increase the risk for development of the metabolic syndrome, hypertension and diabetes mellitus.

Some study findings on the association of air pollution with surrogate markers of atherosclerotic cardiovascular diseases in children and adolescents is presented in Table2.

Reference	Location	Population studied	Aims	Findings
Inflammation, Coagulation, Oxidative stress				
Poursafa & Kelishadi (2010)	Review	Review on the effects of air pollution on platelets	The increase of platelet count and platelet hyper-reactivity towards agonists are emerging as markers of hematologic and hemostatic changes in response to the exposure to air pollutants. The systemic pro-inflammatory and pro-thrombotic response to the inhalation of fine and ultrafine particulate matters is seemingly associated with platelet activation.	It is of particular relevance to further study the significance of platelet activation and anti-platelet therapies in primordial/primary preventive measures in children and adolescents at risk of accelerated atherosclerosis.
Kelishadi et al.(2009)	Isfahan,Iran	A population-based sample of children aged 10-18 years (n=374)	To determine the association of air pollution as well as dietary and physical activity habits with markers of inflammation, oxidative stress and insulin resistance	The Pollutant Standard Index (PSI) and the level of fine particulate matter had significant independent association with all biomarkers studied.

Reference	Location	Population studied	Aims	Findings
Yang et al.(2008)	Review	Review of studies on air pollution and chronic obstructive pulmonary diseases, cardiovascular diseases, asthma, and cancer	To provide some insight about the health problems associated with various air pollutants and their relationship in promoting chronic diseases through changes in oxidative stress and modulation of gene expression	Byproducts of oxidative stress found in air pollutants are common initiators or promoters of the damage produced in chronic diseases.
Chuang et al., (2007)	Taipei, Taiwan	Young healthy university students (n=76)	To investigate whether biological mechanisms linking air pollution to cardiovascular events occurred concurrently in human subjects exposed to urban air pollutants	Air pollution is associated with inflammation, oxidative stress and blood coagulation in healthy young humans.
Endothelial dysfunction				
Poursafa et al.(2011)	Iran	Healthy children	To assess the relationship of air pollution and plasma surrogate markers of endothelial dysfunction in the pediatric age group	The independent relationship of air pollutants with endothelial dysfunction and a pro-coagulant state can be an important factor in atherosclerosis development from early life.

Reference	Location	Population studied	Aims	Findings
Brook (2008)	Review	Review of studies on air pollution and cardiovascular diseases	To address the cardiovascular effects of air pollution and related mechanisms	Air particle exposure may both trigger acute events as well as prompt the chronic development of cardiovascular diseases, one of the mechanisms is by triggering acute endothelial dysfunction.
Nadadur et al., (2007)	USA	Differential gene expression and transcription factor activation profiles in human vascular endothelial cells exposed to a non-cytotoxic dose of fly ash or V following semi-global gene expression profiling of approximately 8000 genes.	To explore potential biomarkers for PM-induced endothelial dysfunction	Cardiovascular effects associated with exposure to PM may be mediated by perturbations in endothelial cell permeability, Membrane integrity; and ultimately endothelial dysfunction.

Table 2. Summary of studies assessing the effects of criteria air pollutants on inflammation, coagulation, oxidative stress and endothelial dysfunction among children and young adults

6. Environmental factors, lifestyle behaviors and chronic diseases

Usually improper lifestyle habits and low educational levels have been considered as the underlying process of the role of low socio-economic position in early life as a predisposing factor for future chronic diseases (Power, et al., 2007) and mortality (Strand & Kunst, 2007), the exposure to air pollutants and its effects on low birth weight and premature birth might have an additional role in this regard.

Lifestyle modifications and strengthening primary care in health system are suggested as the main strategies to prevent and control chronic diseases in low- and middle-income countries (Miranda et al., 2008).

The association between air pollution and chronic diseases may be mediated through systemic inflammatory responses (Brook et al., 2004; Holgate et al. 2003). Generating reactive oxygen species is considered to be linked to a variety of environmental factors. The association of air pollution and inflammation/oxidative stress has been demonstrated (Huang et al., 2003; Ruckerl et al. 2006; Chuang et al., 2007), even among healthy children (Kelishadi et al., 2009) who might have the early stages of atherosclerosis. Such association is also confirmed for air pollutants, notably particulate matters and surrogate markers of endothelial dysfunction and markers of vascular injury (Poursafa et al., 2011). The effects of air pollution on inflammation, coagulation, oxidative stress and endothelial dysfunction from early life confirm the necessity of implications of these findings in relation to public health and regulatory policies for primordial/primary prevention and control of adult chronic diseases from childhood.

The prevalence of malignancies are rapidly accelerating worldwide. Although lifestyle behaviors as smoking (Dominguez et al., 2006), as well as unhealthy dietary and physical activity habits leading to obesity and diabetes are known as a major contributing factor in this regard (Hjartåker et al.,2008), air pollution should be considered as another potential risk factor for developing countries(Nejjari et al.,2003) especially Asian countries , where cancer has become an emerging health threat (Park et al.,2008).This issue is particularly important for children who are susceptible to short-term and long-term effects of air pollutants.

7. Conclusion

Air pollution is a global health issue with serious public health implications particularly for children. Usually respiratory effects of air pollutants are being considered, the importance of other health hazards should be highlighted. In addition to short-term effects, exposure to criteria air pollutants from early life might have long-term hazards principally on chronic non-communicable diseases as cardiovascular diseases and cancers. In view of the emerging epidemic of chronic disease in low- and middle- income countries, the vicious cycle of rapid urbanization in such communities resulting in increasing levels of air pollution and its consequent effects on chronic diseases, as well as the limited financial resources of these countries for planning effective air pollution control programs, public health and regulatory policies for air quality protection should be integrated into the main priorities of primary health care system and into the educational curriculum of health professionals.

We suggest that environmental protection activities, particularly for reducing the emission of criteria air pollutants, should be considered for public health measures taken into account for primordial/primary prevention of chronic diseases especially in developing countries.

8. References

Adar SD, Klein R, Klein BE, Szpiro AA, Cotch MF, Wong TY, O'Neill MS, Shrager S, Barr RG, Siscovick DS, Daviglus ML, Sampson PD, Kaufman JD. (2010).Air Pollution and the microvasculature: a cross-sectional assessment of in vivo retinal images in the population-based multi-ethnic study of atherosclerosis (MESA). *PLoS Med.*7(11):e1000372.

Balakrishnan K, Dhaliwal RS, Shah B. (2010). Integrated urban-rural frameworks for air pollution and health-related research in India: the way forward. *Environ Health Perspect*, 119(1), a12-14

Brook RD, Jerrett M, Brook JR, Bard RL, Finkelstein MM.(2008). The relationship between diabetes mellitus and traffic-related air pollution. *J Occup Environ Med.* 50:32-38.

Brook RD, Franklin B,Cascio W. (2004).Expert Panel on Population and Prevention Science of the American Heart Association. Air pollution and cardiovascular disease: a statement for healthcare professionals from the Expert Panel on Population and Prevention Science of the American Heart Association. *Circulation.*109:2655-2671.

Brook RD, Rajagopalan S, Pope CA 3rd, Brook JR, Bhatnagar A, Diez-Roux AV,Holguin F, Hong Y, Luepker RV, Mittleman MA, Peters A, Siscovick D, Smith SC Jr, Whitsel L, Kaufman JD; American Heart Association Council on Epidemiology and Prevention, Council on the Kidney in Cardiovascular Disease, and Council on Nutrition, Physical Activity and Metabolism. (2010).Particulate matter air pollution and cardiovascular disease: An update to the scientific statement from the American Heart Association. *Circulation.* 121(21):2331-2378.

Brunekreef B, Beelen R, Hoek G, Schouten L, Bausch-Goldbohm S, Fischer P.(2009).Effects of long-term exposure to traffic-related air pollution on respiratory and cardiovascular mortality in the Netherlands: the NLCS-AIR study. *Res Rep Health Eff Inst.* (139):5-71.

Cao J, Yang C, Li J, Chen R, Chen B, Gu D, Kan H. (2010).Association between long-term exposure to outdoor air pollution and mortality in China: A cohort study. *J Hazard Mater.*[Epub,ahead of print]

Carmichael GR, Adhikary B, Kulkarni S, D'Allura A, Tang Y, Streets D, Zhang Q, Bond TC, Ramanathan V, Jamroensan A, Marrapu P. (2009).Asian aerosols: current and year 2030 distributions and implications to human health and regional climate change. *Environ Sci Technol.* 43(15):5811-5817.

Chuang KJ, Chan CC, SuTC. (2007).The effect of urban air pollution on inflammation, oxidative stress, coagulation, and autonomic dysfunction in young adults. *Am J Respir Crit Care Med.*176,370-376.

Dennekamp M, Carey M.(2010). Air quality and chronic disease: why action on climate change is also good for health. *N S W Public Health Bull.* 21(5-6):115-121.

Dominguez LJ, Galioto A, Ferlisi A, Pineo A, Putignano E, Belvedere M, Costanza G, Barbagallo M. (2006).Ageing, lifestyle modifications, and cardiovascular disease in developing countries. *J Nutr Health Aging.*10,143-149.

Evensen KA, Steinshamn S, Tjønna AE, Stølen T, Høydal MA, Wisløff U, Brubakk AM, Vik T. (2008).Effects of preterm birth and fetal growth retardation on cardiovascular risk factors in young adulthood. *Early Hum Dev.*85,239-245.

Hjartåker A, Langseth H, Weiderpass E. (2008).Obesity and diabetes epidemics: cancer repercussions. *Adv Exp Med Biol.* 630,72-93.

Holgate ST, Devlin RB, Wilson SJ, Frew A J. (2003). Health effects of acute exposure to air pollution. Part II: Healthy subjects exposed to concentrated ambient particles. *Res Rep Health Eff Inst.*112:31-50.

Huang SL, Hsu MK, Chan CC. (2003).Effects of submicrometer particle compositions on cytokine production and lipid peroxidation of human bronchial epithelial cells. *Environ Health Perspect.*111:478-482.

Kaplan C. (2010). Indoor air pollution from unprocessed solid fuels in developing countries. *Rev Environ Health;* 25(3),221-242.

Kelishadi R, Mirghaffari N, Poursafa P, Gidding SS. (2009). Lifestyle and environmental factors associated with inflammation , oxidative stress and insulin resistance in children. *Atherosclerosis.* 203:311-319.

Kim JJ; American Academy of Pediatrics Committee on Environmental Health. Ambient air pollution: health hazards to children. (2004). *Pediatrics.*114:1699-1707.

Layshock J, Simonich SM, Anderson KA.(2010). Effect of dibenzopyrene measurement on assessing air quality in Beijing air and possible implications for human health. *J Environ Monit.* 12(12),2290-2298.

Lewis PR, Hensley MJ, Wlodarczyk J, Toneguzzi RC, Westley-Wise VJ, Dunn T. (1998). Outdoor air pollution and children's respiratory symptoms in the steel cities of New South Wales. *Med J Aust.*169,459-463.

Mendez LB, Oldham MJ. (2010). New developments in aerosol dosimetry. *Inhal Toxicol.* 22 Suppl 2:6-14.

Miranda JJ, Kinra S, Casas JP, Davey Smith G, Ebrahim S.(2008). Non-communicable diseases in low- and middle-income countries: context, determinants and health policy. *Trop Med Int Health,*13,1225-1234.

Nandasena YL, Wickremasinghe AR, Sathiakumar N. (2010).Air pollution and health in Sri Lanka: a review of epidemiologic studies.BMC Public Health. *Biol Pharm Bull.*33,780- 783.

Nandi PK, Gorain GC. (2010).Effect of traffic pollution on health of the people at Durgapur (India). *J Environ Sci Eng.* 52,167-172.

Nejjari C, Filleul L, Zidouni N, Laid Y, Atek M, El Meziane A, Tessier JF. (2003). Air pollution: a new respiratory risk for cities in low-income countries. *Int JTuberc Lung Dis.* 7,223-231.

Novaes P, Saldiva PH, Matsuda M, Macchione M, Rangel MP, Kara-José N, Berra A. (2010).The effects of chronic exposure to traffic derived air pollution on the ocularsurface. *Environ Res.*110:372-374.

Nurkiewicz T, PorterD, Hubbs A, Cumpston J, Chen B, Frazer D, Castranova A.(2008). Nanoparticle inhalation augments particle-dependent systemic microvascular dysfunction. *Part Fibre Toxicol.* 9 ,1

Park S, Bae J, Nam BH,Yoo KY. (2008). Aetiology of cancer in Asia. *Asian Pac J Cancer Prev.* 9,371- 380.

Pope CA 3rd, Burnett RT, Thun MJ, Calle EE, Krewski D, Ito K, Thurston GD. (2002).Lung cancer, cardiopulmonary mortality, and long-term exposure to fine particulate air pollution. *JAMA.*287,1132-1141.

Poursafa P, Kelishadi R.(2010). Air pollution, platelet activation and atherosclerosis. *Inflamm Allergy Drug Targets.*9,387-92.

Poursafa P, Kelishadi R, Lahijanzadeh A, Modaresi M, Javanmard SH, Assari R, Amin MM, Moattar F, Amini A, Sadeghian B. (2011).The relationship of air pollution and surrogate markers of endothelial dysfunction in a population-based sample of children. *BMC Public Health.* 11(1):115. [Epub ahead of print]

Power C, Atherton K, Strachan DP, Shepherd P, Fuller E, Davis A, Gibb I, Kumari M, Lowe G, Macfarlane GJ, Rahi J, Rodgers B, Stansfeld S. (2007).Life-course influences on health in British adults: effects of socio-economic position in childhood and adulthood. *Int J Epidemiol,* 36,532-9.

Power MC, Weisskopf MG, Alexeeff SE, Coull BA, Spiro Iii A, Schwartz J. (2010).Traffic-related air pollution and cognitive function in a cohort of older men. *Environ Health Perspect.* [Epub ahead of print]

Puett RC, Hart JE, Schwartz J, Hu FB, Liese AD, Laden F. (2010). Are Particulate Matter Exposures Associated with Risk of Type 2 Diabetes? *Environ Health Perspect.* [Epub ahead of print]

Rankin J, Rushton S, Pless-Mulloli T,Dadvand P. (2011).Association between maternal exposure to ambient air pollution and congenital heart disease: A register-based spatiotemporal analysis. *Am J Epidemiol.* 173,171-182.

Ruckerl R, Ibald-Mulli A, KoenigW. (2006). Air pollution and markers of inflammation and coagulation in patients with coronary heart disease. *Am J Respir Crit Care Med.*173,432-441.

Siddique S, Banerjee M, Ray MR, Lahiri T. (2010).Attention-deficit hyperactivity disorder in children chronically exposed to high level of vehicular pollution. Eur J Pediatr. [Epub ahead of print]

Simpson R, Williams G, Petroeschevsky A, Best T, Morgan G,Denison L. (2005).The short-term effects of air pollution on daily mortality in four Australian cities. *Aust N Z J Public Health.* 29,205-212.

Sinclair KD, Lea RG, Rees WD, Young LE. (2007).The developmental origins of health and disease: current theories and epigenetic mechanisms. *Soc Reprod Fertil* Suppl. ,64,425- 443.

Strand BH, Kunst A. (2007).Childhood socioeconomic position and cause-specific mortality in early adulthood. *Am J Epidemiol.* 165,85-93.

Szyszkowicz M, Willey JB, Grafstein E, Rowe BH, Colman I. (2010). Air pollution and emergency department visits for suicide attempts in vancouver, Canada. *Environ Health Insights.* 4,79-86.

Tung KY, Tsai CH, Lee YL. (2010). Microsomal Epoxide Hydroxylase Genotypes/Diplotypes,Traffic Air Pollution and Childhood Asthma. *Chest.* [Epub ahead of print]

Yoshida T, Yoshioka Y, Fujimura M, Kayamuro H, Yamashita K, Higashisaka K. (2010).Urban aerosols induce pro-inflammatory cytokine production in macrophages and cause airway inflammation in vivo. *Biol Pharm Bull.*33, 780-783.

Zhou Y, Li XY, Chen K, Ye XJ, Shen Y. (2010).Association between air particulate matter and stroke attack or mortality: a Meta-analysis. *Zhonghua Liu Xing Bing Xue Za Zhi.* 31,1300-1305.

Zhuang Y, Sun XM, Wang XY, Shi HY, Zhang ZG, Wang Q. (2010). [The influence of ambient air pollutants on outpatient visits for allergic disease and pollinosis.]. *Zhonghua Yu Fang Yi Xue Za Zhi.* 44,1121-1127.

6

Respiratory Health Effects of Air Pollution and Climate Parameters in the Population of Drobeta Turnu-Severin, Romania

Cristina Petrescu[1], Oana Suciu[1], Romanita Ionovici[2], Olf Herbarth[3], Ulrich Franck[4] and Uwe Schlink[4]
[1]*"Victor Babes" University of Medicine and Pharmacy Timisoara,*
[2]*Public Health Authority Mehedinti,*
[3]*Faculty of Medicine, University Leipzig,*
[4]*Helmholtz Centre for Environmental Research – UFZ Leipzig,*
[1,2]*Romania*
[3,4]*Germany*

1. Introduction

Air pollution represents a problem for the population's health in Drobeta Turnu-Severin, south-west Romania since 1970, when heavy industry started developing fast. Political openings to the western world created opportunities for great investments and development of industry. The impact of air pollution on health became a major concern, especially when a Heavy Water Plant was built on the northern side of the city. Some acute accidents occurred in this Plant and, in order to prevent them, an efficient and modern monitoring system and special protection perimeters were created. Severe accidents disappeared and the values of H_2S (hydrogen sulphide) were kept at very low levels. The population's reaction to the air pollution risk decreased in time. Instead, there were air pollution sources that emitted for a long period of time particulate matter (PM) at very high levels, and gases. A Paper Plant and numerous Power Plants necessary to sustain the heavy industry became significant sources of air pollution. Coal and oil were the main fuels used. The local authorities decided at that time to pay attention to particles and gases, carefully monitoring their levels in the air. Thus, very high levels of deposited particles and sulphates and not so high levels of total suspended particles (TSP) were registered; the maximal TSP concentration admitted daily being 150 $\mu g/m^3$. After 1989, when the Revolution changed the economical system, a great part of the heavy industry was diminished, the industrial sources of air pollution became weaker, and the capacity of monitoring pollution decreased slowly but continuously. A good monitoring system was preserved at the Heavy Water Plant, and an alternative monitoring system was created, to investigate air pollution in the city (Environmental Protection Agency). Unfortunately, there were not created any conditions to establish the levels of PM_{10} (with mass median aerodynamic diameter - MMAD less than 10 µm), coarse particles or $PM_{10-2.5}$ (MMAD between 10 and 2.5 µm) and fine particles (MMAD of 2.5 µm) [Monn, 2001] yet. Observations of photochemical oxidants were performed only in special situations. Stationary air pollution sources diminished in importance, and traffic became a major one. Second hand cars released high concentrations of gases and particles. However, stationary points fixed near the industries or inside the city remained the main sites to monitor the air pollution in Drobeta Turnu-Severin.

The impact of air pollution on human health, especially on respiratory or cardiovascular systems, is well known. Many studies found significant association between air pollution and health effects (Abbey et al., 1999; Linn et al. 2000; Kelly et al., 2003). Particles proved to have a very complex structure (association between particles and gases was found) and they were measured in various ways. In literature, values between 30% and 90% are given for the share of PM_{10} (aerodynamic diameter less than 10 µm) in TSP (Total Suspended Particles) (Chen & Mao, 1998; Fang et al., 1999). Air pollution with gases was considered as having a direct effect on health or was considered as an associated and confounding factor of air pollution with particles. This complexity and ubiquity of particles made them a major concern on human health.

The HEAPS (Health Effect of Air Pollution on Susceptible Population) study performed in Europe (Health Effect of Air Pollution on Susceptible Population [HEAPS], 2003) and the CESAR approach in Central Eastern Europe (Leonardi et al., 2002) offered and extended an accurate analysis of the effects of air pollution on health.

Respiratory diseases tend to increase in incidence and prevalence, but it was difficult to establish a relationship with air pollution due to the multi-factorial aetiology (Zielinski et al., 1997). People with chronic respiratory diseases especially Chronic Pulmonary Obstructive Diseases (COPD), Chronic Bronchitis (Chronic Bronchitis) and Asthma proved to be a susceptible sub-population exposed to air pollution (Dennis et al., 1996; Lindgren et al., 2009; Kurmi et al., 2010). Many studies analysed the existence of a connection between air pollution exposure and the aggravating or triggering of these chronic respiratory diseases in population (Wong et al., 1999). The young and the elderly were considered an especially vulnerable population (Ritz et al., 2009; Simoni et al., 2003).

Climate factors (temperature, humidity, and wind velocity or wind direction) also have an impact on chronic respiratory diseases. Variations of air temperature and humidity were demonstrated to be associated with asthma (Weiland et al., 2004). Acute accidents of air pollution happened in specific conditions of climate (Choukiker, 2005). Seasonality was demonstrated to be an important confounding factor on the relation air pollution - health. Increased variation of climate factors has a well-known direct influence on human health and on air pollution levels and variation to the same extent. Numerous present day studies demonstrate the variation of climate factors, the tendency of global warming, and its effects on human health and not only (Kowats et al., 1999; Van Vjingaarden & Vincent, 2004).

Keeping in mind all additional factors that determine or aggravate respiratory pathology makes us more aware of the complexity of the relation between air pollution and health.

In the study performed, we aimed at examining long- and short-term respiratory health effects of air pollution with particles and gases in a vulnerable population group, and its modification due to climate parameters (air temperature and air humidity). Their changes in time were studied using a longitudinal retrospective inquiry and two time-series analyses of data, obtained in two successive periods and in the same area.

2. Material and methods

In order to assess the long and short-term effects of air pollution on chronic respiratory diseases and its modification by climate parameters we used the following material and methods.

2.1 Description of population and area under study

Drobeta Turnu-Severin is a historic and industrial city situated in south-western Romania, on the north side of the Danube. Its population is about 100,000 inhabitants (Romanian Statistical Yearbook, 2006). This city used to be highly industrialized between 1970 and 1995, with an economical decrease afterwards. Heavy industry, a large carriage manufacturer, a paper plant, and numerous power plants polluted the environment intensively. The climate is temperate with Mediterranean influences. On the south-western side of the city there is a great accumulation lake built over the Danube and used by the Iron Gates Hydropower Plant. This accumulation lake changes the climate factors of the neighbouring area.

2.2 Material

The material consisted in studies on health, air pollution, and meteorological data.

2.2.1 Health data

Health data were provided by the County Hospital Drobeta Turnu-Severin. As we performed this study using two successive study periods, we used different data. In the first study period (1.01.1990-31.12.1997), we used incidence (number of new cases per 100,000 inhabitants) of COPD (ICD-10, J 44.0, 1, 8), of chronic bronchitis (ICD-10, J 41.0, 1, 8), and of asthma (ICD-10, J 45.0, 1, 8), counted in year of study and age groups, in the exposed population from Drobeta Turnu-Severin (study group), and 2 control groups from other urban and rural areas in Mehedinti County, representing the non-exposed population. Choosing diagnosed patients, we took into consideration the main definition of the three clinical endpoints.

In the second (1.01.2000-31.12.2003) study period, daily hospital admissions according to sex and age groups were recorded in the exposed area of Drobeta Turnu-Severin. During these two study periods, acute infectious respiratory disease cases were provided and used as a confounding factor.

2.2.2 Air pollution data

Air pollution data were provided by the Public Health Direction Mehedinti and comprised the maximum, minimum, and average daily concentration ($\mu g/m^3$ air) of total suspended particles (TSP) for the two study-periods. The Environmental Protection Agency (EPA) Drobeta Turnu-Severin provided daily concentrations of SO2 and NO2 only for the period 23.01.2001 - 31.08.2002, included in the second study period. These pollutants were measured at 4 fixed sampling sites inside the city, 1 point being situated in a well known polluted area (near the Paper Plant). The measurements are based on the gravimetric method for TSP and automatic monitoring stations for SO_2 and NO_2.

2.2.3 Meteorological data

Meteorological data were provided by the Romanian Regional Meteorological Centre Craiova, Oltenia and comprised daily maximum, minimum, and average air temperature (Celsius degrees), relative air humidity (%), wind velocity (m/s) and wind direction. The meteorological station is situated in the western part of Drobeta Turnu-Severin. For the analysis of the second study period, we calculated the absolute humidity that gives the amount of water in air, using the relative humidity and the air temperature (Stull, 2000).

2.3 Methods

A statistical analysis of all data gathered in the two study periods was performed.

2.3.1 First study period

A case-control study was performed for the relative risk of air pollution with particles (TSP) on respiratory health. For that purpose, three groups were considered: one population-group exposed to air pollution, and two control groups of non-exposed population, in other urban areas and the rural area of the region. Meteorological factors were considered as confounding factors. Spearman correlation was used to calculate associations of air pollution with TSP - meteorological factors and meteorological factors - respiratory health. Relative risk was measured considering the ratio of the exposed population with a specific chronic respiratory disease to the non-exposed population with a specific chronic respiratory disease. We considered relative risk taking into account an important confounding factor: different age and density distribution of population between urban and rural areas. Spearman's rank correlation coefficient assesses how well an arbitrary monotonic function can describe the relationship between two variables, without making any other assumption about the particular nature of the relationship between variables. The analysis based on rank correlations is insufficient, however, for days having zero hospital admissions due to respiratory ailments.

2.3.2 Second study period

Two time-series analyses were performed in the second study period in order to establish the short-term impact of air pollutants on health, one for TSP and the other considering gases: NO_2 and SO_2.

In the first time-series analysis, the Poisson regression, a method that belongs to the group of generalized linear models (GLM, Freeware R) was used to assess the risk. Missing TSP data were imputed by means of a special Kalman filter, which acts as a low pass for periods longer than 100 days. Absolute humidity was used, although relative humidity was recorded. The values of the current day and the values of the previous day (time lag effect) of the environmental factors were used considering a possible lagged influence on the hospital admissions. The adverse effect of TSP may be possibly modified by air temperature and/or air humidity. In the first time-series analysis, we used 2 approaches.

In the first approach, models of the following type were examined:

$$\ln(\mu_i) \approx \beta_o + \beta_1 dow_i + \beta_2 TSP_i + \beta_3 absHum_i + \beta_4 Tave_i \,,\ i = 1, 2, \ldots, n \,, \tag{1}$$

with μ_i - the expected number of patients in day i, dow_i - the day of the week and $absHum_i$ the absolute humidity.

In the second modelling approach, the annual variation was included. The variable "day" was introduced to represent the Julian day number and modelled with a spline (s):

$$\ln(\mu_i) \approx \beta_o + \beta_1 dow_i + \beta_2 TSP_i + \beta_3 absHum_i + \beta_4 Tave_i + s(day) \tag{2}$$

Hospital admission data and risk factors follow seasonal variation. Generalized additive models (GAM - Hastie & Tibshirani, 1987) and generalized linear models with natural splines were applied. All models were compared by the criterion AIC and explained deviance (the fraction of the deviance in the data explained by the model).

In the second time-series analysis, we applied the additive models and the "Air Pollution and Health: a European Approach - 2" protocol (APHEA-2) (Atkinson et al., 2001). First, a core model without pollutants was built and, in the second stage, the core model was extended by air pollutants. Seasonal patterns and a dummy variable for the day of the week were always included and daily of mean, minimum and maximum air temperature and relative humidity as meteorological confounders were used as in the second study. Time delayed effects were moved up to seven days. Non-linear effects were modelled by thin plate regression splines with automatic smoothness selection by the mgcv-software-package (Wood, 2006). In the extended model, a PDLMs (Zanobetti et al., 2003) with a polynomial degree of three and a maximal lag effect of 10 were used for evaluation of the association and only for interpolated time-series. A threshold analysis was done for all significant effects of values greater than the threshold value, and all the concentration values below the threshold were set to zero. The modifying effect of air humidity for all observed adverse significant effects, including an interaction term (the product of the lagged effect of absolute humidity with the pollutant effect) in the model was investigated.

3. Results obtained

3.1 First period study results

3.1.1 Descriptive statistics

Chronic respiratory diseases (COPD, chronic bronchitis - CB and asthma) incidence **in the exposed population** (who live in Drobeta Turnu-Severin) was the following (Table 1), with a not so high mean of incidence for the entire first study period.

We registered an increased incidence of COPD in the third, fourth, and seventh years of the first study period, respectively in 1993 (330 cases/ $\%_{000}$ inhabitants), 1992 and 1996. Chronic bronchitis incidence was increased in the years 1993 (253 cases/ $\%_{000}$ inhabitants) and 1996, and asthma incidence was increased in the same years, and 1997 (with the highest value - 74 cases/ $\%_{000}$ inhabitants).

Respiratory diseases	Years of first study								Mean
	1	2	3	4	5	6	7	8	
COPD	61	74	**155**	**330**	181	172	**263**	173	176
CB	43	49	116	**253**	140	114	**185**	81	122
Asthma	10	16	30	**53**	36	46	**59**	**74**	40

Table 1. Distribution of chronic respiratory diseases (COPD, CB and asthma) incidence (cases/ $\%_{000}$ inhabitants) in the exposed population (Drobeta Turnu-Severin) during the first study period

Distribution of chronic respiratory diseases incidence according to age groups proved to be very different, with the highest values in the age group 15-64 years for COPD and CB, and in the age's group 1-14 years for asthma (Table 2).

Chronic respiratory diseases (COPD, chronic bronchitis - CB and asthma) incidence **in the 1st control group** (other urban areas) can be seen in Table 3.

Age group distribution of incidence in the 1st control group (other urban areas) can be seen in Table 4.

Respiratory diseases	Incidences on age groups (years)		
	1-14	15-64	>65
COPD	165	**720**	524
CB	6	**532**	445
Asthma	**159**	130	35

Table 2. Distribution of chronic respiratory diseases (COPD, CB and asthma) incidence (cases/ %000 inhabitants) in the exposed population (Drobeta Turnu-Severin) depending on age groups, in the first study period

Respiratory diseases	Years of first study								Mean
	1	2	3	4	5	6	7	8	
COPD	188	75	71	14	286	70	299	101	138
CB	122	75	64	14	56	40	257	37	83
Asthma	66	0	7	0	230	30	42	62	54

Table 3. Distribution of chronic respiratory diseases (COPD, CB and asthma) incidences (cases/ %000 inhabitants) in the 1st control group (other urban areas) during the first study period

Respiratory diseases	Incidence according to age groups (years)		
	1-14	15-64	>65
COPD	253	351	498
CB	37	168	460
Asthma	216	183	38

Table 4. Distribution of chronic respiratory diseases (COPD, CB and asthma) incidence (cases/ %000 inhabitants) in the 1st control group (other urban areas) depending on age group, in the first study period

Chronic respiratory diseases (COPD, chronic bronchitis - CB and asthma) incidence **in the 2nd control group** (rural areas) can be seen in Table 5.

Respiratory diseases	Years of first study								Mean
	1	2	3	4	5	6	7	8	
COPD	139	198	154	116	123	261	174	143	164
CB	92	185	123	78	92	202	134	107	126
Asthma	22	12	21	20	22	29	27	32	23

Table 5. Distribution of chronic respiratory diseases (COPD, CB and asthma) incidence (cases/ %000 inhabitants) in the 2nd control group (rural areas) during the first study period

Age groups distribution of incidences in the 2nd control group (rural areas) can be seen in Table 6.

Respiratory diseases	Incidences on age groups (years)		
	1-14	15-64	>65
COPD	63	798	452
CB	37	616	360
Asthma	25	119	42

Table 6. Distribution of chronic respiratory diseases (COPD, CB, and asthma) incidence (cases/ %000 inhabitants) in the 2nd control group (rural areas) depending on age group, in the first study period

We noticed large variation of chronic respiratory diseases incidence depending on the area and age group.

In the first study period (1.01.1990-31.12.1997) we registered the following levels of **air pollution with TSP** considering the mean, the maximum and minimum values, measured yearly and for the entire first study period (Table 7).

TSP	Years of first study								Mean	Median	SD
	1	2	3	4	5	6	7	8			
M	189	104	71	129	137	161	174	110	**134**	133	39
Max	440	270	120	80	330	360	330	270	275	300	121
Min	50	40	30	2	40	20	60	30	34	35	17

Table 7. Yearly distribution of TSP values (μg/m³ air) in the first study period

Season	Years of first study								Mean	Median	Standard deviation
	1	2	3	4	5	6	7	8			
Winter	175	167	71	80	168	161	165	164	143	**164**	42
Spring	171	141	70	105	101	143	162	096	123	123	35
Summer	194	93	-	131	132	167	174	93	123	131	61
Autumn	217	88	-	163	138	176	203	96	135	**150**	71

Table 8. Seasonal distribution of TSP values (μg/m³ air) in the first study period

The annual means and entire study mean exceeds the maximum admitted concentration in United States for TSP (75 μg/m³ air) or maximum admitted concentration for PM10 in Europe (40 μg/m³ air).

Air pollution with particles (TSP) was also analysed depending on season, for each year and for entire first study period (Table 8).

The season means and medians calculated in each year of the first study and for the total study period indicate frequent surpassing of the annual maximum admitted concentration

for TSP and even daily maximum admitted concentration of TSP (150 µg/m^3 air) in winter and autumn. A possible explanation of increased air pollution in these seasons could be the use of heat sources based on burning fuel.
Climate factors measured yearly and for the entire first study period (Table 9) revealed high relative humidity in the area (over 70%), temperature higher than 10 Celsius degrees and low wind velocity (1.97 m/s). Wind direction was North/North-West.

Climate factors	Years of first study								Mean
	1	2	3	4	5	6	7	8	
T (°C)	12.50	10.65	12.50	11.78	12.92	11.53	11.46	11.24	**11.82**
Hu (%)	75.66	80.58	72.50	74.25	75.50	74.33	74.25	71.00	**74.75**
w.v (m/s)	2.00	1.96	2.19	2.22	1.95	1.81	1.85	1.85	**1.97**

Table 9. Yearly mean values of climate factors in the first study period

The seasonal evolution of climate factors (Table 10) indicates that the highest value of temperature was registered in summer (21.63 °C) and the lowest in winter (3.03 °C), the highest relative humidity in autumn (82.29%) and the lowest in summer (68.83%), the highest wind velocity in winter (2.22 m/s) and the lowest in autumn (1.65 m/s). We noticed that in winter there was low temperature, high relative humidity (78.24%) and increased wind velocity. The investigated area is situated near the Danube and not far from an accumulation lake built on the Danube: the Iron Gates Hydropower plant, fact that explains increased air humidity in all seasons (over 65%).

Climate factors	Seasons				Mean
	Winter	Spring	Summer	Autumn	
T (°C)	**3.03**	16.59	**21.63**	6.20	11.82
Hu (%)	**78.24**	70.49	**68.83**	**82.29**	**74.75**
w.v (m/s)	**2.22**	2.06	**2.03**	1.65	1.97

Table 10. Seasonal mean values of climate factors in the first study period

3.1.2 Relative risk

A significant relative risk of TSP was found for COPD in exposed population in comparison with the 1st control group (other urban areas) and the 2nd control group (rural areas, see Figure 1).
The relative risk of TSP on chronic respiratory diseases in the exposed population measured in comparison with the non-exposed populations (1st and 2nd control groups) presented as main confounding factors different age group distribution and density of population between urban and rural areas. We excluded the differences between the two control groups comparing them. We considered the registered differences when we measured the relative risk for exposed population considering the 2nd control group (rural area). The registered values for the measured relative risk of TSP on chronic respiratory diseases (COPD, CB, asthma) can be seen in Tables 11 and 12.

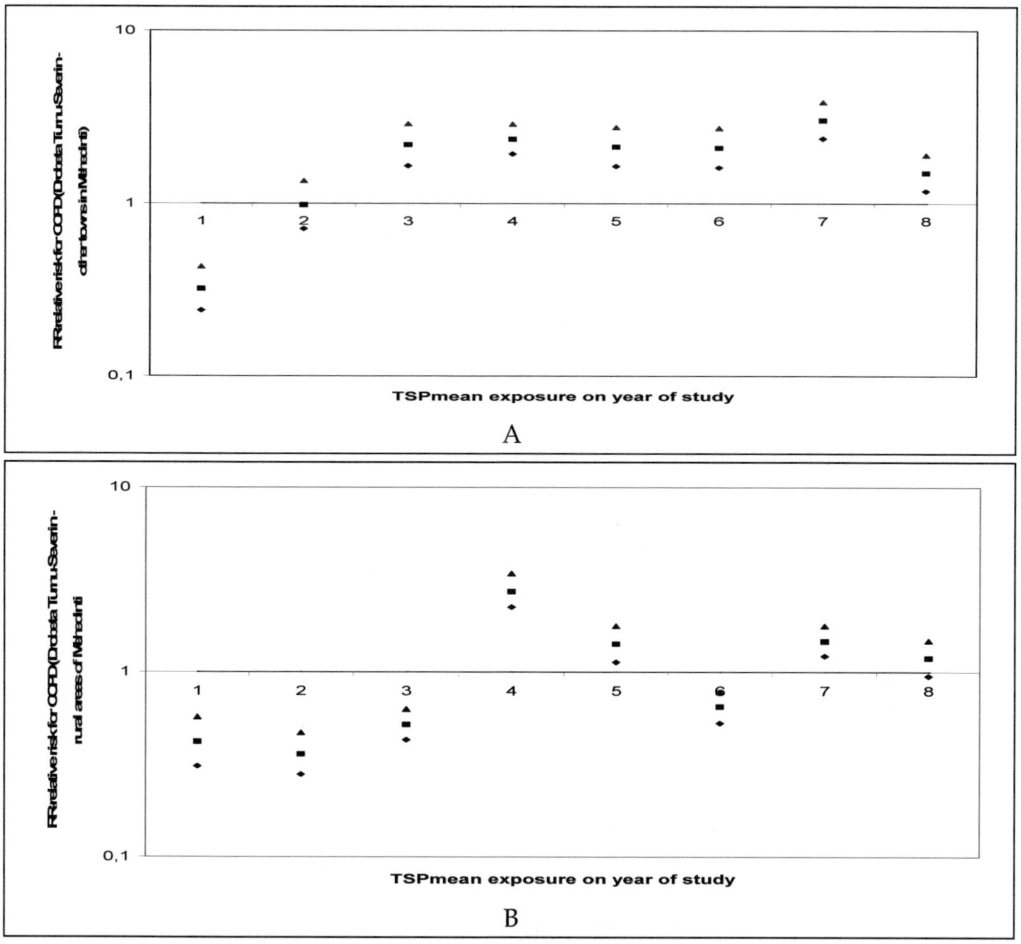

Fig. 1. Relative risk (RR) values for COPD in the population exposed to TSP related to the 1st control group - other urban areas (A) and the 2nd control group - rural area (B)

Chronic respiratory diseases	Relative risk according to age group		
	1-14 years	15-64 years	> 65 years
COPD (RR) CI	0.65 0.53-0.79	**2.05** 1.8-2.32	**1.05** 0.93-1.18
CB (RR) CI	0.16 0.07-0.37	**3.16** 2.66-3.76	0.96 0.85-1.09
Asthma (RR) CI	0.73 0.6-0.9	0.71 0.56-0.88	0.92 0.58-1.45

Confidence interval Nurminen-Miettinen

Table 11. Relative risk of TSP on chronic respiratory diseases considering age groups and 1st control group (other urban areas)

There is an observable increase in the relative risk of TSP on **chronic bronchitis** in the age group **15-64 years** and on **COPD** in the age groups **16-64 years** and **over 65 years**, and in the exposed population as compared to non-exposed population from other urban areas.

Chronic respiratory diseases	Relative risk according to age groups		
	1-14 years	15-64 years	> 65 years
COPD (RR) CI	0.65 0.64-0.66	**2.08** 2.02-2.12	**2.94** 2.94-2.95
CB (RR) CI	0.15 0.1-0.23	**3.18** 3.03-3.44	0.96 0.96-0.97
Asthma (RR) CI	0.73 0.73-0.74	0.71 0.69-0.72	0.89 0.75-1.04

Confidence interval Nurminen-Miettinen

Table 12. Relative risk of TSP on chronic respiratory diseases considering age group and 2nd control group (rural areas) with exclusion of confounding factors between urban and rural areas

Comparing the exposed with the non-exposed population from rural areas, after excluding the differences urban - rural, we noticed the same increased risk of TSP on **chronic bronchitis** in the age group **15-64 years** and on **COPD** in the age groups over **65 years** and **15-64 years.**

3.1.3 Spearman correlation of variables

We applied the following Spearman correlations: chronic respiratory diseases - air pollution with TSP, air pollution with TSP - climate factors, and chronic respiratory diseases – climate factors. Significant results were obtained only for the Spearman correlation asthma – relative air humidity (Table 13).

		Asthma incidence 15-64 years
Relative air humidity (%)	rho	**- 0850****
CI	p value	0.007

**Correlation is significant at 0.01 level (2-tailed)

Table 13. Spearman correlation (rho) asthma incidence – relative air humidity

This result indicates a protective effect of elevated air humidity on asthma, especially in the age group 15-64 years.

3.2 Second period study results

3.2.1 Descriptive statistics

Descriptive analysis performed in the second study period (1.01.2000-31.12.2003) can be seen in Table 14. The values for all variables were available for 1461 days, but for TSP only for 820 days. The patients were classified according to sex.

	Minimum	First quartile	Average	Median	Third quartile	Maximum	Standard Error
COPDm (cases)	0	0	0.91	1	1	7	1.17
COPDf	0	0	0.43	0	1	4	0.70
CBm	0	0	0.21	0	0	4	0.48
CBf	0	0	0.16	0	0	3	0.41
Asthma m	0	0	0.24	0	0	4	0.52
Asthma f	0	0	0.46	0	1	5	0.80
TSP (µg/m^{3}air)	18	75	111.42	105	142	289	47.39
T min	-14	1	7.78	8.2	14.7	25.8	8.22
Tave (^{0}C)	-9.2	5.1	12.78	12.8	20.8	31.7	9.42
Tmax	-6	9.6	18.61	19.4	27.6	42.6	10.9
Rel. hu (%)	29	58	67.96	68	79	100	14.55
w.v. (m/s)	0	3	5.64	5	8	16	3.18

m-male, f-female

Table 14. Descriptive analysis in the second study period

We realised that the day of the week (dow) is an important confounding factor (Figure 2).

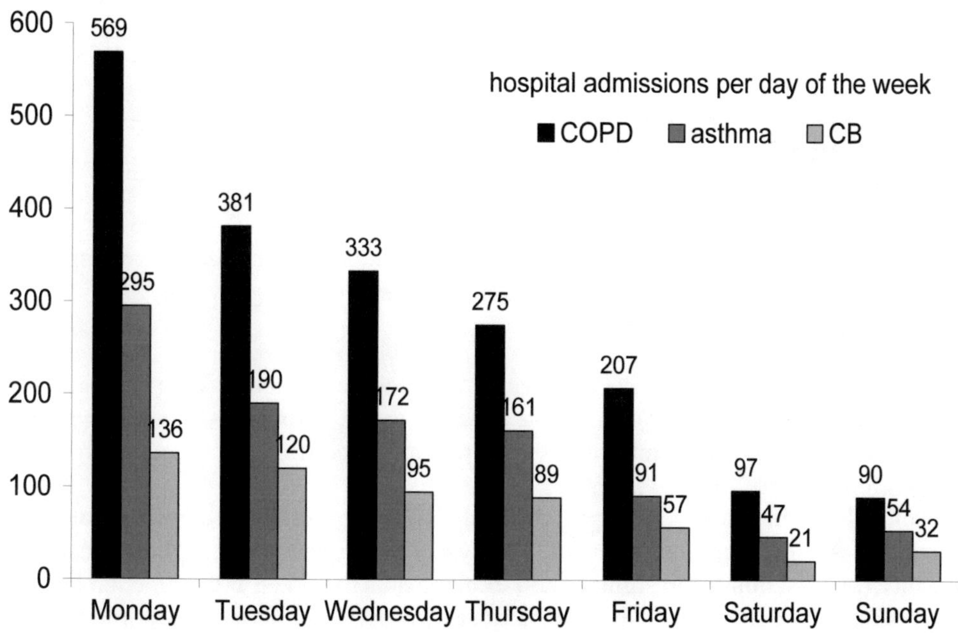

Fig. 2. Number of patients with chronic respiratory diseases for each day of the week

The number of patients is maximal on Monday and minimal on Sunday. One explanation could be the admission of patients on Monday and the discharge from hospital on Friday. The average period of hospitalisation of a patient is 5 days.

3.2.2 Risk evaluation - first time-series analysis results

COPD (Chronic Obstructive Pulmonary Diseases)

In the **first approach** (Table 15), a protective effect of absolute humidity on COPD (abs Hum coeff. = - 0.0347) was noticed. The protective effect of air humidity was modified by TSP (abs Hum: TSP = - 0.0011). The relative risk of this modification was rather small with low statistical significance (RR = exp(0.001) ~1.001, p=0.008).

Variable 1	Coefficient	s.e.	P value
Intercept	1.1870	0.0640	<0.0001
Tuesday	-0.4496	0.0670	<0.0001
Wednesday	-0.5425	0.0693	<0.0001
Thursday	-0.7590	0.0743	<0.0001
Friday	-0.9780	0.0803	<0.0001
Saturday	-1.7410	0.1084	<0.0001
Sunday	-1.8440	0.1134	<0.0001
absHum	**-0.0347**	0.0088	<0.0001
absHum:TSP	0.0011	0.0006	0.0833

Table 15. Results of the first model approach for COPD

This change of the humidity effect suggested a high influence of the season on respiratory diseases (especially COPD), the annual trend, and temperature being involved.

The second approach

The yearly variation and the effect of temperature were modelled with a spline. The absolute humidity has a significant adverse effect on the age group 18-44 years. A positive difference between absolute humidity of the current and preceding day increases the risk in three age groups (Table 16).

The absolute humidity varies direct proportionally with the number of admitted patients in the hospital. A significant influence of TSP was not observed. The spline function of seasonality has local minima in summer and local maxima in winter. If the function becomes 0, there is no influence on the expected number of hospital admissions. The function has positive values in winter and this indicates an increased risk on hospital admissions, the negative one is in summer and the risk is reduced.

Variable 1	Coefficient	s.e.	P value	RR
Intercept	-2.9672	0.2133	<0.0001	-
Tuesday	-0.4002	0.0480	<0.0001	0.6702
Wednesday	-0.5390	0.0500	<0.0001	0.5833
Thursday	-0.7336	0.0533	<0.0001	0.4802
Friday	-1.0021	0.0588	<0.0001	0.3671
Saturday	-1.7597	0.0795	<0.0001	0.1721
Sunday	-1.8520	0.0821	<0.0001	0.1569
absHum 18-44	**0.0600**	0.0229	0.0088	**1.0619**
Dif absHum 18-44	**0.1352**	0.0682	0.0473	**1.1448**
Dif absHum 45-64	**0.0645**	0.0221	0.0035	**1.0667**
Dif absHum 65-74	**0.0653**	0.0262	0.0126	**1.0678**

Table 16. Results of the second model approach for COPD

Chronic Bronchitis

In the **first approach,** we did not find a significant effect of TSP on chronic bronchitis. The absolute humidity has a protective effect at very high value (more than 6g/m^3 air).
In the **second approach,** we found a TSP risk on chronic bronchitis and a protective effect of absolute humidity on the same disease (Table 17).

Variable 1	Coefficient	s.e.	P value	RR
Intercept	-1.3895	0.1754	<0.0001	-
Tuesday	-0.1534	0.0643	0.0170	0.8578
Wednesday	-0.3731	0.0684	<0.0001	0.6886
Thursday	-0.4016	0.0698	<0.0001	0.6692
Friday	-0.8759	0.0806	<0.0001	0.4165
Saturday	-1.8801	0.1196	<0.0001	0.1526
Sunday	-1.4482	0.1002	<0.0001	0.2350
TSP lag 1 18-44	**0.0071**	0.0015	<0.0001	**1.0071**
absHum	**-0.0648**	0.0171	<0.0001	**0.9372**

Table 17. Results of second model approach for chronic bronchitis

Asthma

In the **first approach** (Table 18), TSP had an adverse effect on asthma.

Variable 1	Coefficient	s.e.	P value
Intercept	0.3169	0.1053	0.0026
Tuesday	-0.3226	0.0911	0.0004
Wednesday	-0.5215	0.0972	<0.0001
Thursday	-0.5972	0.0993	<0.0001
Friday	-1.1167	0.1194	<0.0001
Saturday	-1.7127	0.1507	<0.0001
Sunday	-1.7060	0.1507	<0.0001
TSP	**0.0167**	0.00836	0.0453
absHum:TSP	**-0.0022**	0.00073	0.0023

Table 18. Results of first model approach for asthma

Variable 1	Coefficient	s.e.	P value	RR
Intercept	-1.9800	0.1986	<0.0001	-
Tuesday	-0.4420	0.0562	<0.0001	0.6427
Wednesday	-0.5402	0.0578	<0.0001	0.5826
Thursday	-0.6037	0.0592	<0.0001	0.5468
Friday	-1.1739	0.0722	<0.0001	0.3092
Saturday	-1.8348	0.0945	<0.0001	0.1596
Sunday	-1.6989	0.0890	<0.0001	0.1829
TSP lag 1 0-17	**0.0068**	0.0014	<0.0001	1.0068
absHum 45-65	**-0.0341**	0.0149	0.0222	0.9665
absHum >66	**-0.0463**	0.0227	0.0412	0.9548

Table 19. Results of second model approach for asthma

Absolute humidity modified this effect and was protective. The relative risk of this modification was exp(-0.0022) ~0.008, for a fixed TSP value and a rise of 1 g water/m³.
In the **second approach,** TSP has an adverse effect on asthma in age group 0-17 years. Absolute humidity has a protective effect on asthma in older people (Table 19).

3.2.3 Descriptive statistics – considering gases

TSP, gases (NO_2 and SO_2), number of cases of chronic respiratory diseases and climate factors were analysed during the time interval 23.01.2001 - 31.08.2002 (Table 20) in the second study period.

	Missing (%)	Cases/ year	Mean	Median	Minim	Maxim	Standard Deviation
COPD (cases)	0	511	1.4	1	0	10	1.55
CB (cases)	0	160	0.44	0	0	3	0.71
Asthma (cases)	0	282	0.77	0	0	7	1.08
TSP (μg/m³air)	42	-	123.4	122	22	289	44.12
SO2 (μg/m³air)	16	-	4.68	4,6	1.7	10.4	1.55
NO2 (μg/m³air)	16	-	11.8	11	1	33.5	3.84
T high (⁰C)	0	-	20.23	21.5	-4	39.5	10.08
Tmean (⁰C)	0	-	14.06	14.9	-8.2	31.7	9.03
Tlow (⁰C)	0	-	8.81	9.6	-13.2	25.8	8.19
Rel. hu (%)	0	-	67.76	68	29	98	13.93
AbsHum (g/m³)	0	-	8.79	8.23	2.05	19.2	4.03

Table 20. Descriptive analysis during the interval 23.01.2001 - 31.08.2002 in the second study period

Hospital admissions showed a weekly and a seasonal cycle. The maximum of average hospital admissions occurred on Monday and the minimum on Sunday (Figure 3) during this time interval, this result being similar with the situation of the second study period. Moreover, in this time interval we studied the seasonal variation. The maximum average monthly admission was in January and the minimum in August for COPD and asthma, and September for chronic bronchitis (Figure 4).

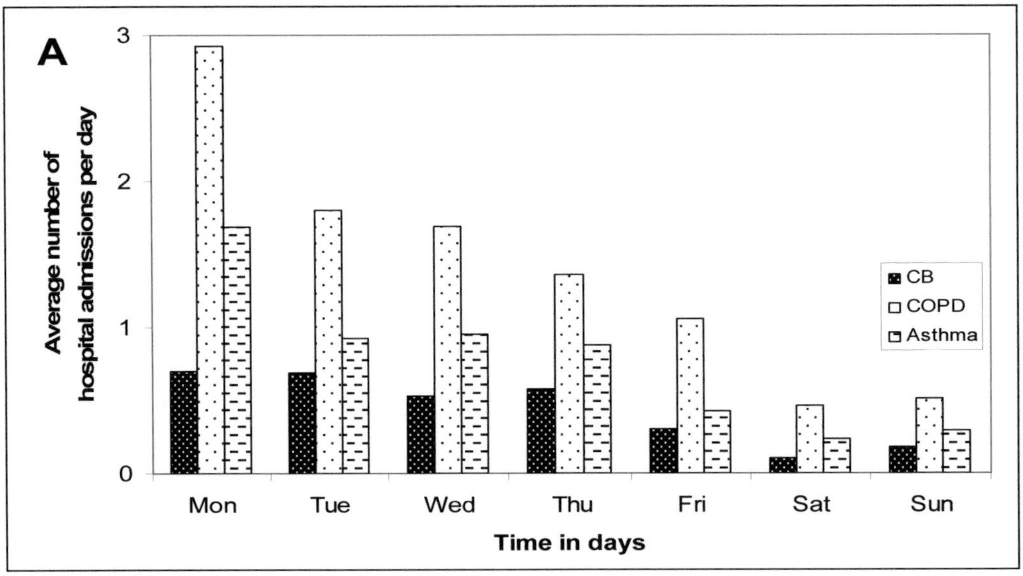

Fig. 3. The weekly cycle of daily hospital admission

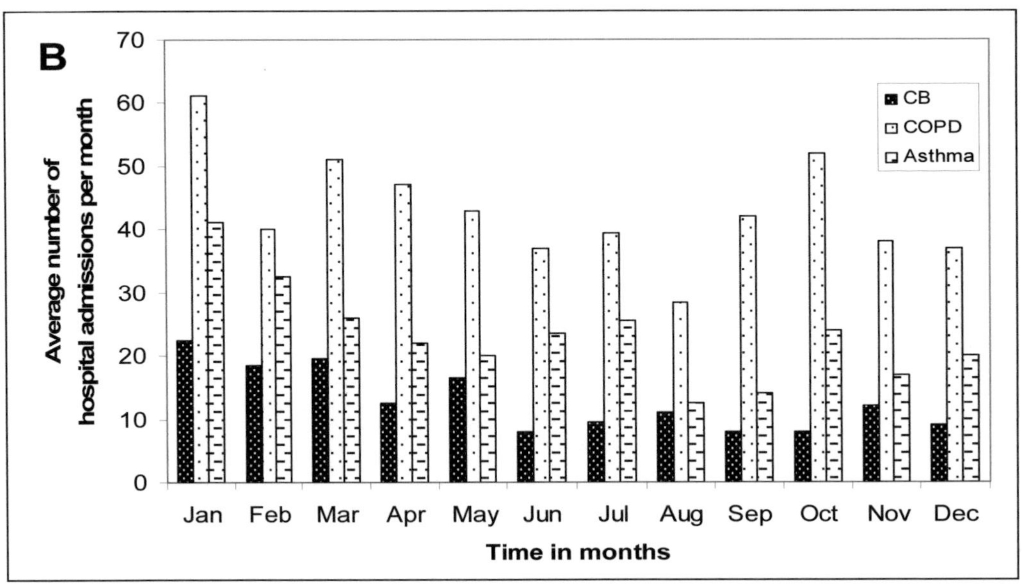

Fig. 4. The seasonal cycle of monthly hospital admission

3.2.4 Risk evaluation - second time-series analysis results considering gases

Effects of TSP on chronic bronchitis

Significant results of applying this model were found only for chronic bronchitis.
The core model for chronic bronchitis consisted of factor variables for workdays/holidays and nonlinear effects of time, of temperature with a lag of 6 days, of change of humidity over the last three days.

For the single lag model, an increase in hospital admissions for chronic bronchitis of $_{0.3}3.3_{6.4}\%$ with a lag of 1 day and $_{0.12}2.8_{5.7}\%$ with a lag of 4 day for an elevation of TSP with 10 µg/m³ air was registered (Figure 5). Displaying the estimates with confidence intervals as a triple of percentiles $_{2.5}50.0_{97.5}$ is recommended (Louis & Zeger, 2009). The effects were slightly smaller for the time-series with missing values. For PDLMs the association was significant with an increase of $_{0.01}1.6_{3.1}\%$ in hospital admissions for chronic bronchitis for an elevation of 10 µg/m³ TSP with a lag of three days.

A threshold analysis revealed there was no level, below which no effect of TSP was present for both lagged TSP effect levels (Figure 6).

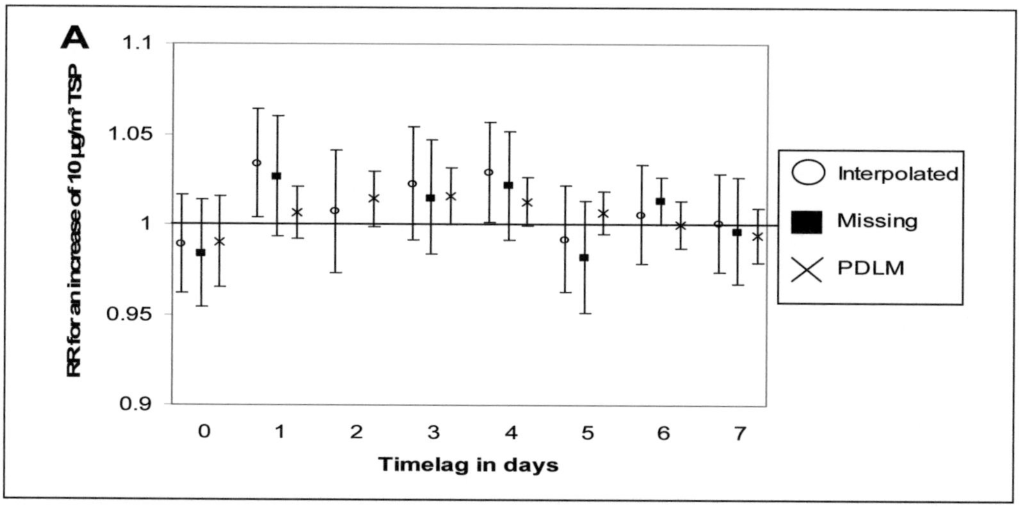

Fig. 5. Relative risk (CI 95%) for an increase of TSP with 10 µg/m³ air in single lag model with original and interpolated data and PDLMs with interpolated data

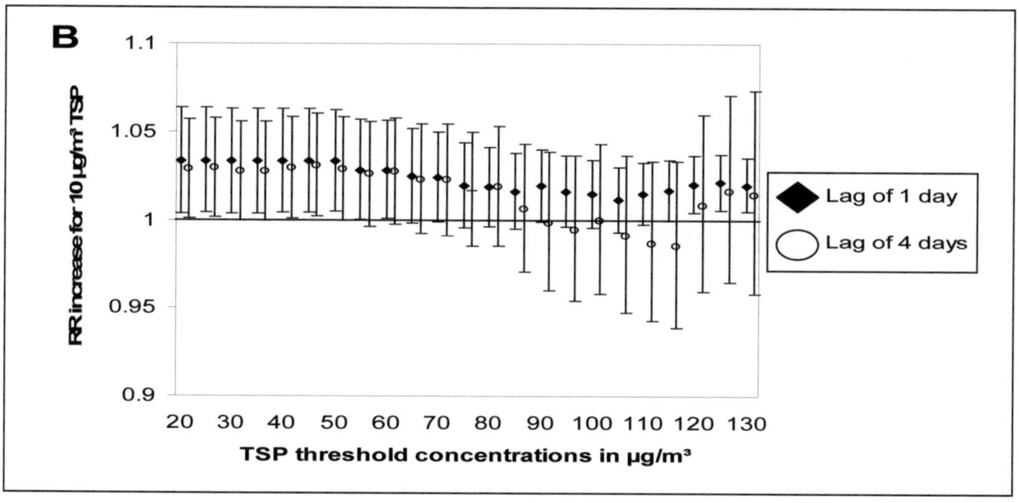

Fig. 6. Relative risk (CI 95%) for an increase of TSP with 10 µg/m³ air using the threshold of 1 day and 4 days

When we investigated subgroups depending on gender the effect was significant, the number of male hospital admissions increased to $_{1.6}4.09_{7.22}$ % with a lag of 4 days and the female hospital admissions increased to $_{0.434}4.44_{8.60}$% with a lag of one day TSP exposure (Figure 7).
The modification of the significant effects of TSP by absolute humidity was analysed adding an interaction term between the two variables (Leitte et al., 2009). For low concentration of TSP (less than 60 µg/m³), increased humidity reduces the risk of hospital admission due to chronic bronchitis (Table 21). Concentrations of TSP of more than 60 µg/m³ cancel the antagonistic effect.

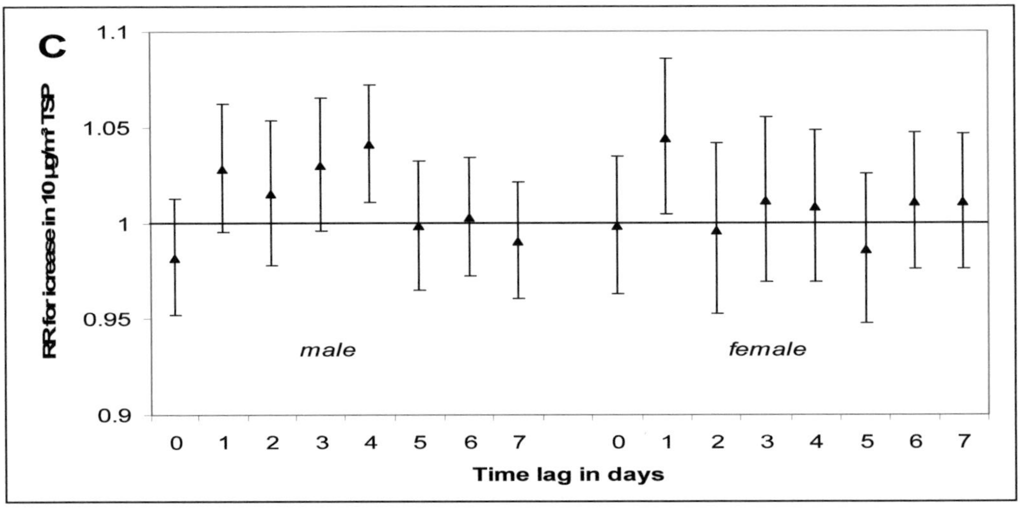

Fig. 7. Relative risk (CI 95%) for an increase of TSP with 10 µg/m³ air for gender subgroups

	Change of hospital admissions for chronic bronchitis (%)	p-value
TSP with lag of 1 day	6.97	0.002
Modification by absolute humidity, moving the average on the same and last day	-0.41	0.05
TSP with a lag of 4 days	6.02	0.009
Modification by absolute humidity with a lag of 4 days	-0.37	0.092

Table 21. Modifying effect of TSP (increment of 10 µg/m³) by absolute humidity (increment of 1 g/ m³) on hospital admissions for chronic bronchitis

Effects of SO_2 on chronic bronchitis

We observed a significant increase of 15% and 9% hospital admissions due to chronic bronchitis for an increment of 1 µg/m³ SO_2 with a delay of two or seven days when we considered a time-series with interpolated values.

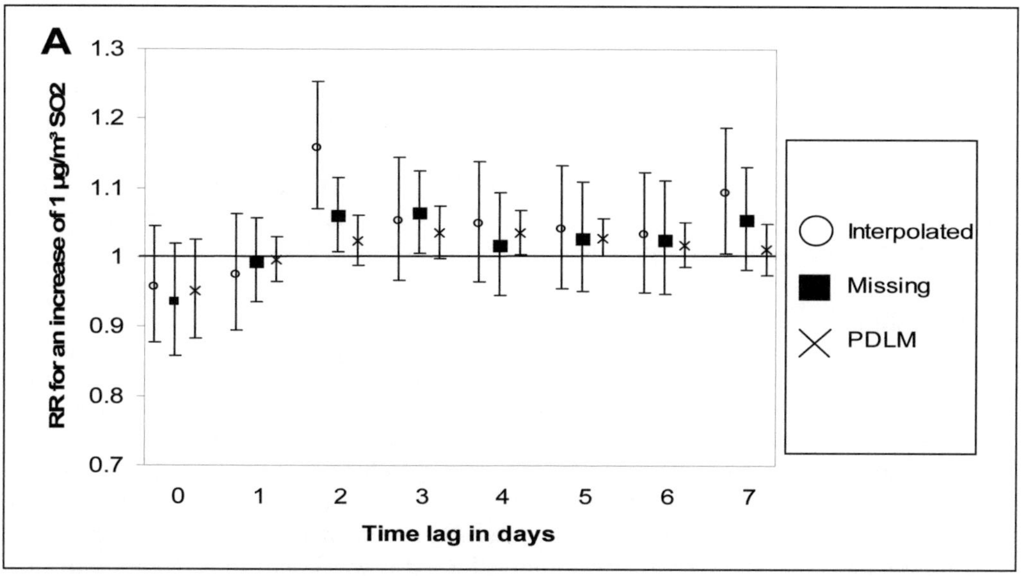

Fig. 8. Relative risk (RR - 95% CI) of an increase of 1 µg/m³ SO_2 (single lag model with interpolated, original data and PDLMs for interpolated data) for hospital admissions for chronic bronchitis and different time lags

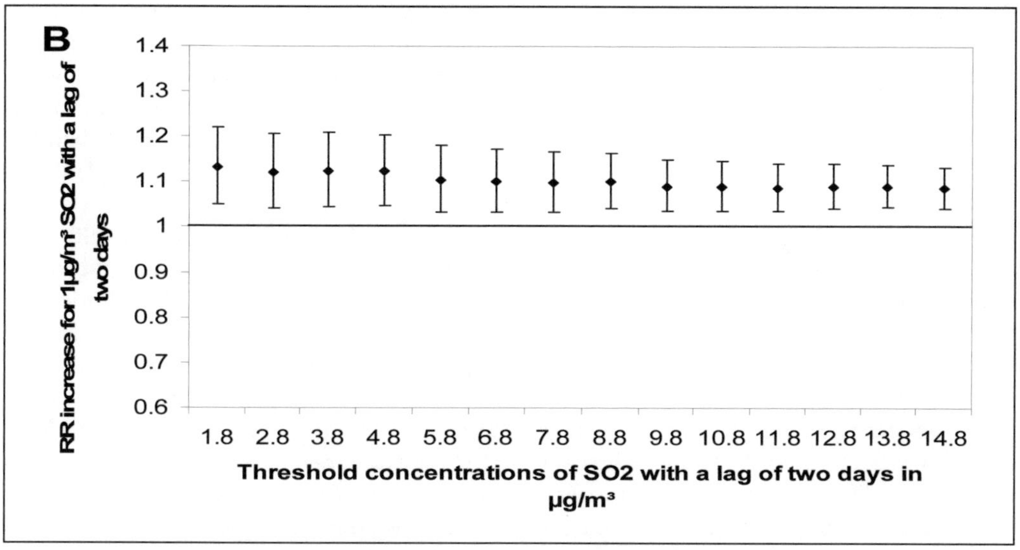

Fig. 9. Relative risk (RR - 95% CI) of an increase of 1 µg/m³ SO_2 for hospital admissions for chronic bronchitis using thresholds for a lag effect of two days

When we considered a time-series with missing values the association was reduced to a $_{1}6_{12}$% and $_{1}6_{13}$% increase of hospital admissions for chronic bronchitis for the same increment with 1 µg/m³ SO_2 with a lag of two and three days. For PDLMs an increment of 1 µg/m³ SO_2 determined an increase of $_{0.40}3.6_{6.81}$% and $_{0.11}2.8_{5.58}$% hospital admissions for chronic bronchitis (Figure 8).

There was no threshold, under which no adverse effect of SO_2 on chronic bronchitis existed (Figure 9).
All significant single pollutant effects were included in a multi pollutant model. In this model the effect of SO_2 with a delay of two days remained significant (RR=13% for a increase of 1 µg/m^3 SO_2) and the effect of TSP with a delay of four days became weakly significant (RR=2.5% for an increase of 10 µg/m^3 TSP, p=0.09). Adverse influences of TSP, SO_2, NO_2 on total respiratory admissions, COPD and asthma, and of NO_2 on chronic bronchitis were found, but without statistical significance.

4. Discussion

Air pollution with particulate matter and gases is well known in south-eastern Europe, as it is reported in numerous studies performed in this area. High levels of black smoke were found in Serbia (Milutinović et al., 2009), increasing concentrations of SO_2 were noticed in Prague (Duben, 1988). In Greece, a study that investigated the gas concentration evolution along the year indicated the powerful influence of the season on the air pollutant levels, being minimal in summer and maximal in winter (Papaioannou et al., 2010). A "Convention on Long-Range Trans-boundary Air Pollution (CLRTAP)" was opened for signature in 1979 and came into force in 1983. This Convention has now 51 parties and priority activities include review and possible revision of its most recent protocols with special focus on Eastern Europe and South Eastern Europe.
In our study, we found an increased level of air pollution with particles (134 µg/m^3 air in the first study period and 111.42 µg/m^3 air or 123.45 µg/m^3 air in the second study period) and gases (4.68 µg SO_2/m^3 air and 11.8 µg NO_2/m^3 air - in the second study period - 1 year), which surpasses the maximal admitted concentrations of TSP for the United States (75 µg/m^3 air) and the maximal admitted concentration of PM10 specified by the European Commission (40 µg/m^3 air corresponding to 57 - 133 µg/m^3 air TSP) and are lower than the maximal admitted concentration of gases (SO_2 - 20 µg/m^3 air and NO_2 - 40 µg/m^3 air) (2008/50/EC).
Climate change and climate factors influence both air pollution and human health (Mc Michael et al., 2008) and is also a largely investigated topic. Climate change was not found to have high impact on the distribution of the deposition, nor did regional air pollution (in the form of sulphate aerosols) (Alcamo et al., 2002). In our study, we also did not find an association between climatic factors and air pollutants. Variations in meteorological parameters were evident between the two periods under study: average temperature increased from 11.82^{0}C to 12.78^{0}C; average relative air humidity decreased from 74.5% to 67.96% and wind velocity from 1.97m/s to 5.64 m/s. These results are similar with other results from literature (Kovats et al., 1999; Wjingaarden & Vincent, 1995). In literature an association climate factors - respiratory health was found, especially between air temperature, humidity and asthma (Weiland et al., 2004). In our study, the first period, we identified a protective effect of air relative humidity on asthma (Spearman correlation), and in the second period of study first time-series analysis was revealed a protective effect of absolute air humidity on chronic bronchitis and asthma. Analysing the evolution of results between the two approaches of the first time-series analysis, we noticed seasonality as a powerful confounding factor, this being considered only in the second approach. The greatest confusion created by seasonality was present especially in the case of COPD (chronic obstructive pulmonary diseases) when humidity transformed itself from a

protective factor on COPD into an adverse one, between the two approaches. The seasonal influence on COPD admission in hospital was modelled by a spline function of seasonality with local minima in summer and local maxima in winter.

Respiratory health appeared as being frequently impacted by air pollution. Many epidemiological studies described the long term effects of air pollutants on the population with chronic respiratory diseases, such as the impact on symptoms investigated in Switzerland (Zemp et al., 1999), symptoms which occurred in children exposed to air pollution in the Netherlands (Brauer et al., 2002; Abbey et al., 1995), or the impact of air pollution on other chronic diseases as carcinoma (Chen & Goldberg, 2009). Respiratory health evolution was difficult to analyse between the two study periods, because of different approaches. Only a qualitative comparison between the two study periods was possible. A similar approach was considered necessary in China, where short-term effects were studied (numerous time-series analyses). Long-term effects of air pollutants on health are less studied so far (Kan et al., 2009).

Case-control studies proved to be useful for the investigation of long-term effects of air pollution on health, investigating especially the respiratory symptoms (Burr et al., 2004). Some authors analysed the study performed in South-Eastern Europe, exposing the bad management of resources in this area and the great levels of air pollution (especially sulphur dioxide) (Jedrychowski, 1995). The case-control study performed in the first period of study offered results about long-term effects of air pollution with TSP on new cases of hospitalisation for chronic respiratory diseases. A similar study on hospital attendance due to respiratory diseases as a result of air pollution effects was made in Turin, Italy (Migliareti et al., 2007). A significant result of the first study period was the impact of outdoor air pollution on COPD at different ages, in the active population and the elderly. Similar results were obtained analysing some epidemiological studies performed in developing and industrialized countries (Viegi et al., 2006).

Short-term effects of air pollution on respiratory health were frequently approached by researchers in the area, revealing the weak effect of particulate matter and the stronger effect of sulphur dioxide on respiratory health in Paris (Dab et al., 1996), the effect of fine particulate matter on cardiovascular disease and carcinoma (Laden et al., 2006), or the association of elderly population exposed to higher risk of mortality (Gouvea & Fletcher, 2000). The two time-series analyses performed in the second study period made possible an evaluation of the short-term risk of the TSP (both analysis) and SO2 and NO2 (the second analysis) on respiratory health.

The impact of air pollution with particles on chronic bronchitis was showed in literature (Herbarth et al., 2001). Although, initially, air pollution was investigated for high levels of exposure, there is also a tendency to investigate air pollution effects on low levels (Pope et al., 1995). This approach is also supported by the results of our study, we did not find a threshold under which no effects occur for TSP and sulphur dioxide. Although the study designs were different, the impact of TSP on chronic bronchitis at a specific age (15-64 years in the first study period, 18-44 years in the second study period) was evident. In our study, active population seems to be more exposed to outdoor air pollution. Similar results were found in literature, a significant number of cases of chronic bronchitis being found in adults by a study in Barcelona (Kunzli & Perez, 2007).

In the second time-series analysis of the second study period, significant results were found only for chronic bronchitis due to air pollution with TSP and gases (SO_2) using an extension

of the time delayed effect from 1 day (in the first time-series analysis) to 7 days, PDLMs for interpolated series and an interaction term in order to observe the modifying effect of absolute humidity. Seasonality as a confounding factor was always considered.
For a single lag model, the hospital admissions for chronic bronchitis increased with 3.3% for a lag of 1 day and with 2.8% for a lag of 4 days due to an increase of **TSP** by 10 μg/m^3 air (interpolated series). The effect was lower for the series with missing values and more realistic for the series with PDLMs. When we included the TSP effect on chronic bronchitis in the multi pollutant model, the relative risk remained weakly significant (RR=2.5% for an increase of 10 μg/m^3 TSP, p=0.09), only for a lag of 4 days. The **modifying effect of absolute humidity** on the effect of **TSP on chronic bronchitis** was present and statistically significant for a time lag of 2 days and TSP concentrations lower than 60 μg/m^3 air, with a decrease of 0.41% admitted cases (p=0.05) to an increase in 1 g/m^3 air absolute humidity. The differences between health effects according to sex were described in literature (Kenedy et al., 2007). In our study, the effect for hospital admissions was more delayed and slightly lower for male admissions, but these differences between sexes are not statistically significant.
The single lag model also registered an increase of chronic bronchitis with 15% for a lag of 2 days and 9% for a lag of 7 days, for an enhancing with 1 μg/m^3 air **SO_2**, considering interpolated series. When we considered the series with PDLMs the increase of chronic bronchitis admissions was lower than 3.6% and 2.8% for a lag of 2 and 3 days, respectively, of SO_2 increment with 1 μg/m^3 air. The series with PDLMs indicated more realistic results, similar with the data found in literature (Wilson et al., 2005). When we included SO_2 effect on chronic bronchitis on multi pollutant model, the relative risk remained statistically significant (RR=13% for a increase of 1 μg/m^3 SO_2), only for a lag of 2 days. In this multi pollutant model there was no correlation between SO_2 and TSP and multicolinearity should be negligible. In our study, we did not find a threshold for TSP, neither for SO_2. This result indicates the necessity of considering low levels of air pollution as a factor with possible respiratory health impact, too. Air pollution in Drobeta Turnu-Severin, Romania, represents a risk factor for respiratory health of the exposed population, especially because of high levels of SO_2 and TSP in outdoor air. Increased air humidity of this area is a protective factor for people with chronic bronchitis and asthma.

5. Conclusion

Outdoor air pollution is a reality in Drobeta Turnu-Severin Romania, beginning during the communist period and continuing after the revolution. High levels of TSP and sulphates registered in the eighth decade of the past century diminished in time, but not in a sufficient amount to avoid its impact on health. The monitoring system of air pollution has a tendency to diminish its potential. The Danube and the accumulation lake "Iron Gates" change the climate factors and the ecosystem in the area. Humidity is higher than in other areas of the region and it affects the relationship air pollution - respiratory health, acting as a protective factor on specific chronic respiratory diseases as chronic bronchitis and asthma. The different study designs (case-study for the first study - 8 years and time-series analysis for the second successive study - 4 years) indicated similar results regarding the TSP effect on chronic bronchitis in active population. A short period of time extended as procedures time-series analysis (1 year) included in the second study indicated similar results for TSP and SO_2 on chronic bronchitis, also revealing the delayed effects of 4 days or 2 days respectively for both pollutants. Air humidity appeared as modifying the effect of TSP on chronic bronchitis for a delayed effect of 2 days, revealing itself as being protective. A limitation of the study results consists regards the action of TSP and humidity on COPD. In the first study period, the long-term exposure of the population to TSP acted as a trigger factor for COPD, in active and elderly population. In the second study period, TSP and humidity seemed to have an antagonistic effect on COPD. Moreover, seasonality determined a specific evolution of this disease during the year. A powerful impact of seasonality referred to the action of TSP on asthma, the differences of results between the two approaches of the second study period, in the first time-series analysis being huge. New studies are necessary in order to clarify these aspects.

6. Acknowledgments

This study was performed with the financial support of the BMBF (Bundesministerium fur Bildung und Forschung), MECT (Ministerul Educatiei, Cercetarii si Tineretului), 6th Framework Program of the European Commission, Project NoMiracle (Contract No. 003956).

7. References

2008/50/EC. (2008). Directive 2008/50/EC of the European Parliament and of the Council of 21 May.

Abbey, D.; Nishino, N.; McDonnell, W.; Burchette, R.; Knutsen, S.; Beeson, L. & Yang, J. (1999). Long-term inhalable particles and other air pollutants related to mortality in nonsmokers, *Am. J. Respir. Crit. Care Med.*, Volume 159, No. 2, pp. 373-382.

Alcamo, J,; Meyerhofer, P.; Guardans, R.; Van Harmelen, T.; van Minnen, J.; Onigkeit, J.; Posch, M. & de Vries, B. (2002). An integrated assessment of regional air pollution and climate change in Europe: findings of the AIR-CLIM Project, *Environental Scienc & Policy*, Volume 5, No. 4, pp. 257-272, doi: 10.1016/SI 462-9011(02)00037-0.

Atkinson, R. & al. (2001). Acute effects of particulate air pollution on respiratory admissions - results from APHEA 2 project, *Am J Respir Cell Mol Biol*, No. 164, pp. 1860-1866.

Brauer, M. & al. (2002). Air pollution from traffic and the development of respiratory infections and asthmatic and allergic symptoms in children, *American Journal of Respiratory and Critical Care Medicine*, Volume 166, pp. 1092-1098.

Burr, M.; Karani, G.; Davies, B.; Holmes, B. & Williams, K. (2004). Effects on respiratory health of a reduction in air pollution from vehicle exhaust emissions, *Occup Environ Med.*, Volume 61, No. 3, pp. 212-218.

Chen, H. & Goldberg, S. (2009). The effects of outdoor air pollution on chronic illnesses, *McGill J Med*, Volume 12, No. 1, pp. 58-64.

Chen, M. & Mao, I. (1998). Spatial variations of airborne particles in metropolitan Taipei, *Sci Total Environ*, No. 209, pp. 226-231.

Choukiker, S. (2005). Major air pollution episodes: Environmental distortions that kill. An account of major air pollution episodes and their repercurssions arguing for a responsible environmental awareness, *Environment: Disaster Management*, http://www.visionriviewpoint.com/print.ap?articleid=26.

Dab, W.; Medina, S.; Quénel, P.; Le Moullec, Y.; Le Tertre, A. ; Thelot, B. ; Monteil, C. ; Lameloise, P. ; Pirard, P. ; Momas, I. ; Ferry, R. & Festy, B. (1996). Short term respiratory health effects of ambient air pollution : results of the APHEA project in Paris, *J. Epidemiol Community Health.*, Volume 50, Suppl. 1, s42-s46.

Dennis, R.; Maldonado, D.; Baena, E. & Martinez, G. (1996). Woodsmoke exposure and risk for obstructive airway disease among women, *CHEST*, Volume 109, No. 1, pp. 115-119.

Duben (transl) Giustino, K. (1988). Air quality in Prague, *Lidove Noveni*, pp. 7.

Fang, G.; Chang, C.; Wu, Y.; Fu, P.; Chang, K. & Yang, D. (1999). The characteristic study of TSP, PM10 and PM2.5 in the rural site of central Taiwan, *Sci Total Environ*, No. 232, pp. 177-184.

Gouvea, N. & Fletcher, T. (2000). Time-series analysis of air pollution and mortality effects by cause, age and socioeconomic status, *J Epidemiol Community Health*, Volume 54, pp. 750-755.

Hastie, T. & Tibshirani, R. (1987). Generalized Additive Models: some applications (with discussion), *JASA*, No. 82, pp. 371-386.

Herbarth, O. & al. (2001). Effect of sulfur dioxide and particulate pollutants on bronchitis in children - a risk analysis, *Environ Toxicol*, Volume 16, No. 3, pp. 269-276.

Jedrychowski, W. (1995). Review of recent studies from central and Eastern Europe associating respiratory health effects with high levels of exposure to "traditional" air pollutants, *Environ Healh Perspect.*, Volume 103, Suppl. 2, pp. 15-21.

Kan, H.; Chen, B. & Hong, C. (2009). Health Impact of Outdoor air pollution in China: current knowledge and future research needs, *Environ Health Perspect*, Volume 117, No. 5, pp. A187.

Kelly, F.; Dunster, C. & Mudway, I. (2003). Air pollution in the elderly: oxidant/antioxidant issues worth consideration, *Eur. Respir. J. Suppl.*, No. 40, 70s-75s.

Kennedy, S.; Chambers, R.; Du, W. & Dimich-Ward, H. (2007). Environmental and occupational exposures: do they affect chronic obstructive pulmonary disease differently in women and men?, *Proc Am Thorac Soc*, Volume 4, No. 8, pp. 692-694.

Kowats, R.; Haines, A.; Stanwell-Smith, R.; Martens, P.; Menne, B. & Bertolini, R. (1999). Climate change and human health in Europe, *BMJ*, No. 318, pp. 1682-1685.

Kunzli, N. & Perez, L. (2007). The public health benefits of reducing air pollution in Barcelona metropolitan area, www.creal.cat/media/upload/arxius/.../informe-contaminacio-eng.pdf.

Kurmi, O.; Semple, S.; Simkhada, P.; Smith, W. & Ayres, J. (2009). COPD and chronic bronchitis risk of indoor air pollution from solid fuel: a systematic review and meta-analysis, *Thorax*, Volume 65, pp. 221-228, doi: 10.1136/thx.2009.124644.

Laden, F.; Schwartz, J.; Speizer, F. & Dockery, D. (2006). Reduction in fine particulate air pollution and mortality, *American Journal of Respiratory and Critical Care Medicine*, Vol. 173.

Leitte, A.; Petrescu, C.; Franck, U.; Richter, M.; Suciu, O.; Ionovici, R.; Herbarth, O. & Schlink, U. (2009). Respiratory health, effects of ambient air pollution and its modification by air humidity in Drobeta-Turnu Severin, Romania. *Science of the Total Environment*, No. 407, 4004 - 4011.

Leonardi, G.; Houthuijs, D.; Nikiforov, B.; Volf, J.; Rudnai, P. & Zejda, J. (2002). Respiratory symptoms, bronchitis and asthma in children of Central and Eastern Europe. *Eur. Respir. J.*, Volume 20, No. 4, pp. 890-898.

Lindgren, A.; Stroh, E.; Montnemery, P.; Nihlen, U.; Jakobson, K. & Axmon, A. (2009). Traffic-related air pollution associated with prevalence of asthma and COPD/chronic bronchitis. A cross-sectional study in Southern Sweden, *Int J Health Geogr.*, Volume 8, No. 2, doi: 10.1186/1476-072X-8-2.

Linn, W.; Szlachcic, Y.; Gong, H. & Berhane, K. (2000). Air pollution and daily hospital admissions in metropolitan Los Angeles, *Environ.Health Perspect.*, Volume 108, No. 5, pp. 427-434.

Louis, T.A.; Zeger, S.L. (2009). Effective communication of standard errors and confidence intervals, *Biostatistics,* Volume 10, No. 1, pp. 1-2.

McMichael, A. & al. (2008). International study of temperature, heat and urban mortality: the ISOTHURM project, *Int J Epidemol,* Volume 37, No. 5, pp. 1121-1131.

Migliaretti, G.; Damalso, P. & Gregori, D. (2007). Air pollution effects on the respiratory health of the resident adult population in Turin, Italy, *International Journal of Environmental Health Research,* Volume 17, No. 5, pp. 369-379.

Milutinović, S.; Nikić, D.; Stosić, L.; Stanković, A. & Bodanović, D. (2009). Short-term association between air pollution and emergency-room admissions for chronic obstructive pulmonary disease in Nis, Serbia, *Cent Eur J Public health,* Volume 17, No. 1, pp. 8-13.

Monn, C. (2001). Exposure assessment of air pollutants: a review on spatial heterogeneity and indoor/outdoor/personal exposure to suspended particulate matter, nitrogen dioxide and ozone, *Atmospheric Environment,* Volume 35, No. 1, pp. 1-32.

National Institute of Statistics of Romania, Romanian Statistical Yearbook. (2006). http://www. insse.ro/cms/rw/pages/index.en.do [acesed 30 september 2008].

Papaioanou, A.; Viras, L.; Nastos, P. & Paliatsos, A. (2010). Temoral evolution of sulfur dioxide and nitrogen oxides in the city of Volos, Greece, *Environ Monit Assess.,* Volume 161, No. 1, pp. 485-494.

Pope, A.; Bates, D.& Raizenne, M. (1995). Health effects of particulate air pollution: time for reassessment?, *Environmental Health,* Volume 103, pp. 472-480.

Ritz, T.; Meuret, A.; Wilhelm, F. & Roth, W. (2009). Changes in pCO2 symptoms, and lung function of asthma patients during capnometry - assisted breathing training, *Applied Psychophysiology and Biofeedback,* No. 34, pp. 1-6.

Simoni, M.; Jaakkola, M.; Carrozzi, L.; Baldacci, S.; Di Pede, F.; Yiegi, G. (2003). Indoor air pollution and respiratory health in the elderly, *ERJ,* Volume 21, No. 40 suppl. 15s-20s.

Stull, R. (2000). Meteorology for scientists and engineers, In: *Pacific Grove California,* Brooks Cole.

The HEAPS project Health Effect of Air pollution on Susceptible Subpopulation, study design. (2003). Air pollution and recruitment of Myocardial Infarction patients, *Epidemiology,* Volume 14, No. 5, pp. 591-592.

Van Wjingaarden, W. & Vincent, L. (2005). Trends in relative humidity n Canada from 1953-2003, *J. Geophys. Res.,* 110, D22102, doi: 10.1029/2005JD05925, http://www.yorku.ca/wlaser/publications/Papers%20(PDF%20Format)/2006/2006_3.pdf.

Viegi, G.; Maio, S.; Pistelli, F.; Baldacci, S. & Carrozzi, L. (2006). Epidemology of chronic obstructive pulmonary disease: health effects of air pollution, *Respirology,* pp. 523-532, doi: 10.1111/j1400-1843.2006.00886.x.

Weiland, S.; Husing, A.; Strachan, D.; Rzehac, P. & Pearce, N. (2004). Climate and the prevalence of symptoms of asthma, allergic rhinitis, and atopic eczema in children, *Occup.Envion. Med.*, No. 61, pp. 609-615.

Wilson, A. & al. (2005). Air pollution, weather, and respiratory emergency room visits in two northern New England cities: an ecological time-series study, *Environ Res*, Volume 97, pp. 312-321.

Wong, T.; Lau, T.; Yu, S.; Neller, A.; Wong, S.; Tam, W. & Pang, S. (1999). Air pollution an hospital admissions for respiratory and cardiovascular diseases in Hong Kong, *Occup. Environ. Med.*, Volume 56, No. 10, pp. 679-683.

Wood, S. (2006). Generalized Additive Models: an introduction with R. In: *Chapman and Hill/CRC*, New York.

Zanobetti, A. & Schwatz, J. (2003). Air pollution and emergency admissions in Boston, MA, *J Epidemiol Community Health*, No. 60, pp. 890-895.

Zemp, E. & al. (1999). Long-term ambient air pollution and respiratory symptoms in adults (SAPALDIA study), *Am J Respir Crit Care Med.*, Volume 159, No. 4, pp. 1257-1266.

Zielinski, J.; Macnee, W.; Wedzicha, J.; Ambrosino, N.; Braghiroli, A.; Dolensky, J.; Howard, P.; Gorzelak, K.; Lahdensuo, A.; Strom, K.; Tobiasz, M. & Weitzenblum, E. (1997). Causes of death in patients with COPD and chronic respiratory failure, *Monaldi Arch Chest Dis.*, Volume 52, No. 1, pp. 43-47.

7

Observation and Research on the Typical Atmosphere of Beijing Olympic Games by Optical Remote Sensing

Wenqing Liu, Wei Huang, Pinhua Xie and Jianguo Liu
State Environmental Protection Key Laboratory of Optical Monitoring Technology, Ministry of Environmental Protection of the People's Republic of China, Anhui Institute of Optics & Fine Mechanics, Chinese Academy of Sciences, Hefei, Anhui P.R. China

1. Introduction

The 29th Olympic and Paralympics Games were held from August 8th to September 17th, 2008 in China's capital - Beijing. Air quality is a critical concern to China and the world. To ensure good air quality during the Olympic Games, Beijing Municipal Government and the Environmental Protection Agency of China launched a program called "Air Quality Guarantee Plan for the 29th Olympics in Beijing (AQGP)". The period starting from July 1st, 2008 marked an intensified endeavour to improve Beijing's air quality by shutting down heavy polluters. 3.3 million motor vehicles were removed from city streets on alternate days, depending on whether the license plate ended in an odd or even number during the period of July 20th to Sep. 20th, 2008 (the alternative day-driving scheme). At the same time, the adjacent cities and provinces also did their best to close major polluters that may have influence on the air quality of Beijing. The exceptional strict pollution control measures have hugely reduced the source emissions, making a green Olympic Games achievable. Beijing's air quality has witnessed significant improvements due to enhanced pollution control efforts and new environment criteria.

Beijing's air quality was intensively monitored by environment protection authorities and research institutions before, during, and after the Olympic Games. The rich scientific data collected during the period have become solid evidences for judging the impact of exposure to air pollution on human health. To evaluate the effectiveness of the AQGP program, Beijing Municipal Environmental Protection Agency led a plan named "Assessment of AQGP and Air quality Monitoring, Forecasting and Early Warning".

There are numerous publications (Cermak & Knutti, 2009; Huang *et al.*, 2010; Y. Li *et al.*, 2010; X. G. Liu *et al.*, 2009; Mijling *et al.*, 2009; Shao *et al.*, 2009; Shen *et al.*, 2011; Shou-bin *et al.*, 2009; Simonich, 2009; Streets *et al.*, 2007; X. Y. Tang *et al.*, 2009; W. T. Wang *et al.*, 2009; Westerdahl *et al.*, 2009; Yao *et al.*, 2009) to now on evaluation of these pollution measures. Hou. et. al. (Hou *et al.*, 2010) measured the levels of Beijing residents' exposure to PM_{10} during three different time periods. Their study showed that during the Olympic Games, population-weighted PM_{10} exposure came down by 46% and 19%, respectively, compared with the pre-OG and the post-OG periods, indicating that in addition to favourable weather conditions, enhanced traffic and emission control policies and measures have produced a noticeable effect on PM_{10} reduction.

For persistent organic pollutants (POPs), Zhang et. al. (L. F. Zhang *et al.*, 2010) systematic studied hexachlorobenzene in the ambient air before and after the Olympic Games. Hexachlorobenzene concentration was found decreased sharply in the winter of 2008 comparing with that of 2007 due to the implementation of a series of "Green Olympic" policies. Gas-particle partitioning shows that the increase of hexachlorobenzene levels in winter time was mainly contributed by the high total suspended particulate from combustion processes such as coal-burning and traffic emission.

Xin et. al. (Xin *et al.*, 2010) used Beijing-Tianjin-Hebei Atmospheric Environment Monitoring Network to monitor and provide warnings of the atmospheric quality in Beijing and its surrounding area during the Beijing 2008 Olympic Games. During the Olympic Games, the mean concentration of SO_2, $PM_{2.5}$, NO_2, O_3_8h max, and O_x were 12.5±4, 56±28, 23±4, 114±29, 95±17 μg/m^3 in the region, respectively, and fell by 51.0%, 43.7%, 13%, 20.2%, and 18.9%, respectively, compared to the prophase mean concentration before the Olympic Games. After the Olympic Games, SO_2, $PM_{2.5}$ and NO_x increased significantly as the temporary atmospheric pollution control measures was terminated. Wang et. al's study (S. X. Wang *et al.*, 2010) showed that the mobile source NO_X and non-methane volatile organic compounds (NMVOC) reduced by 46% and 57%, respectively, during the Olympic Games period due to strict pollution policies.

Zhou et. al. investigated the effects of pollution measures on urban motor vehicle emissions in Beijing. (Zhou *et al.*, 2010) Comparison between the emission intensity before and during the 2008 Olympics shows a reduction of 44.5% and 49.0% in daily CO and NOx emission from motor vehicles. Their results suggest that reasonable traffic system improvement strategies along with vehicle technology improvements can contribute to controlling total motor vehicle emissions in Beijing after the Olympic Games. Wang et. al. (Wang & Xie, 2009) investigated the traffic-related air pollution in the urban streets. The average reduction of PM_{10}, CO, NO_2 and O_3 are 28%, 19.3%, 12.3%, and -25.2%, respectively. Another study (X. Wang *et al.*, 2009) evaluated on-road emission factors and black carbon profiles. Diesel trucks are believed to be a major source of summertime BC in Beijing. A mobile laboratory(M. Wang *et al.*, 2009) also was employed to evaluate changes in on-road air pollutants.

Liu et. al. (J. F. Liu *et al.*, 2009) measured the hourly concentrations of BTEX (Benzene, Toluene, Ethylbenzene, m,p-Xylene and o-Xylene) in the urban area of Beijing. They found that during the games, the mean daytime concentrations of benzene, toluene, ethylbenzene, m,p-xylene and o-xylene were 2.37, 3.97, 1.92, 3.51 and 1.90 μg/m^3, respectively, and were 52.8%, 63.9%, 56.4%, 56.8%, and 46.9% lower than those after the games. Secondary pollutants and regional impact also were studied. (T. Wang *et al.*, 2010)

He et. al. (S. Z. He *et al.*, 2010) measured atmospheric H_2O_2 and organic peroxides in urban Beijing (at the Peking University campus), from 12th July to 30th September, before and during the Beijing Olympic Games full-scale control. They concluded that the heterogeneous removal of H_2O_2 is faster than that of methyl hydroperoxide (MHP), as indicated by the strong negative correlation between the H_2O_2-to-MHP ratio and the aerosol surface area.

Wang. et. al. (B. Wang *et al.*, 2010) assessed variation of ambient non-methane hydrocarbons by gas chromatography equipped with a quadrupole mass spectrometer and a flame ionization detector (GC/MSD/FID) as parts of the field Campaign for the Beijing Olympic Games Air Quality program (CareBeijing). Their findings demonstrate the effectiveness of the air quality control measures enacted for the 2008 Olympics and indicate that controlling vehicular emissions could be the most important measure to improve the air quality of Beijing.

The abovementioned studies used all kinds of methods to evaluate the air quality during Beijing Olympic Games, but nearly none used optical remote sensing methods. Here we will introduce our investigations on the air quality of Beijing Olympic Games using all kinds of optical remote sensing techniques. Due to the limited room in this chapter, we will only introduce part of our results here.

The emission flux of important pollution sources was measured with mobile differential optical absorption spectroscopy (DOAS) and light detection and ranging (LIDAR). The sources are Capital Iron and Steel (CIS), Yanshan Petrochemical (YP), and Capital Airport (CA), respectively. It was found that CIS contributed most to emissions of NO_2, SO_2, and particulate matter, but the overall emission flux from June to September was decreasing, representing the effectiveness of air pollution control policy. It was also found that the emission flux came from YP declined as well, while the NO_2 emission of CA only had a slight drop.

The MAX-DOS system located in the region of pollution sources monitored the tropospheric column density of SO_2 and NO_2, which reflected the whole layer concentration of SO_2 and NO_2 around pollution sources. The results showed that the column density of SO_2 and NO_2 of CIS, YP and SO_2 decreased significantly, where the column density of NO_2 of CIS and YP decreased by 30-40%, whereas that of CA reduced less.

LIDAR measurements proved that emissions of particulate matter can be reduced to some extent by the alternative day-driving scheme and plant production restriction. Weather is the major factor for cleaning of particulate matters, implying that the emission control policy can only reduce the accumulation of pollutant in some extent, but the decisive factor of pollutant elimination is still weather.

The observation to 4 main pollutants (SO_2, NO_2, CO, PM_{10}) in all stations showed their concentration reached the China National Level II Standard and the guideline values of WHO. The concentration of NH_3 was found to be higher in the daytime while lower during the night-time. Varieties of optical remote sensing techniques were used to monitor the air quality of 2008 Beijing Olympic Games, and were proved to be critical in air quality monitoring system.

The tropospheric nitrogen dioxide over the Beijing Olympic Games venues was measured by ground-based Multi Axis Differential Optical Absorption Spectroscopy (MAX-DOAS), and was compared with the result of Ozone Monitoring Instrument (OMI) (Duncan *et al.*, 2010). It showed that the results of MAX-DOAS were higher than that of OMI, and can up to 2.4 times the result of OMI; both instruments had good correlation (r=0.64) when it is clear. There was more difference between the results of both instruments for the existing of cloud when it was rainy or cloudy, with a correlation coefficient of 0.19, but the results of MAX-DOAS had better consistency with LP-DOAS, with a correlation coefficient of 0.92.

2. Experimental methods - optical remote sensing

An integrated spatiotemporal monitoring system (Fig. 1) for regional complex air pollution was constructed which comprises a variety of optical remote sensing techniques, such as differential optical absorption spectroscopy (DOAS)(Winer & Biermann, 1994), light detection and ranging (LIDAR)(Vierling *et al.*, 2008), Fourier transform infrared spectroscopy (FTIR)(Movasaghi *et al.*, 2008), tunable diode laser absorption spectroscopy (TDLAS) (Lackner, 2007), tapered element oscillating microbalance (TEOM) (Patashnick & Rupprecht, 1991). It was worth pointing out that all the instruments mentioned here were developed in our institute (Chen *et al.*, 2009; He *et al.*, 2009; Y. B. He *et al.*, 2010; Jin *et al.*, 2010; S. W. Li *et al.*, 2009; S. W. Li *et al.*, 2008; Lu *et al.*, 2010a, 2010b; Luo *et al.*, 2011; Si *et al.*, 2010; Y. Y. Tang *et al.*, 2010; J. Xu *et al.*, 2010; Z. Y. Xu *et al.*, 2010; S. A. Zhang *et al.*, 2009; Zhu *et al.*, 2010). We established 11 monitoring stations including 3 super monitoring stations and 2 mobile monitoring vehicles. From June 10th to September 20th of 2008, the air quality of Beijing area was monitored continuously and in real-time at these stations. Enormous data on all kinds of pollutants were acquired to assess the consequences of air pollution source control policies implied during the Olympic Games.

3. Experimental results and discussions

The integrated spatiotemporal monitoring system consists of 11 monitoring stations equipped with a variety of aforementioned optical remote sensing instruments. There are 3 super monitoring stations and 2 mobile monitoring vehicles. Air quality of Beijing area was monitored continuously and in real-time in these stations from June 10th to September 20th, Enormous data on all kinds of pollutants were acquired to assess the consequences of air pollution source control policies implied during the Olympic Games. Here we will choose some typical data to demonstrate applications of optical remote sensing techniques.

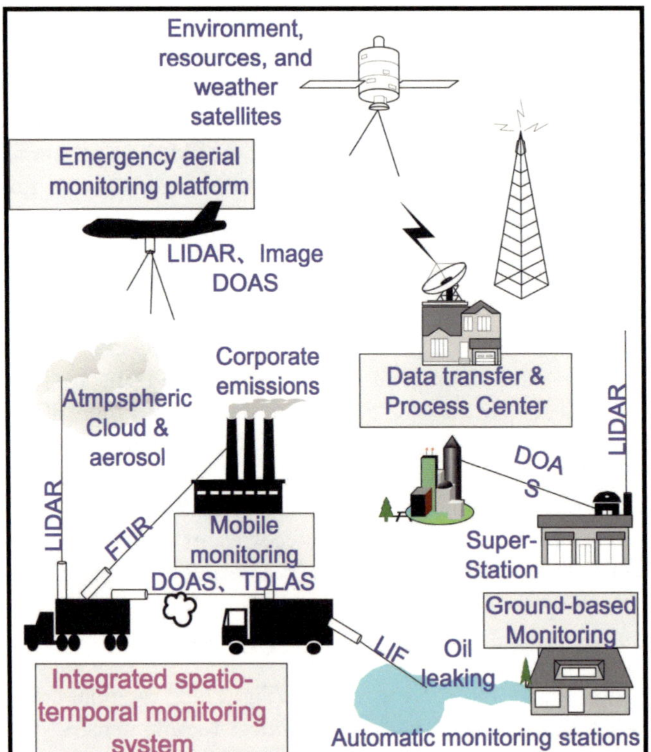

Fig. 1. Integrated spatiotemporal monitoring system for regional complex air pollution

3.1 Aerosol extinction & backscattering coefficients

We've done continuously measurements to air quality over the Olympic Games main sports center - "Bird's Nest" using double wavelength Raman LIDAR with three channels. The Raman LIDAR was installed at the Institute of Remote Sensing Applications (IRSA), Chinese Academy of Sciences, which is located about 500 meters north of the "Bird's Nest". The data acquired from Raman LIDAR are used to analyze the optical characteristics of representative weather during the period of Olympic Games.

Fig. 2 shows the aerosol extinction coefficient spatial and temporal distributions during Aug. 7th - Aug. 25th, 2010. Obviously, one can find that the aerosol extinction coefficients on Aug. 8th and Aug. 9th are considerably large, suggesting pretty heavy air pollution, and thereafter the atmosphere became clear due to several rainfalls.

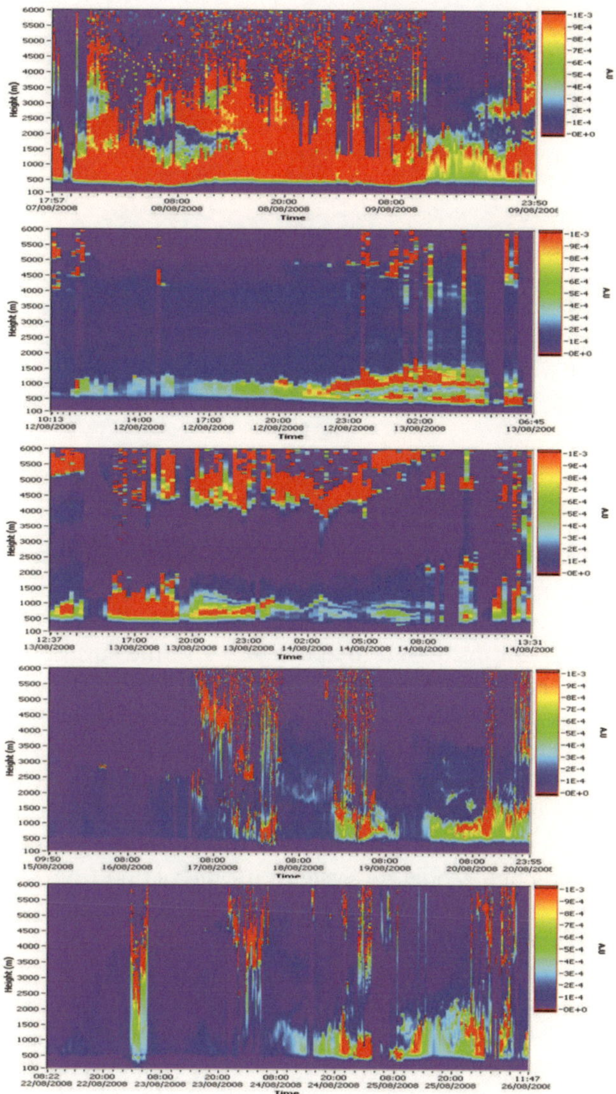

Fig. 2. Aerosol extinction coefficient vertical distributions during Aug. 7th ~ 25th, 2008

The aerosol extinction coefficient at some wavelength can give the distribution of pollutants, but it cannot provide the particle size information. Fig. 3 shows the spatial and temporal distribution of backscatter signal and extinction coefficient at 355 nm & 532 nm from Aug. 7th to 9th, 2010. As can be seen from the figure, the backscatter signals at 355 nm and 532 nm illustrate the spatial distribution of small aerosol particles clearly, and the heights of most of these small particles are lower than 600 meters. The most abundant distribution is at ~400 meters, and this is verified by the ratio of aerosol extinction coefficients at two different wavelengths. The height distribution and diurnal change of boundary layer are clearly demonstrated in the spatial -temporal distribution (Fig. 3).

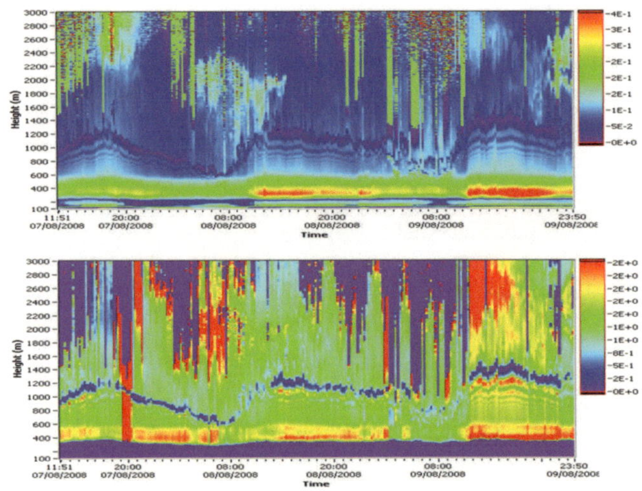

Fig. 3. Ratio vertical distribution of backscatter signal (upper pannel) and extinction coefficient (lower pannel) of 355/532 nm from Aug. 7th to 9th, 2008

It is worth noting that Yang (Yang *et al.*, 2010) also used a Dual-wavelength Polarized LIDAR to analyzes the variation of aerosol extinction coefficients during the Beijing Olympic Games. Their results showed that (1) during the Beijing Olympic Games, the aerosol extinction coefficient decreased to about 42.3% in the surface layer (below 250 m) compared with that in 2007 under almost the same meteorological conditions, indicating the effectiveness of local air pollution control measures in Beijing areas; (2) the analysis of the aerosol extinction coefficient obtained before (Jul. 20th–Aug. 7th, 2008) and during the Beijing Olympic Games (Aug. 8th–Aug. 24th, 2008) shows a maximum decrease of daily aerosol extinction coefficient in the layer of 0.5−1.5 km. To elucidate such a phenomenon, the transport of PM_{10} from surrounding areas to Beijing was simulated by NAQPMS (the Nested Air Quality Prediction Modeling System) for the period from Jul. 20th to Aug. 24th, 2008. Analysis of PM_{10} transport from neighboring areas indicated a decrease of 36.6% during the Olympic Games.

Aerosol optical depth (AOD) is a manifestation of air pollution to some extent, and angstrom wavelength exponent (AE) is an indicator of the relative size of aerosol particles. Fig. 4 illustrates the distribution of aerosol optical depth and angstrom exponent between Aug. 7th and Aug. 25th, 2010. One can clearly see that the AODs in the first 3 days are large, and the AOD of Aug. 8th is much larger. The data of Aug. 10th, Aug. 11th, Aug 14th, and Aug. 21st are missed due to rains in these days. According to the variations of AOD, we found that the AOD rose significantly after rainfalls, hinting a procedure of pollutants accumulation. The average Angstrom exponents calculated from Fig. 4 is 2.64. Fig. 4 shows that the heavy pollution on Aug 8th-9th is caused by large aerosol particles. The AEs are very small after the rainfalls on Aug 14th and Aug. 21st, meaning the average aerosol particle size is small, and this agrees well with other atmospheric fine particle size measurements.

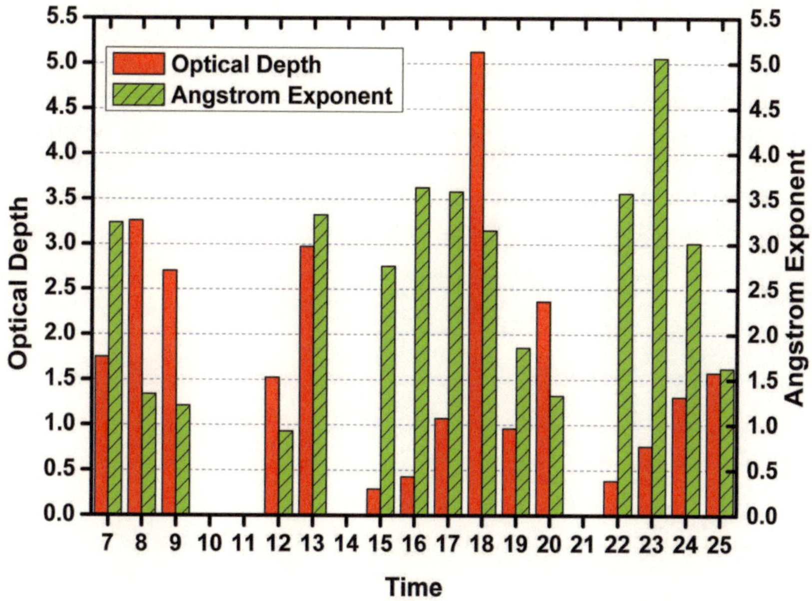

Fig. 4. Distribution of aerosol optical depth and Angstrom exponent during 20:00~23:59 during Aug. 7th ~ 25th, 2008

From the measurements of Raman LIDAR, it was concluded that: (a) the distribution of pollutants on Aug 7-9th exhibits period variation with time, and the heavy pollution on Aug. 8-9th is attributed to large aerosol particles; (b) The echo signals of LIDAR at different wavelengths clearly depicts the spatial distribution of aerosol particle size. The altitude of most small particles is about 400 meters. The AOD become much larger after rains, suggesting an accumulation process of pollutants; (c) the average AE is 2.64 during the period of Olympic Games, and its diurnal changes are in very good agreement with the data of ground-based atmospheric fine particle measurement.

3.2 Measurement of SO_2, NO_2, O_3 and PM_{10}

5 of the 11 monitor stations are used to measure the conventional pollutants in Beijing and surrounding areas. These stations are located in Yanshan Petrochemical (YP), Yugang Town (YT), Institute of Remote Sensing Applications, Chinese Academy of Sciences (IRSA), Capital Airport (CA), and Yongledian Town (YLD), respectively. The concentration of gaseous pollutants, i.e., SO_2, NO_2, and O_3, was measured using DOAS. We tested the performance of the DOAS system and calibrateed the lamp spectra every day. The comparison of these data with other instruments in IRSA and YP was used to ensure the accuracy and reliability of field measurements. TEOM was used to get the PM_{10} data. The

measurements started from Jun. 20th of 2008, and ended in September, 2008. We will introduce the time series, average diurnal change, regional transport, and comparison of pollutant concentration in YP, YT, IRSA, and CA.

3.2.1 Time series of pollutant concentration

Fig. 5 is the hourly average time series of pollutants. The PM_{10} concentration of IRSA station is much higher than that of YT because IRSA is very close to the Bird's Nest, and there were more human activities near this station, while YT is in suburb area. The maximum concentration of PM_{10} in IRSA station occurred at 13:00, July 4th, and is 470.6 μg/m^3. The maximum concentration of PM_{10} in YT station occurred at 22:00, July 27th, and is 354.0 μg/m^3. The sources of PM_{10} are very complicated, e.g., coal and dust pollution, industrial emissions, vehicle emissions, and pollutants transport. The weak southerly wind from July 4th to Aug. 10th hindered the diffusion of pollutants, and made the concentrations of PM_{10} and O_3 measured in each station are pretty high.

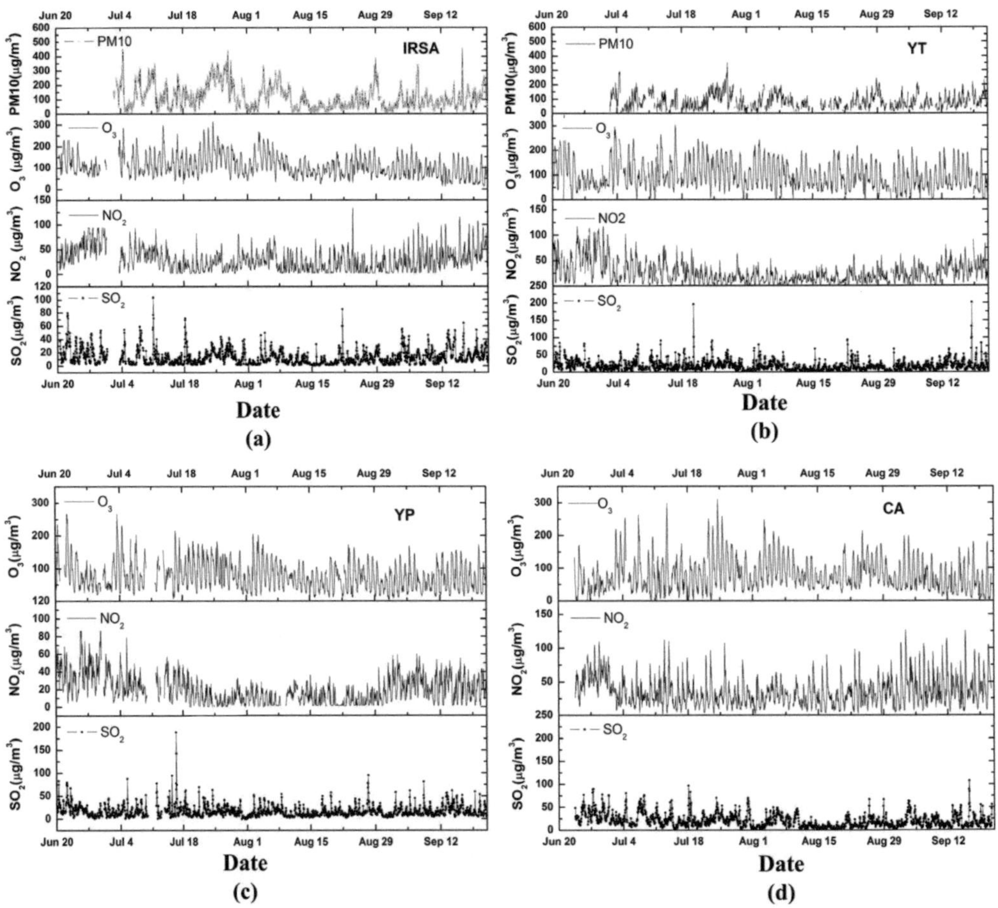

Fig. 5. Time series of observed SO2, NO2, O3 and PM10

O_3 came from the photochemical reactions during the daytime, and exhibited apparent diurnal variation trend, i.e., the peak value occurred in the afternoon, as the valley happened in the night. The maximum hourly average O_3 concentration occurred at 13:00, Jun. 22nd in IRSA station, and the value is 324 μg/m^3. The concentrations of SO_2 and NO_2 were under the requirement of National Atmosphere Environment Quality Level II, which are 500 μg/m^3 and 120 μg/m^3. The main source of SO_2 came from combustion of sulfur compounds. The peak values of SO_2 concentration were more frequently found in YP and YT stations. NO_2 was found to have a contrary trend with O_3 in concentration change, i.e., the concentration was high during the night, but became lower in the day due to low photolysis concentration. There are lots of rainy days before July due to Mongolian cold- and low-swirl. In addition, the emission reduction policy started to effect since July. These two factors caused the concentration of NO_2 generally was high as a result of lacking of photolysis. The concentrations of primary pollutants, SO_2 and NO_2, have significant falling with the implementation of emission reduction policy, particularly during the Olympic Games period.

3.2.2 Diurnal variation of pollutants in different monitoring sites

Fig. 6 presents the average diurnal variations of SO_2, NO_2, O_3, and PM_{10} at different monitoring sites in different stages. The average diurnal variation curves of urban site (Fig. 6a, IRSA) and suburban site (Fig. 6b, YT) show that the concentration of PM_{10} only changed slightly, and the PM_{10} concentration of IRSA site is higher than that of YT site. The PM_{10} concentration measured in IRSA site was found decreasing continuously from Jul. 20th (The alternative day-driving scheme starting date) to September. The sharpest decreasing occurred between 01:00 ~ 12:00, and the PM_{10} concentration average diurnal variations at other stages were nearly equal. The PM_{10} concentration measured at YT site had a significant decline in September than before, and the main difference happened between 02:00 ~ 10:00. The average difference between urban and suburban changed from 50% of Jul. 1st - Jul. 19th to 25% of Sept. 6th - Sept. 17th.

The measured NO_2 concentration in all the monitor sites exhibited lower value from Jul. 20th to Aug. 24th. The diurnal variation has an obvious double-peak. The first peak appeared at 05:00 ~ 09:00, suggesting that it may be related with local traffic emission and human activity. The second peak came out at 19:00 ~ 22:00, and this is originated from concentration accumulation due to atmospheric inversion effects of near-ground atmosphere and lack of photolysis in the night.

The comparison of NO_2 diurnal variations in different monitoring sites showed that the concentration of NO_2 in Capital Airport (CA) changed very little. This is reasonable when considering the large number of air flights every day, and the effect of alternative day-driving scheme is relative small. The measured NO_2 concentration in July is much lower than that of June, and there is no clear peak value in the morning. The NO_2 concentration of all sites during Sept. 6th -Sept. 16th increased a little bit compare with previous measurements, especially at around 19:00 in Capital Airport and 21:00 in IRSA.

The NO_2 concentrations were found to decrease most in IRSA and CA sits from the average diurnal variation of SO_2. The intermittent SO_2 peaks appeared in YT site (16:00) before applying alternative day-driving scheme and YP site (10:00) during Jul. 1st - Jul. 19th. The NO_2 average diurnal variations generally showed single peak, and only YP and YT sites showed double-peak. The NO_2 concentration is higher during the night-time in urban sites, while high NO_2 concentration was found in other sites, e.g., YT and YP sites were found to have peak value between 07:00 ~ 09:00.

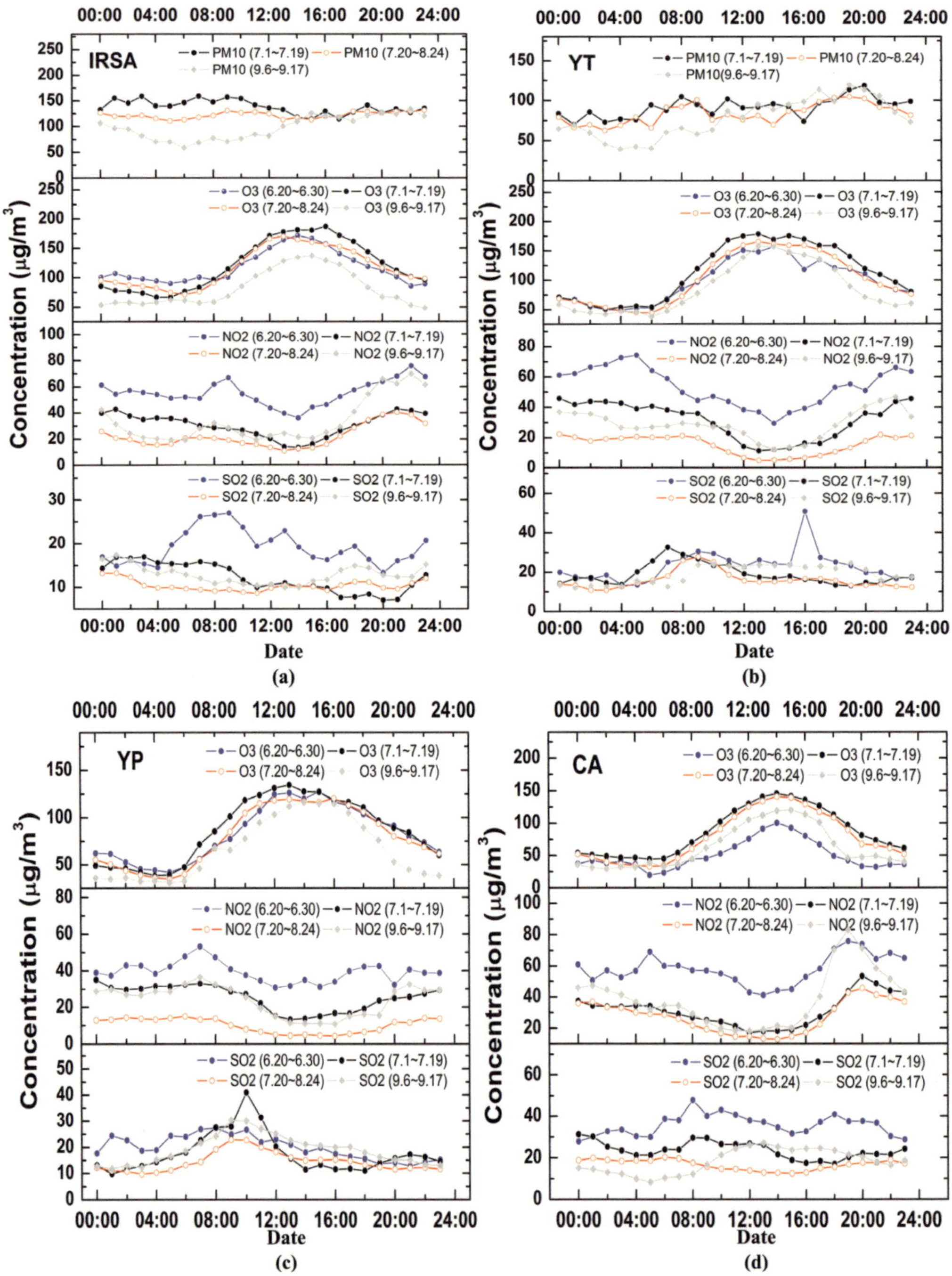

Fig. 6. Average diurnal patterns of SO_2, NO_2, O_3 and PM_{10} in 4 monitoring stations

In contrast with NO_2 and SO_2, the measured O_3 concentration increased somewhat at the beginning of applying alternative day-driving scheme. This can be explained by that O_3's originate in Beijing is basically controlled by VOCs. The decreasing of NOx resulted in local

O_3's consumption declination too, thus caused the increasing of O_3 concentration. The statistic data after Jul. 20th demonstrated that the O_3 concentration decreased continuously. O_3 concentration exhibits typical diurnal variation pattern, that is, the peak value appeared in the daytime, while the concentration became lower due to the reaction with NO. Measurements in urban sites showed that the O_3 concentrations in Jul. 20th - Aug. 24th are comparable to that of June, but slightly lower. The peak O_3 concentration appeared earlier from 16:00 of Jul. 1st - Jul. 19th to 13:00 of Jul. 20th - Aug. 24th, while the peak appeared time from Sept. 6th – Sept. 17th is almost the same as that the beginning of alternative day-driving scheme and the difference between peak and valley values became smaller and smaller. O_3 concentrations measured in YP and YT sites gave apparent double-peak features, where the first peak appeared around 13:00, and the second peak appeared around 16:00. The second peak probably is related with O_3's regional transport. Although YT site is far from urban area, O_3 concentration was found to increase at the beginning stage of alternative day-driving scheme, but it became much lower afterwards, and the second peak became not so obvious. We found the same trend for O_3 in Capital Airport, whose peak value appeared at around 14:00. The maximum concentration of O_3 during Jul. 1st - Jul. 19th increased by nearly 40 $\mu g/m^3$ compare with that of June, and become a little lower afterwards.

3.2.3 Observation of pollutants regional transport

We will take the measured data of Aug. 8th in four sites as an example to investigate the effect of the regional transport in southwest direction. SO_2 and O_3 were chosen as tracer pollutants. Fig. 7 shows the air mass backward trajectory at 200 m, 500 m, and 1000 m heights in 24 hours using HYSPLIT model, where the star labeled the location of IRSA. One can find from Fig. 7 that the main wind direction is southwest, and the air mass transportation is from southwest (YP) to northeast (IRSA).

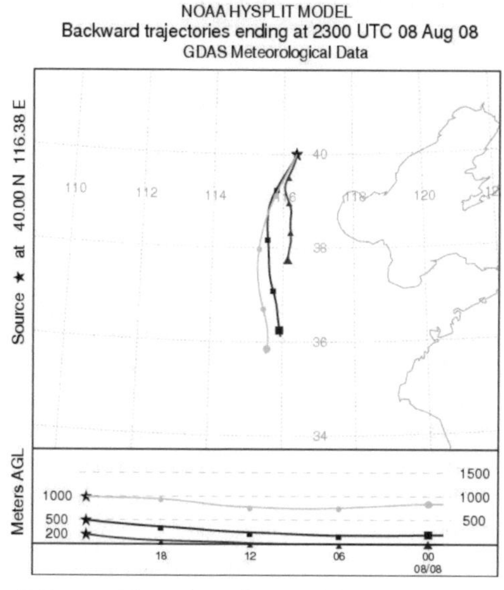

Fig. 7. Air trajectories for different altitudes of Aug. 8th, 2008

Fig. 8 presents the diurnal variations of SO_2 and O_3 on Aug. 8th, 2008. The YP monitor site got SO_2 peak value at 12:00 first, and then the peak values were found in YT, IRSA, and CA sites at 13:00, 14:00, and 15:00, respectively. The time gap between the peak values is about 1 hour. SO_2 was found to move from southwest (YP) to northwest (IRSA, CA). Same as SO_2, the O_3 peak value was found in YP site (12:00), and then YT, IRSA, and CA sites. The delay between peaks is about 1 hour as well. The O_3 concentration of YP site was found to be the lowest, and keep at 125-128 μg/m^3 level during 12:00-16:00, and lowered a little bit at night-time, but the second clear peak was found after 21:00. YT and IRSA sites then found O_3 concentration peak values during night-time at 22:00 and 23:00, respectively. These peak values may be related with pollutants regional transport. Such findings are consistent with the SO_2 vertical column density measurements of MAX-DOAS.

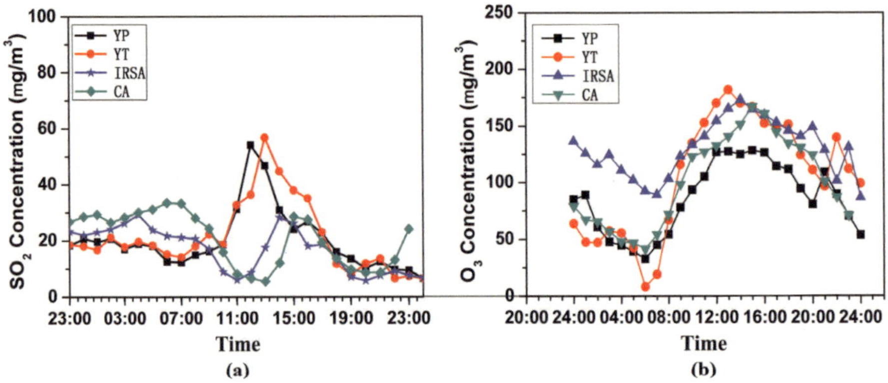

Fig. 8. Diurnal variation of SO_2 (a) and O_3 (b) measured at 4 monitoring stations on Aug. 8th, 2008

3.2.4 Comparison of pollutants concentration in different monitoring sites

Here we present the diurnal averages of PM_{10}, SO_2, and NO_2 concentration; the maximum hourly average concentration of O_3 measured during Jun. 20th - Sept. 22nd in every monitoring site. Table 1 gives the concentrations of conventional pollutants, PM_{10}, SO_2, NO_2 and O_3, measured at different stages between June and September, 2008, where diurnal averages value were used for PM_{10}, SO_2, and NO_2, while maximum hourly average values were used for O_3.

PM10: Fig. 9 presents the comparison of PM_{10} daily average concentration and statistical diurnal average concentration measured in IRSA and YT sites. Since IRSA site is close to the Bird's Nest and there are more human activities around the site, the concentration measured here was about 40% higher than that of YT site. The PM_{10} variation patterns in two sites are similar, and high concentration was found in two periods (Jul. 24th - Jul. 28th, Aug. 4th -Aug. 7th). The respirable particulate matters accumulate due to the static stable weather and resulted in the PM_{10} concentration in IRSA exceeded the National Atmospheric Environmental Level II standard (150 μg/m^3). We can see from Table 1 that the PM_{10} concentration of IRSA which located in urban area was larger than that of YT, and the reductions were 12.1% and 9.9%, respectively. The measurements at urban sites showed that of PM_{10} concentration had a big growth in September.

SO_2: Fig. 10 presents the variation of NO_2 daily average concentration and the contrast of statistical average concentration. No concentration over standard was found during the whole campaign. Since there were lots of precipitations between Jun. 28th and Jul. 4th, the concentrations of pollutants like NO_2 were relatively low due to wet sedimentation. The NO_2 concentration of CA was higher than other sites, and the highest concentration

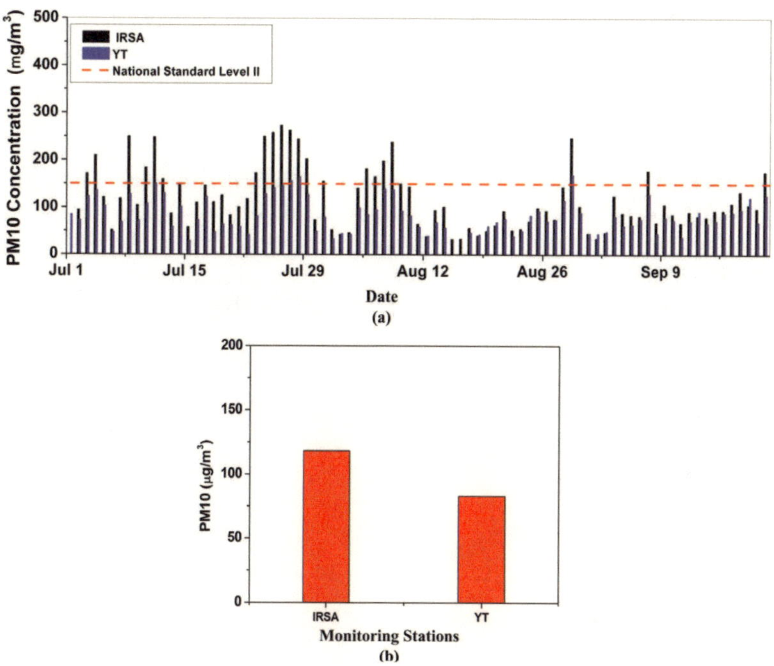

Fig. 9. Diurnal mean variation (a) and statistical average daily mean of PM_{10} concentration (b) during the campaign

Site	Pollutants (μg/m³)	Measurement Periods							
		June	7.1-7.19	7.20-8.24	8.8-8.24	9.6-9.17	7.1-7.19/June Decease	7.20-8.24/7.1-7.19 Decrease	9.6-9.17/8.8-8.24 Decrease
YT	NO_2	51.7	32.0	15.6	13.9	28.6	38.1%	51.3%	-105.6%
	SO_2	22.7	18.2	15.6	11.4	18.5	19.8%	14.3%	-61.7%
	O_3	176.1	200.4	177.8	155.9	166.1	-13.8%	11.3%	-4.03%
	PM10		90.0	81.3	66.3	75.9		9.6%	-14.5%
IRSA	NO_2	54.5	33.2	21.9	18.7	32.8	39.1%	34.0%	-75.8%
	SO_2	19.4	12.3	10.2	7.7	15.1	36.6%	17.1%	-97.0%
	O_3	185.0	208.3	179.7	136.8	144.7	-12.6%	13.7%	-2.5%
	PM10		139.0	122.2	80.8	94.4		12.1%	-16.9%
YP	NO_2	37.8	26.6	10.3	10.1	24.1	29.6%	61.3%	-142.5%
	SO_2	19.1	18.4	14.4	12.6	18.8	3.7%	21.7%	-50.0%
	O_3	145.4	157.4	135.7	111.6	125.0	-8.3%	13.8%	-8.9%
CA	NO_2	57.8	32.5	27.5	28.4	20.5	43.8%	15.4%	-35.8%
	SO_2	37.2	25.1	16.8	10.7	18.1	32.5%	33.1%	-68.3%
	O_3	106.0	157.6	153.7	124.1	129.7	-48.7%	2.5%	-3.3%

Table 1. Concentrations of routine pollutants at 4 monitoring stations from June to September, 2008

measured on Jun. 27th was 56.8 μg/m³. The average concentrations of each site was found to decrease more than that before implementation of alternative day-driving scheme, especially for IRSA site where the reduction up to 35%. Due to the continuous rain showers between Aug. 8th and Aug. 15th, the NO_2 concentrations in several sites were found decreased considerably because of NO_2 purge. CA's SO_2 concentration decreased significantly during the period of Jul. 20th - Aug. 24th compare with the beginning of the alternative day-driving scheme. After the Olympic Games, the SO_2 concentrations measured in all sites were higher than that of August. Specifically, the concentration of SO_2 was found to increase by 97% (Table 2). One can find the CA statistical diurnal average concentration is the highest in all the sites, followed by YT, YP and IRSA. This can be explained by there are more flights in CA area.

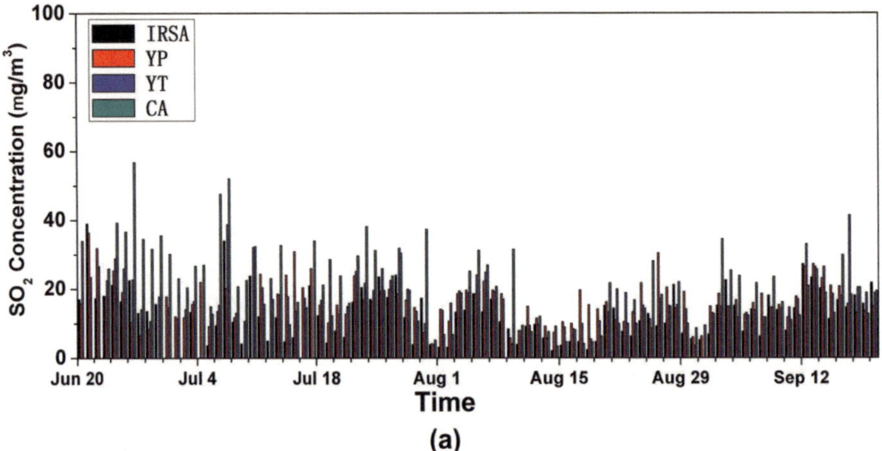

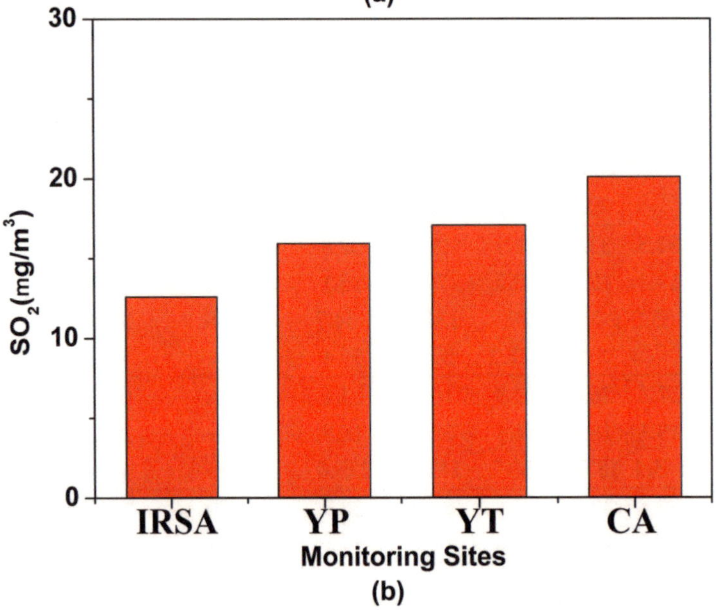

Fig. 10. Daily mean variation and statistical average daily mean of SO_2 during the whole campaign

NO_2: Fig. 11 gives the NO_2 diurnal average concentration and the comparison of statistical average concentration. Same as SO_2, there was no concentration over standard was found. There were lots of rains between Jun. 25th and Jul. 1st. NO_2 accumulated due to lack of photolysis in this period, and resulted in high concentration during Jun. 27th and Jun. 29th. As the weather changed and photochemical activity was enhanced after Jul. 1st, the NO_2 was decomposed through photolysis and NO_2 concentrations at all sites decreased. The static weather pollutants accumulated on Jul. 29th and Aug. 1st-Aug. 8th resulted in high NO_2 concentration. With the coming of Olympic Games opening ceremony, the NO_2 concentration in CA and IRSA increased gradually, but decreased abruptly on Aug. 8th. Same as that of SO_2, NO_2 concentration in BA site was higher than other sites. With the implementation of emission reduction policy, NO_2 concentration in YP and YT decreased noticeably during Jul. 20th to Aug. 24th. The concentration in all sites increased to some extent in September, especially in YP and YT sites, and the growths are 142% and 105%, respectively. Judged from statistical diurnal average concentration, CA and IRSA had a higher NO_2 concentration due to human activity, then followed by YT and YP.

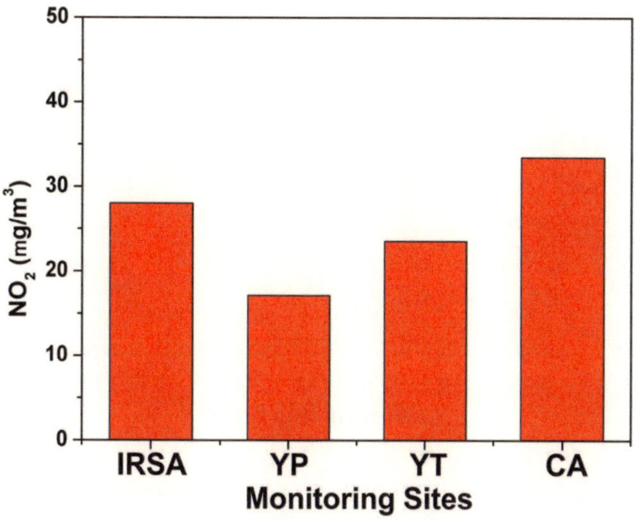

Fig. 11. Daily mean variation and statistical average daily mean of NO_2 during the whole campaign

O_3: Fig. 12 gives the O_3 diurnal hourly concentration and the comparison of statistical hourly maximum average concentration. As a result of rainy weather, the low photochemical activities led to low O_3 concentration. The concentration exceeding standard happened mainly during the static stable weather period, and they are Jul. 11th - Jul. 25th, and Aug. 2nd - Aug. 6th. In particular, the O_3 concentration in several sites increased at a rate of about 40% from Aug. 1st to Aug. 6th with the maximum happened on Aug. 3rd, and then decreased at a rate of 7%/day. From Aug. 8th to the end of Olympic Games, O_3 concentration measured in 4 sites meet the National Air Quality Standard Level II overall and were kept in a relative low level except on some days. From the statistical hourly average maximum concentration, YT site had the highest O_3 concentration, followed by IRSA and CA, and YP had the lowest O_3 concentration.

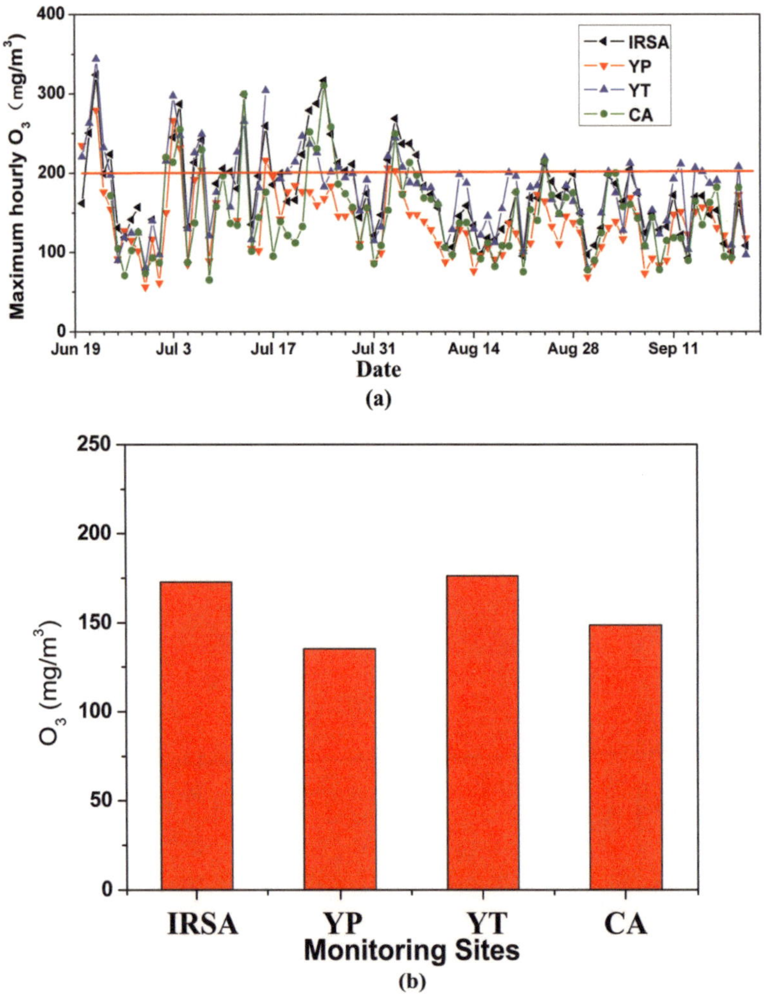

Fig. 12. Maximum hourly ozone concentrations on a daily basis and statistical average of the maximum hourly ozone during the whole campaign

The exceeding standard hourage of each site is shown in Fig. 20. One can find that O_3 concentrations go beyond the standard on Jun. 22nd, Jul. 3rd - Jul. 4th, Jul. 7th, and Jul. 16th. It is worth to mention that the period over standard in YP, YT and IRSA on Jun. 22nd was found to last up to 9 hours. With the implementation of pollution reduction policy, O_3 concentration in YP decreased significantly. The O_3 concentration in all sites meet the requirement of National Air Quality Standard Level II except O_3 concentration of YT site exceeded the standard occasionally.

SO_2 and NO_2 concentrations were found to meet the National Air Quality Level II standard during the whole campaign. The SO_2 and NO_2 concentration declined in the first stage of alternative day-driving scheme (Jul. 1st – Jul. 19th). The reduction of NO_2 was far larger than that of SO_2, and reduced up to 30% in all the sites, especially in CA site. All the pollutants concentration continue to decrease in the second stage (Jul. 20th – Aug. 24th), and NO_2 concentration in YP site which is in suburban decreased most significantly and up to 60%, then followed by YT site. The SO_2 concentration in CA site decreased most drastically and up to 33%.

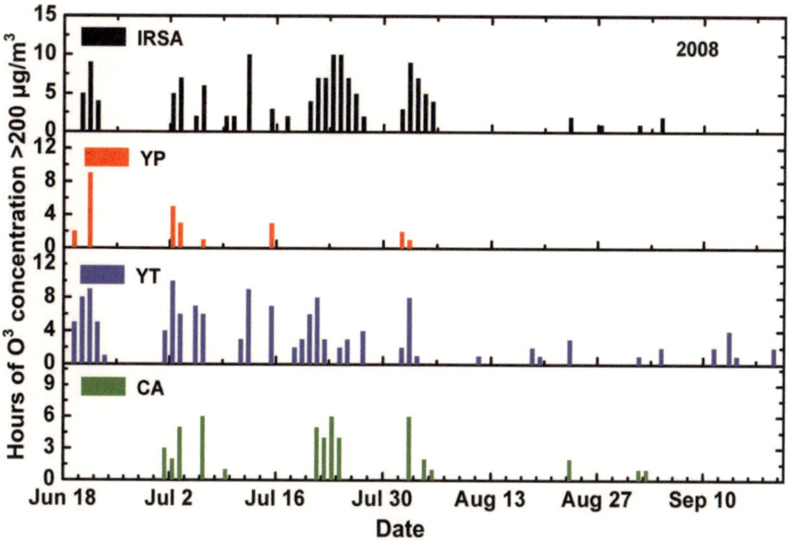

Fig. 13. Hourage of O_3 exceeding standard

The O_3 concentration was found to go up then go down at the beginning of the alternative day-driving scheme. This can be explained by that Beijing is located in the VOCs control area generated by O_3. The reduction of NOx due to alternative day-driving scheme will cause O_3 concentration to increase. Compare with June, O_3 in CA site increase most significant and up to 48.7%, and that of YT and IRSA increase by about 13%. The regional transport existed due to the influence of the southwest wind. The YT and IRSA sites located in the downwind of YP site can find peak O_3 in turn.

Since PM_{10} is more closely affected by human activities, the PM_{10} concentrations in monitoring sites in urban area were higher. The PM_{10} concentrations of IRSA which is in urban area were found to have more reduction than that of YT site during the Olympic Games period compare to the first stage, and decreased by 12.1% and 9.6%, respectively.

When comparing the Olympic Games period with the first stage of emission reduction policy, SO_2, NO_2, O_3, and PM_{10} concentrations in all monitoring sites decreased by 14%-33%, 15%-6%, 2.5%-14%, 10%-12%, respectively. But the concentrations of these pollutants increased during the Paralympic Games period.

3.3 Comparison of NO_2 measured by MAX-DOAS and OMI

Fig. 14 presents the comparison of the tropospheric NO_2 colmun density of MAX-DOAS and OMI from Jun. 10th to Sept. 14th. The MAX-DOAS setup was installed close to the Bird's Nest. One can easily find the results of ground-basd MAX-DOAS are higher than that of OMI, and can up to 2.4 times on some days (Fig. 14a). This can be explained by spatial resolution difference between MAX-DOAS and OMI. The OMI's spatial resolution is 13×24 km^2 whereas MAX-DOAS has much higher resolution. This means the results of OMI is the average of a large area including the urban and suburban area. We know that the NO_2 of suburban area is much lower than that of urban area. This was validated by the average

NO_2 column density measured in 3 sites (Fig. 14b). Fig. 14b showed the column density ratio of MAX-DOAS to OMI decrease to 1.5 - 1.8 (Table 2). One can image that with more sites, the MAX-DOAS average will be close to OMI measurements.

	OMI (10^{15}molec./cm^2)	MAX-DOAS (10^{15}molec./cm^2)	Average (10^{15}molec./cm^2)
June	8.54	12.46	13.13
August	7.81	15.91	14.36
September	9.05	19.91	15.36

Table 2. Comparison of monthly average between OMI and MAX-DOAS

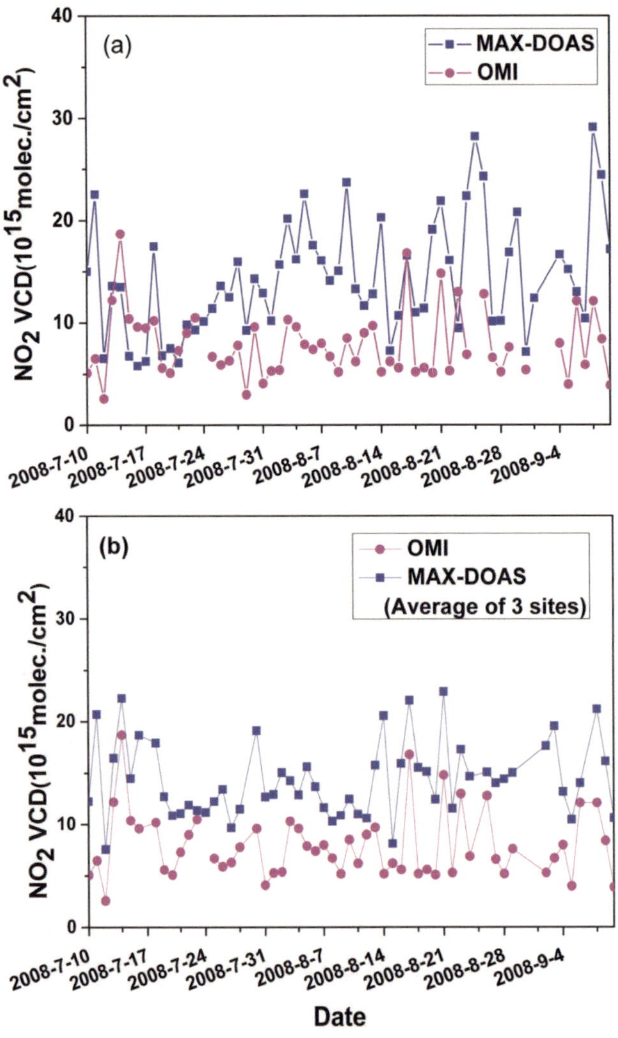

Fig. 14. Comparison between the results of OMI and MAX-DOAS,(a) is the result of comparison between OMI and MAX-DOAS near the Olympic venues,(b) is comparison between OMI and the average of three sites, IRSA, CIS and YT, which are in the same pixel of OMI

It is worth to mention that Yu (Yu *et al.*, 2010) also assessed the effect of the air quality ensuring measures during the game period using the tropospheric NO_2 column density retrieved from Ozone Monitoring Instrument (OMI) onboard AURA satellite. About 40% reduction in tropospheric NO_2 column density over the Beijing area is obtained from the assessment during July to August, 2008, a key period of air quality ensuring measures for the Beijing 2008 Olympic Games.

4. Conclusion

China hosted XXIX Olympic and Paralympics Games in its capital, Beijing, from August to September of 2008. To improve the air quality during this period, extreme stringent pollutions control policies were applied to Beijing and its neighbouring provinces. A variety of optical remote sensing techniques were employed on the measurements of the air quality of Beijing at different stages of Olympic Games. The aerosol extinction coefficients and aerosol optical depth between August 7th and August 25th were obtained using Raman LIDAR. Differential Optical Absorption Spectroscopy (DOAS) was utilized to evaluate the density of SO_2, NO_2, O_3. The density of SO_2, NO_2, O_3 from July 20th to August 24th were reduced by 14%-33%, 15%-61%, and 2.5%-14%, than that of July 1st to July 19th, respectively. The troposphere NO_2 column density acquired using multi-axis differential DOAS (MAX-DOAS) was found to be much higher than that from OMI with a maximum factor of 2.4. PM_{10} was found to decline by 10% - 12% during the Olympic Game period using TEOM. The results were compared with data from other groups using similar methods or different methods like satellite observations or air quality modelling system, and demonstrated the efficiency of pollution control measures to improve the air quality during the XXIX Olympic Games in 2008.

Using the optical remote techniques aforementioned along with other techniques like satellite observation, an integrated spatiotemporal monitoring system for regional complex air pollution was constructed. The results demonstrated the efficiency of pollution control measures in surrounding areas in improving air quality in Beijing. Optical remote sensing techniques were proved to be critical in air quality monitoring system.

5. Acknowledgment

The experimental and theoretical work done at AIOFM were supported by grants from National Science Foundation of China (40905010, 40805015), National High Technology Research and Development Program (2009AA063006), and Special Fund of Environmental Protection and Public Service (201109007).

6. References

Cermak, J., & Knutti, R. (2009). Beijing olympics as an aerosol field experiment. *Geophysical Research Letters*, 36, L10806.

Chen, Z. Y., Liu, W. Q., Zhang, Y. J., Zhao, N. J., He, J. F., & Ruan, J. (2009). Measurements of aerosol distribution by an elastic-backscatter lidar in summer 2008 in beijing. *Chinese Optics Letters*, 7, 753-755.

Duncan, B. N., Yoshida, Y., Olson, J. R., Sillman, S., Martin, R. V., Lamsal, L., et al. (2010). Application of omi observations to a space-based indicator of nox and voc controls on surface ozone formation. *Atmospheric Environment*, 44, 2213-2223.

He, S. Z., Chen, Z. M., Zhang, X., Zhao, Y., Huang, D. M., Zhao, J. N., et al. (2010). Measurement of atmospheric hydrogen peroxide and organic peroxides in beijing before and during the 2008 olympic games: Chemical and physical factors influencing their concentrations. *Journal of Geophysical Research-Atmospheres*, 115.

He, Y., Zhang, Y. J., Kan, R. F., Xia, H., Geng, H., Ruan, J., et al. (2009). Open-path online monitoring of ambient atmospheric co2 based on laser absorption spectrum. *Spectroscopy and Spectral Analysis*, 29, 10-13.

He, Y. B., Kan, R. F., Englich, F. V., Liu, W. Q., & Orr, B. J. (2010). Simultaneous multi-laser, multi-species trace-level sensing of gas mixtures by rapidly swept continuous-wave cavity-ringdown spectroscopy. *Optics Express*, 18, 20059-20071.

Hou, Q., An, X. Q., Wang, Y., & Guo, J. P. (2010). An evaluation of resident exposure to respirable particulate matter and health economic loss in beijing during beijing 2008 olympic games. *Science of the Total Environment*, 408, 4026-4032.

Huang, X. F., He, L. Y., Hu, M., Canagaratna, M. R., Sun, Y., Zhang, Q., et al. (2010). Highly time-resolved chemical characterization of atmospheric submicron particles during 2008 beijing olympic games using an aerodyne high-resolution aerosol mass spectrometer. *Atmospheric Chemistry and Physics*, 10, 8933-8945.

Jin, L., Gao, M. G., Liu, W. Q., Lu, Y. H., Zhang, Y. J., Wang, Y. P., et al. (2010). Application of sof-ftir method to measuring ammonia emission flux of chemical plant. *Spectroscopy and Spectral Analysis*, 30, 1478-1481.

Lackner, M. (2007). Tunable diode laser absorption spectroscopy (tdlas) in the process industries - a review. *Reviews in Chemical Engineering*, 23, 65-147.

Li, S. W., Liu, W. Q., Xie, P. H., Wang, F. S., & Yang, Y. J. (2009). Real-time forecasting model for monitoring pollutant with differential optical absorption spectroscopy. *Spectroscopy and Spectral Analysis*, 29, 3057-3060.

Li, S. W., Liu, W. Q., Xie, P. N., Li, A., Qin, M., Peng, F. M., et al. (2008). Observation of the nighttime nitrate radical in hefei, china. *Journal of Environmental Sciences-China*, 20, 45-49.

Li, Y., Shao, M., Lu, S. H., Chang, C. C., & Dasgupta, P. K. (2010). Variations and sources of ambient formaldehyde for the 2008 beijing olympic games. *Atmospheric Environment*, 44, 2632-2639.

Liu, J. F., Mu, Y. J., Zhang, Y. J., Zhang, Z. M., Wang, X. K., Liu, Y. J., et al. (2009). Atmospheric levels of btex compounds during the 2008 olympic games in the urban area of beijing. *Science of the Total Environment*, 408, 109-116.

Liu, X. G., Zhang, Y. H., Jung, J. S., Gu, J. W., Li, Y. P., Guo, S., et al. (2009). Research on the hygroscopic properties of aerosols by measurement and modeling during carebeijing-2006. *Journal of Geophysical Research-Atmospheres*, 114.

Lu, C. P., Liu, W. Q., Zhao, N. J., Liu, L. T., Chen, D., Zhang, Y. J., et al. (2010a). Influence of humidity on characteristic of laser-induced soil plasmas. *Spectroscopy and Spectral Analysis*, 30, 2885-2888.

Lu, C. P., Liu, W. Q., Zhao, N. J., Liu, L. T., Chen, D., Zhang, Y. J., et al. (2010b). Measurement and analysis of copper in soil using laser-induced breakdown spectroscopy. *Spectroscopy and Spectral Analysis*, 30, 3132-3135.

Luo, Y. H., Liu, W. Q., Bian, L. G., Lu, C. G., Xie, P. H., Si, F. Q., et al. (2011). The retrieval of ozone column densities by passive differential optical absorption spectroscopy during summer at zhongshan station, antarctic. *Spectroscopy and Spectral Analysis*, 31, 456-460.

Mijling, B., van der A, R. J., Boersma, K. F., Van Roozendael, M., De Smedt, I., & Kelder, H. M. (2009). Reductions of no2 detected from space during the 2008 beijing olympic games. *Geophysical Research Letters*, 36.

Movasaghi, Z., Rehman, S., & Rehman, I. U. (2008). Fourier transform infrared (ftir) spectroscopy of biological tissues. *Applied Spectroscopy Reviews*, 43, 134-179.

Patashnick, H., & Rupprecht, E. G. (1991). Continuous pm-10 measurements using the tapered element oscillating microbalance. *Journal of the Air & Waste Management Association*, 41, 1079-1083.

Shao, M., Lu, S. H., Liu, Y., Xie, X., Chang, C. C., Huang, S., et al. (2009). Volatile organic compounds measured in summer in beijing and their role in ground-level ozone formation. *Journal of Geophysical Research-Atmospheres*, 114.

Shen, J. L., Tang, A. H., Liu, X. J., Kopsch, J., Fangmeier, A., Goulding, K., et al. (2011). Impacts of pollution controls on air quality in beijing during the 2008 olympic games. *Journal of Environmental Quality*, 40, 37-45.

Shou-bin, F., Gang, T., Gang, L., Yu-hu, H., Jian-ping, Q., & Shui-yuan, C. (2009). Road fugitive dust emission characteristics in beijing during olympics game 2008 in beijing, china. *Atmospheric Environment*, 43, 6003-6010.

Si, F. Q., Xie, P. H., Dou, K., Zhan, K., Liu, Y., Xu, J., et al. (2010). Determination of the atmospheric aerosol optical density by multi axis differential optical absorption spectroscopy. *Acta Physica Sinica*, 59, 2867-2872.

Simonich, S. L. M. (2009). Response to comments on "atmospheric particulate matter pollution during the 2008 beijing olympics". *Environmental Science & Technology*, 43, 7590-7591.

Streets, D. G., Fu, J. S., Jang, C. J., Hao, J. M., He, K. B., Tang, X. Y., et al. (2007). Air quality during the 2008 beijing olympic games. *Atmospheric Environment*, 41, 480-492.

Tang, X. Y., Shao, M., Hu, M., Wang, Z. F., & Zhang, J. F. (2009). Comment on "atmospheric particulate matter pollution during the 2008 beijing olympics". *Environmental Science & Technology*, 43, 7588-7588.

Tang, Y. Y., Liu, W. Q., Kan, R. F., Zhang, Y. J., Liu, J. G., Xu, Z. Y., et al. (2010). Spectroscopy processing for the NO measurement based on the room-temperature pulsed quantum cascade laser. *Acta Physica Sinica*, 59, 2364-2368.

Vierling, K. T., Vierling, L. A., Gould, W. A., Martinuzzi, S., & Clawges, R. M. (2008). Lidar: Shedding new light on habitat characterization and modeling. *Frontiers in Ecology and the Environment*, 6, 90-98.

Wang, B., Shao, M., Lu, S. H., Yuan, B., Zhao, Y., Wang, M., et al. (2010). Variation of ambient non-methane hydrocarbons in beijing city in summer 2008. *Atmospheric Chemistry and Physics*, 10, 5911-5923.

Wang, M., Zhu, T., Zheng, J., Zhang, R. Y., Zhang, S. Q., Xie, X. X., et al. (2009). Use of a mobile laboratory to evaluate changes in on-road air pollutants during the beijing 2008 summer olympics. *Atmospheric Chemistry and Physics*, 9, 8247-8263.

Wang, S. X., Zhao, M., Xing, J., Wu, Y., Zhou, Y., Lei, Y., et al. (2010). Quantifying the air pollutants emission reduction during the 2008 olympic games in beijing. *Environmental Science & Technology*, 44, 2490-2496.

Wang, T., Nie, W., Gao, J., Xue, L. K., Gao, X. M., Wang, X. F., et al. (2010). Air quality during the 2008 beijing olympics: Secondary pollutants and regional impact. *Atmospheric Chemistry and Physics*, 10, 7603-7615.

Wang, T., & Xie, S. D. (2009). Assessment of traffic-related air pollution in the urban streets before and during the 2008 beijing olympic games traffic control period. *Atmospheric Environment*, 43, 5682-5690.

Wang, W. T., Primbs, T., Tao, S., & Simonich, S. L. M. (2009). Atmospheric particulate matter pollution during the 2008 beijing olympics. *Environmental Science & Technology*, 43, 5314-5320.

Wang, X., Westerdahl, D., Chen, L. C., Wu, Y., Hao, J. M., Pan, X. C., et al. (2009). Evaluating the air quality impacts of the 2008 beijing olympic games: On-road emission factors and black carbon profiles. *Atmospheric Environment*, 43, 4535-4543.

Westerdahl, D., Wang, X., Pan, X. C., & Zhang, K. M. (2009). Characterization of on-road vehicle emission factors and microenvironmental air quality in beijing, china. *Atmospheric Environment*, 43, 697-705.

Winer, A. M., & Biermann, H. W. (1994). Long pathlength differential optical-absorption spectroscopy (doas) measurements of gaseous hono, no2 and hcho in the california south coast air basin. *Research on Chemical Intermediates*, 20, 423-445.

Xin, J. Y., Wang, Y. S., Tang, G. Q., Wang, L. L., Sun, Y., Wang, Y. H., et al. (2010). Variability and reduction of atmospheric pollutants in beijing and its surrounding area during the beijing 2008 olympic games. *Chinese Science Bulletin*, 55, 1937-1944.

Xu, J., Xie, P. H., Si, F. Q., Dou, K., Li, A., Liu, Y., et al. (2010). Retrieval of tropospheric no2 by multi axis differential optical absorption spectroscopy. *Spectroscopy and Spectral Analysis*, 30, 2464-2469.

Xu, Z. Y., Liu, W. Q., Kan, R. F., Zhang, Y. J., Liu, J. G., Zhang, S. A., et al. (2010). Study on the arithmetic of absorbance inversion based on tunable diode-laser absorption spectroscopy. *Spectroscopy and Spectral Analysis*, 30, 2201-2204.

Yang, T., Wang, Z. F., Zhang, B., Wang, X. Q., Wang, W., Gbauidi, A., et al. (2010). Evaluation of the effect of air pollution control during the beijing 2008 olympic games using lidar data. *Chinese Science Bulletin*, 55, 1311-1316.

Yao, X. H., Xu, X. H., Sabaliauskas, K., & Fang, M. (2009). Comment on "atmospheric particulate matter pollution during the 2008 beijing olympics". *Environmental Science & Technology*, 43, 7589-7589.

Yu, H., Wang, P. C., Zong, X. M., Li, X., & Lu, D. R. (2010). Change of no2 column density over beijing from satellite measurement during the beijing 2008 olympic games. *Chinese Science Bulletin*, 55, 308-313.

Zhang, L. F., Huang, Y. R., Shi, S. X., Zhou, L., Zhang, T., Dong, L. A., et al. (2010). Concentration and gas-particle partitioning of hexachlorobenzene in the ambient air before and after the beijing olympic games. *Bulletin of Environmental Contamination and Toxicology*, 85, 1-4.

Zhang, S. A., Dong, F. Z., Zhang, Z. R., Wang, Y., Kan, R. F., Zhang, Y. J., et al. (2009). Monitoring of oxygen concentration based on tunable diode laser absorption spectroscopy. *Spectroscopy and Spectral Analysis*, 29, 2593-2596.

Zhou, Y., Wu, Y., Yang, L., Fu, L. X., He, K. B., Wang, S. X., et al. (2010). The impact of transportation control measures on emission reductions during the 2008 olympic games in beijing, china. *Atmospheric Environment*, 44, 285-293.

Zhu, Y. W., Liu, W. Q., Xie, P. H., Dou, K., Qin, M., & Si, F. Q. (2010). Monitoring and analysis for vertical profiles of air pollutants in boundary layer of beijing. *Chinese Journal of Geophysics-Chinese Edition*, 53, 1278-1283.

Part 3

Air Pollution and Animals

8

Air Pollution and Domestic Animals

René van den Hoven
Section Equine internal and Infection Medicine, Department of Small Animals and Horses, Vetmeduni Vienna, Vienna, Austria

1. Introduction

The Neolithic revolution, which began some 12,000 years ago in Turkey and in other parts of the Fertile Crescent, caused man to adopt a sedentary lifestyle, which on its turn speeded up the process of domestication of animals. The dog was already domesticated before this revolution and had served man as help in hunting. During hunting, man probably figured out that a few of the hunted species could be tamed easily, so subsequently other species such as the chicken, the duck, the goose, the sheep, the goat, the cow, the pig and the camel were domesticated. This implicated for these animals a life closely to their masters, many of them in stalls or in corrals. In ancient farm types, man and animals shared the same airspace, especially during winter. Alternatively, in some regions herd animals guided by shepherds were still allowed to be in the fields in relative freedom, some of them, however, only for a part of the year.

Interestingly, the horse was not domesticated by the sedentary peoples in the Middle East or those around the Mediterranean See, but from the nomadic people of the Eurasian steppes. Recent excavations in Kazakhstan showed that horses were ridden 5,500 years ago by the Botai people (Outram et al., 2009). About 1000 to 1500 BC the horse then enters the Near, Middle and Far East, mainly as a war animal. In those days, the horse was already an expensive animal that had to be well cared for and therefore was kept in stables. Some of these were really large, such as for example the one Pharaoh Ramses II had built for 460 horses at Piramesse 3300 years ago. According to Xenophon horses had to be stabled always. With the current knowledge this was not really smart from a veterinary point of view.

Compared to horses, cat and dog share much more indoor atmosphere with man, whereby these species become more exposed to harmful events like man. Swine, poultry and to a lesser extend cattle are exposed to natural, man-made and self-made air pollution. Furthermore, they may share their environment with their care takers for a part of the day. Therefore, studying diseases of animals living close with humans, or even sharing the same rooms, could bring clues for better understanding risk factors for human health and the pathophysiology caused by poor air quality.

2. General aspects air pollution on animals

It should be considered that, in the history of the Earth, the composition of the atmosphere has not always been ideal at every moment, yet life has evolved as we know it today. Several huge environmental disasters occurred during the development of the Earth and countless forms of life were lost. From those few species that survived, new species have evolved. About 10 million years after the great Cretaceous-Tertiary extinction, the era of the dinosaurs suddenly had ended, the mammals subsequently entered the scene and prosper so successful that they dominated the life forms of the Eocene, which is about 55-40 million years ago. In the development of modern mammals, from the veterinary point of view also a by-product called man was created. This species managed within a relative short time to disturb the environment by the by-products of those activities that are euphemistically called the cultural development.

It was the increasing global population that caused intensive livestock production practices. The counter trade of the huge production of meat, eggs and milk resulted in the generation, accumulation and disposal of large amounts of wastes around the world. Aerosolization of microbial pathogens, endotoxins, odours, and dust particles are inevitable consequences of the generation and handling of waste material of the food production chain, originating from animals. Next to effects of the outdoor environmental air pollution, animals kept in huge facilities are exposed to and often diseased due to self-made indoor air pollution.

The effects of poor air quality on domestic animals principally can be divided in health damage caused by the in-door environment and by out-door air pollution. Pollutants may enter the system by inhalation or ingestion. In air pollution, mostly inhalation triggers the health problems, but occasionally deposition of particles from industrial exhaust on pasture land may affect health directly. Eventually, this may result in toxic residues in meat, milk or eggs without obvious clinical symptoms displayed by the animals producing these products. Problems with high dioxin levels in milk of dairy cows or zinc-induced arthritis in growing foals are examples of pasture grass contamination by deposits of smoke from nearby industrial activities.

The dog, the cat and the horse are exposed to the same health hazards as their masters regarding air pollution. Reineroa et al., (2009) reviewed the comparative aspects of feline asthma and brought evidence that important similarities between human and feline response to inhaled allergens exist. The role environmental aeroallergens, however, was only shown in a few studies, but evidence suggests that some environmental allergens can cause disease in both cats and humans. Ranivand & Otto (2008) showed in their epidemiology study that the prevalence of asthma had increased over the last 20 years in cats in a large urban city. This seems to have happened in man as well.

Animals may be involuntarily acting as sentinels for detecting potential harmful effect on the organism of indoor air pollution. From the scope of comparative pathology, diseases of domestic animals associated with adverse environmental factors may give clues to the pathophysiology of the health disorders of man caused by air pollution.

3. Effects of air pollution on animals

3.1 Production animals

Pigs, poultry, cattle, goats and to a far lesser extend sheep are kept in indoor facilities for a variable part of their life, often for all of their life. For dairy cattle, goats and sheep these facilities are quite open and air quality is to a certain degree comparable with the outdoor air quality. The quality of this air is still much better than that of the closed facilities for swine and poultry (Wathes et al., 1998). These buildings are rather closed and the natural or mechanically ventilation is via small air inlets and outlets. Indoor temperature is regulated to create optimal growing conditions, whereby heat loss via ventilation is kept to a level that is just on the boundary of what is still physiologically tolerable. The other reasons for closing these types of buildings as much as possible are the strict bio security procedures applied in order to avoid or reduce introduction of potential infectious material via air or fomites. The temperature in the facilities for optimal growth can be quite high. For instance, one day-old broiler chicks are kept at a room temperature of 34°C the first days of the raising period. Thereafter, ambient temperature will be lowered daily by 1 °C. The high temperatures facilitate growth of fungi and bacteria especially around the drinkers where water is spilled by the animals. The most common used litter for broilers is wood shavings. Sometimes alternatives such as shredded paper, chopped straw and pulverised bark or peat may be used. The bird's respiratory tracts are challenged by dust coming off the litter. Up to 40,000 broilers may be raised in a single house, on littered floors. A production cycle of broilers only takes 42 days on average. In this period the chicks will growth from about 60 grams to about 2000 grams. Thus, by the end of the raising period, the houses are well filled with animals and their activities increase dust levels in the air. In laying birds, although stocking density is lower, this beneficial effect on pollution, however, is offset by the longer housing period. The result is a larger accumulation of manure, usually in pits, which are only emptied infrequently (Harry, 1978). Hence, it is not surprising that especially in poultry houses high concentrations of ammonia, airborne dust, endotoxin and micro organisms can be measured (Wathes et al., 1998).

Fattening pigs are kept in grid floored pens and thus are exposed to fumes of their own faeces and urine for their entire existence, which is of not more than 6-7 month. Also in many piggeries high levels ammonia, airborne dust, endotoxin and micro organisms can be found (Wathes et al., 1998).

The indoor atmosphere in swine and poultry confinement buildings thus contains toxic gases, dusts and endotoxin in much higher concentrations than those in outdoor environments. Apart from minimal ventilation, poor stable design leading to poor homogeneity of ventilation causes locally stagnant air pockets. According to Dunham (1991), recommended maximal concentrations of gases or contaminants in piggeries are: 2.4 mg dust /m^3; 7 ppm ammonia, 0.08 mg endotoxin/m^3, 10^5 colony-forming units (cfu) of total microbes/m^3; and 1,540 ppm. carbon dioxide. Concentrations of bacteria up to 1.1 $x10^6$ cfu/m^3, inhalable dust content of 0.26 mg/m^3 and ammonia concentration of 27 ppm have been reported to occur in facilities during winter, while at summer lower concentrations were measured (Scherer & Unshelm, 1995). Less difference between in- and outdoor temperature in summer allows better ventilation of the buildings.

A fraction of the smallest and most respirable particles are manure particles containing enteric bacteria and endotoxin (Pickrell, 1991). The concentration of these airborne bacteria and endotoxin, of course, is related to the level of pen cleanliness. Regarding generated toxic gasses, ammonia concentrations in the air are primarily affected by level of pen hygiene, but also by volume of the building, pig density and pig flow management (Scherer & Unshelm, 1995). Furthermore, season plays a role as well as was shown by Scherer & Unshelm (1995). Similar factors on ammonia levels are known to play a role in farrowing units and poultry houses (Harry, 1978). Ammonia is considered as one of the most important inhaled toxicant in agriculture. Dodd & Gross (1980) reported that 1000 ppm for less than 24 hour caused mucosal damage, impaired ciliary activity, and secondary infections in laboratory animals. Since this level is nearly never achieved, it is rather the long-term, low level exposure to ammonia that seems to be related to its ability to cause mucosal dysfunction with subsequent disrupting of innate immunity to inhaled pathogenic micro organisms (Davis & Foster, 2002). Generally, the toxic effects of chronic ammonia exposure do not extend into the lower respiratory tract (Davis & Foster, 2002).

In pigs this combined effects of ammonia and endotoxin predispose the animals to infections with viruses and bacteria, both primary pathogenic and opportunistic species. Although food producing animals appear to be capable of maintaining a high level of efficient growth in spite of marked degrees of respiratory disease (Wilson et al., 1986), at a certain level of respiratory insufficiency rapid growth can no longer be attained. In that case the production results will be uneconomically. Ventilation is often at a just acceptable level. In their overview, Brockmeier et al., (2002) summarized the facts on porcine respiratory diseases. They are the most important health problem for the industrial pork production today. Data collected from 1990 to 1994 revealed a 58% prevalence of pneumonia at slaughter in pigs kept in high-health herds. These animals originate from better farms and thus incidence of pneumonia in less well managed farms is higher. Respiratory disease in swine is mostly the result of a combination of primary and opportunistic infectious agents, whereby adverse environmental and management conditions are the triggers. Primary respiratory infectious agents can cause serious disease on their own, however, often uncomplicated infections are observed. More serious respiratory disease will occur if these primary infections become complicated with opportunistic bacteria. Common agents are porcine reproductive and respiratory syndrome virus (PRRSV), swine influenza virus (SIV), pseudorabies virus (PRV), possibly porcine respiratory coronavirus (PRCV) and porcine circovirus type 2 (PCV2) and Mycoplasma *hyopneumoniae, Bordetella bronchiseptica,* and *Actinobacillus pleuropneumoniae. Pasteurella multocida,* is the most common opportunistic bacteria, other common opportunists are *Haemophilus parasuis, Streptococcus suis, Actinobacillus suis,* and Arcanobacterium *pyogenes.*

Workers in pig or poultry facilities are exposed to the same increased levels of carbon monoxide, ammonia, hydrogen sulphide, or the dust particles from feed and manure as the animals (Pickrell, 1991). As a result, workers in swine production tend to have higher rates of asthma and respiratory symptoms than any other occupational group. Mc Donnell et al. (2008) studied Irish swine farm workers in concentrated animal feeding operations and measured their occupational exposure to various respiratory hazards. It appeared that swine workers were exposed to high concentrations of inhalable (0.25–7.6 mg/m3) and respirable (0.01–3.4 mg/m3) swine dust and airborne endotoxin (166,660 EU/m3). Furthermore, the 8 hour time weighted average ammonia and peak carbon dioxide exposures ranged from 0.01–3 ppm and 430–4780 ppm, respectively.

Lesions caused by air pollution in production animals mainly include inflammatory processes. Neoplastic diseases are rather uncommon. This holds true for animals such as swine that are mainly kept indoors, as well as for cattle and sheep that are kept a variable part of their lives outdoors. This was shown in an abattoir survey some 5 decades ago performed in 100 abattoirs throughout Great Britain during one year (Anderson et al., 1969). All tumours found in a total of 1.3 million cattle, 4.5 million sheep and 3.7 million pigs were recorded and histologically typed. Just 302 neoplasias were found in cattle, 107 in sheep and 133 in pigs. Lymphosarcoma was the commonest malignancy in all three species. Lymphosacoma was considered as entirely sporadic, since herds with multiple cases were not found in the UK. The other form, a lentivirus infection that causes outbreaks of enzootic bovine leukaemia was not present in the UK at those days. The 25 primary lung carcinomas in cattle were well-differentiated adenocarcinomas of acinar and papillary structure, squamous and oat-cell forms and several anaplastic carcinomas of polygonal-cell and pleomorphic types. They represented only 8.3 % of all neoplasms, occurring at a rate of 19 per million cattle slaughtered. No primary lung cancers were encountered in sheep or pigs.

Outdoor air pollution could affect farm animals kept at pastures in urban and peri-urban areas. In the past (1952), a severe smog disaster in London was reported to have caused respiratory distress of prize cattle that were housed in the city for a cattle exhibition (Catcott, 1961). It was likely the high level of sulphur dioxide that was responsible for acute bronchiolitis and the accompanying emphysema and right-sided heart failure. Since some of the city farms are located rather in the periphery of cities than in the centre, the inhaled concentrations of pollutants by production animals is likely less than the concentrations inhaled by pet animals living in the city centres or close to industrial estates.

3.2 Companion animals

Bukowski & Wartenberg (1997) described clearly the importance of pathological findings in domestic animals with respect to analysis of the effects of indoor air pollution in a review. Radon and tobacco smoke are believed to be the most important respiratory indoor carcinogens. Already 42 years ago Ragland & Gorham (1967) reported that dogs in Philadelphia had an eight times higher risk developing tonsillar carcinoma than dogs from rural areas. Bladder cancer (Hayes et al., 1981), mesothelioma (Harbison & Godleski, 1993), lung and nasal cancer (Reif et al., 1992, 1993) in dogs are strongly associated with carcinogens released by human in-door activities. In cats, passive smoking increased the incidence of malignant lymphoma (Bertone et al., 2002). By measuring urinary cotinine, passive smoking of the cats can be quantified. However, the late Catherine Vondráková (unpublished results) observed that there was no direct association with the amount of cigarettes that were smoked in a household and the level of cotinine in the urine of the family cat. Nevertheless, there was evidence that exposed cats showed reduced lung function. Measurement of lung function in small animals and in cats particularly, is difficult and usually only possible with whole body plethysmography (Hirt et al., 2007). For this purpose the cat is placed in a Perspex plethysmography box. Whether this method has sufficient accuracy is still to be proven (van den Hoven, 2007).

The effect of outdoor air pollution on companion animals, so far, has not been studied extensively. Catcott (1961) however described that in the smog incident of 1954 in Donora, Pennsylvania about 15% of the cities dogs were reported to have experienced illness. A few died. Diseased dog were mostly less than 1 year old. Symptoms were mostly mild respiratory problems lasting for of 3-4 days. Also some cats had been reported ill. Further indirect evidence exsists provided by observations made during the smog disaster of 1950 in Poza Rica Mexico. Many pets were reported ill or died. Especially canary birds appeared sensitive, since 100% of the population died (Catcott, 1961). The cause of mortality in the dogs and cats, however, was not professionally established; the information was merely that what the owners had reported, when asked on the incident.

Recently, Manzo et al. (2010) reported that dogs with chronic bronchitis and cats with airways inflammatory disease are at increased risk of exacerbating their conditions if exposed to prolonged urban air pollutants. In this respect they respond similar to man. The authors advise to suppress ongoing inflammatory processes by medical therapy and avoid exercising pets outdoors in urban areas during peak pollutant periods.

3.3 Horses

The reason for the domestication of the horse must be attributed to its athletic ability. The quieter donkey and the ox had been domesticated earlier as draft animals. The horse is one of the mammals with the highest relative oxygen uptake and therefore capable of covering long distances at high speed. The tidal volume of a 500 kg horse at rest is 6-7 L and at racing gallops 12-15 L. At rest a horse breathes 60-70 L of air per minute, which corresponds to about 100,000 L/day. During a race, the ventilation rate increases up to 1800 L /min. With this huge amount of air moving in and out the respiratory track, large quantities of dust particles are inhaled and may sediment in the airways. This on its term could have adverse consequences for lung function. Any decrease of lung function could affect the horse's performance over any distance that is longer than 400 meters. Respiratory problems have a direct impact on the racing career of racehorses, if not successfully treated. Horses that are submitted to less intensive exercise, however, can perform up to expectation for quite a long time, if they are only affected by a small decrease of lung function. This can easily be understood if one considers the huge capacity of the equine cardiopulmonary system. An overview of the physiological aspects of the sport horse is given by van den Hoven (2006).

Horses are not exposed to the negative effects of tobacco smoke or radiation, because stables and the living rooms of man mostly do not share common air spaces. Yet, this does not automatically imply that there is a healthy atmosphere in a horse stable. In those countries where horses are kept in stalls, subacute and chronic respiratory diseases are serious and common problems. In countries like New Zealand, where horses live almost exclusively outdoors, these diseases are less well known.

Many equestrian enterprises are situated in the periphery of urbanized areas. Thus urban air pollution must be considered next to the health challenge by poor indoor air quality. In the suburban and urban enterprises mostly adult animals are employed. Riding schools, racehorse training yards and fiacre horse enterprises are examples of yards that may be located in or near city parks or urban green zones. Horses on these yards are either housed in barns or in individual open-fronted loose boxes. The latter have top doors that are mostly left open (Jones et al., 1987) in order to optimize air circulation. Nevertheless, in many of these boxes due to their small doors, the minimal air change rate of 4/ hour is hardly attained (Jones et al., 1987).

The younger animals mainly are kept in rural areas, mostly at stud farms. Here they are kept out-doors partly or continuously. In winter and prior to horse auctions the youngsters will be stabled for longer periods, just to the moment that many of them will be shipped to suburban or urban enterprises. Other young animals will remain in the countryside. A special category of animals are the breeding animals. After having served in sporting events for short or longer periods in the (sub)urban environment, these animals return to the countryside. Mares are bred to stallions and are mostly kept at pasture for all day, or at least a part of the day. If housed, stables are not necessarily well designed and are as traditional as those of racehorses. Thus, exposure to poor air quality is not uncommon in broodmares. Breeding stallions, have only limited freedom, and yet remain large parts of the day in the barn. Stallion barns are mostly better designed than those for mares; often the more valuable stallions have open-front boxes.

Principally almost all horses will be exposed during a variable period of their life to air of poor quality. The sports and working horses stabled and exercised in (sub)urban regions are exposed to the air pollution caused by traffic and industrial activities too (Fig.1.). Indoor and outdoor air pollution must have an impact on the lung health of our horses. Therefore it is not unexpected that respiratory disease is a major problem for horse industries worldwide (Bailey et al., 1999).

Fig. 1. Fiacre horses of Vienna waiting for tourist. Horses are daily exposed for at least 12 hours to air of the inner city. Most horses are also housed in the city (Foto by R. van den Hoven)

Traditional stable design for horses is based on non-empirical recommendations extrapolated from studies of other agricultural species (Clarke, 1987), ignoring fundamental differences in requirements of the equine athlete. Even now in 2010, only a fraction of the horses are housed in modern well designed stables. But even in the traditional stables, with a median floor space of about 12 m^2 (Jones et al., 1987) stocking density is much less than with production animals. Moreover, many horses have their individual living area, but often still share a common airspace with poor air quality.

Organic dust in the common or individual air space, released by moving of bedding and hay is the main pollutant in horse stables (Ghio et al., 2006). Sometimes dust levels in stalls are less than 3 mg/m^3, but during mucking out, the amount increased to 10-15 mg/m^3, of which 20 - 60% is of respirable particles. Measured at the level of the breathing zone, during eating of hay, dust levels may be 20-fold higher than those measured in the stable corridor (Woods et al., 1993). Dust concentrations of 10 mg/m^3 are known to be associated with a high prevalence of bronchitis in humans. Apart from hay and bedding, cereal food may contain considerable levels of dust. It has been shown that dry rolled grains may contain 30 - 60-fold more respirable dust than whole grains or grains mixed with molasses (Vandenput et al., 1997). Respirable dust is defined as particles smaller than 7 μm (McGorum et al., 1998). Respirable particles are capable of reaching the alveolar membrane (Clarke, 1987) and interact with alveolar cells and Clara cells. In this respect current findings by Snyder et al., (2011) in chemical and genetic mouse models of Clara cell and Clara cell secretory protein (CCSP) deficiency coupled with Pseudomonas *aeruginosa* LPS elicited inflammation provide new understanding on the pathophysiology of chronic lung damage. In this study, the authors reported evidence for anti-inflammatory roles of the airway epithelium and elucidated a mechanism whereby Clara cells likely regulate this process. Injured airway epithelium and mice deficient in expression of CCSP respond more robustly to inhaled LPS, leading to increased recruitment of PMNs.

Kaup et al. (1990b) mention that their ultrastructural study suggests that Clara cells are the main target for antigens and various mediators of inflammation during bronchial changes that occur in horses with recurrent airway obstruction (RAO).

The main constituents of stable dust are mould spores (Clarke, 1987) and it may contain at least 70 known species of fungi and Actinomycetes. Most of these micro organisms are not considered as primary pathogens. Occasionally infection of the guttural pouch with *Aspergilles fumigatus* may occur (Church et al., 1986). The guttural pouch is a 300 mL diverticulum of the Eustachian tube (Fig 2).

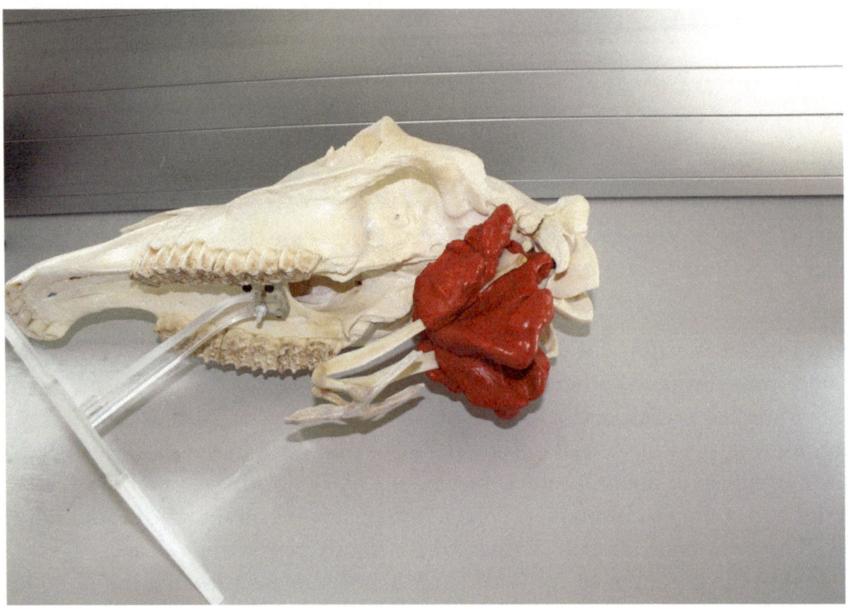

Fig. 2. Horse skull with plastinated guttural pouches. (Preparation kindly provided by Univ-Prof. Dr. Horst König, Section Anatomy, Vetmeduni, Vienna, Austria)

The walls of the guttural pouches are in contact with the base of the skull, some cranial nerves and the internal carotic artery. In case of a fungal infection of the air sac, the fungal plaque is commonly located at the dorsal roof, but may occupy the other walls as well

(Fig.3). The fungus may invade and erode the wall of the adjacent artery. The resulting haemorrhage is not easily controlled and the horse may die due to blood loss.

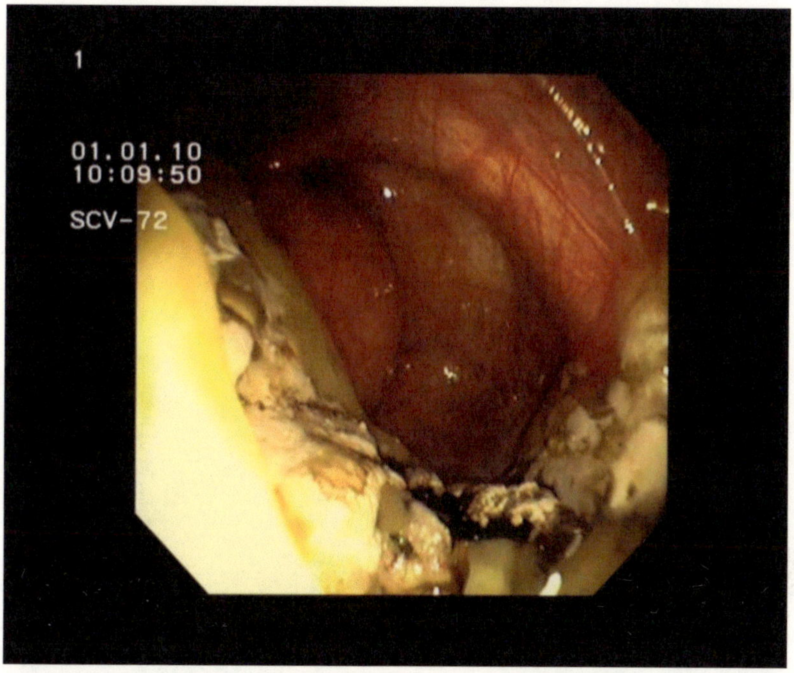

Fig. 3. Endoscopic view of the guttural pouch with a mycotic plaque

A special infection associated with inhalation of bacteria present in the dust generated by dried faeces is the pneumonia caused by *Rhodococcus equi* of young foals (Hillidge, 1986). *R.equi* is a conditional pathogen causing disease in immunologically immature or immune-deficient horses. It can even cause disease in immuno compromised man. The key to the pathogenesis of *R. equi* pneumonia is the ability of the organism to survive and replicate within alveolar macrophages by inhibiting phagosome-lysosome fusion after phagocytosis. Only the virulent strains of *R. equi* having virulence-associated plasmid-encoded 15-17 kDa proteins (VapA) cause the disease in foals (Byrne et al., 2001; Wada et al, 1997). This large plasmid is required for intracellular survival within macrophages. Next to VapA an antigenically related 20-kDa protein, VapB is known. These two proteins however are not expressed by the same *R. equi* isolate. Additional genes carring virulence plasmids e.g. VapC, -D and –E are known. These are co-ordinately regulated by temperature with VapA (Byrne et al., 2001). Expression of the first occurs when *R. equi* is cultured at 37 °C, but not at 30° C. Thus it is plausible that the majority of cases of *R. equi* pneumonia are seen during the summer months. The prevalence of *R. equi* pneumonia is further associated with the airborne burden of virulent *R. equi,* but unexpectedly it seems not directly to be associated with the burden of virulent *R. equi* in the soil (Muscatello et al., 2006). Only under special conditions of the soil, the virulent organisms may be a thread to foals. Dry soil and little grass and holding pens and lanes which are sandy, dry, and lack sufficient grass cover are associated with elevated airborne concentrations of virulent *R. equi.* Hence, Muscatello et al. (2006) consider that environmental management strategies aiming to reduce the level of

exposure of susceptible foals to airborne virulent *R. equi* likely will reduce the impact of *R. equi* pneumonia on endemically affected farms.
If contaminated dust is inhaled by foals of less than 5 month, pulmonary abscesses will develop (Fig. 4). Faecal contamination of pasture and stalls are a prerequisite for the bacteria to establish. Other dust-born bacterial infections are not known in the horse. The non viable components of dust appear to play a major role in the airway diseases of mature horses.

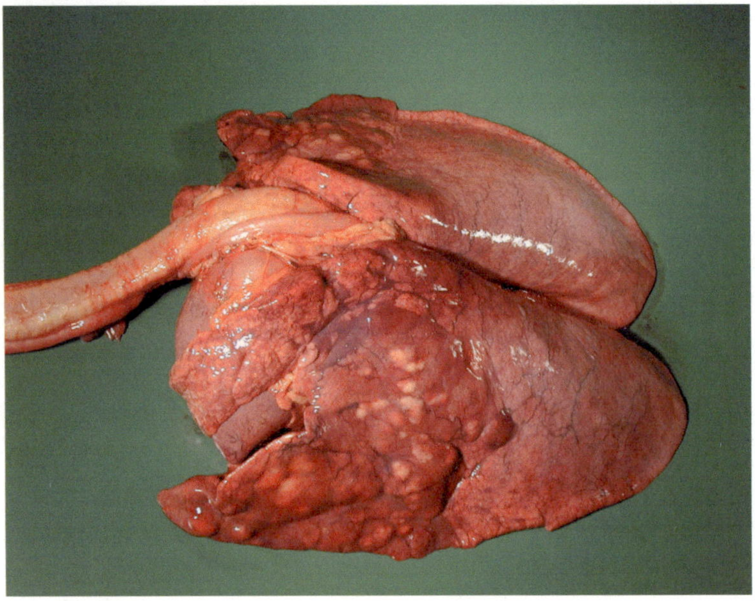

Fig. 4. Pulmonary abscesses

Any threshold limiting value (TLV) for exposure to mould spores or dust are yet not known in horses (Whittaker et al., 2009). In man working for 40 h/week in a dusty environment, the TLV is 10 mg/m^3 (Anonymous, 1972). However, chronical exposure of 5 mg/m^3 caused serious loss of pulmonary function in operators of grain elevators (Enarson et al., 1985). Also Khan & Nachal, 2007 showed that long-term exposure to dust or endotoxin is important for the development of occupational pulmonary diseases in man. In this respect long periods of stabling causing a cumulative exposure effect of dust and endotoxins could result in the development of pulmonary disease in both horses that are susceptible to respiratory disorders and horses that are otherwise healthy (Whittaker et al., 2009).
Generally, horses that are exposed to excess organic dust will develop mild, often subclinical lower airway inflammation. This may contribute to poor performance (see IAD). The symptoms initially seem to share common aspects with the organic dust toxic syndrome in man (van den Hoven, 2006). Some horses could show severe hyperreactivity to organic dust and will display asthma-like attacks after exposure (see RAO). Especially the feeding of mouldy hay is a well-known risk factor for this (McPherson et al., 1979). Commonly incriminated allergens for such sensitive horses are the spores of *Aspergillus fumigatus* and endotoxins. The specific role of β-glucans is still in discussion.
The origin of the moulds may be found in the feedstuff offered to horses. Buckley et al. (2007) analysed Canadian and Irish forage, oats and commercially available equine

concentrate feed and found pathogenic fungi and mycotoxins. The most notable fungal species were *Aspergillus* and *Fusarium*. Fifty per cent of Irish hay, 37% of haylage and 13% of Canadian hay contained pathogenic fungi. Apart from problems by inhalation, these fungi may produce mycotoxins that are rather ingested with the feed than inhaled. T2 and zearalenone appeared to be the most prominent. Twenty-one per cent of Irish hay and 16% of pelleted feed contained zearalenone, while 45% of oats and 54% of pelleted feed contained T2 toxins.

Next to fungal antigens, inhaled endotoxins induce a dose dependent airway inflammatory response in horses (Pirie et al., 2001) and even a systemic response on blood leucocytes can be observed (Pirie et al., 2001; van den Hoven et al., 2006). Inhaled endotoxins in horses suffering RAO are likely not the only determinants of disease severity, but do contribute to the induction of airway inflammation and dysfunction (Pirie et al., 2003).

Whittaker et al. (2009) measured total dust and endotoxin concentrations in the breathing zone of horses in stables. Dust was collected for six hours with an IOM MultiDust Personal Sampler (SKC) positioned within the breathing zone of the horse and linked to a Sidekick sampling pump. The study confirmed earlier studies that forage has a greater effect on the total and respirable dust and endotoxin concentrations in the breathing zone of horses than the type of bedding.

Due to absence of slurry pits under their living area and the low stocking density, noxious gases generated indoors generally play a less important role in development of equine airway disease. Nevertheless, with poor stable hygiene, ammonia released from the urine by urease producing faecal bacteria may contribute to airway disease too.

The effect of air pollution on horses working in the open air has not been extensively studied, but the few studies performed on ozone showed that horses appear less susceptible to the acute effects of ozone compared to humans or laboratory animals (Tyler et al., 1991; Mills et al., 1996). Marlin et al. 2001 found that the anti-oxidant activity of glutathione in the pulmonary lining fluid is likely a highly efficient protective mechanism in the horse. Although it is not likely that ozone is a significant risk factor for the development of respiratory disease in horses, the ability of ozone to act in an either additive or synergistic way with other agents or with already existing disease can not be neglected. Foster (1999) described that this occurs in humans. Diseases associated with poor air quality are follicular pharyngitis, equine inflammatory airway disease and recurrent airway obstruction.

In man exposed to air pollution in large cities, respirable particles and toxic gas levels appear to be associated with acute and subacute cardiopulmonary mortality (Neuberger et al., 2007). Such effects have not been noticed in horse exposed to urban air pollution.

3.3.1 Follicular pharyngitis

Follicular pharyngitis in horses causes narrowing of the pharyngeal diameter and increased upper respiratory airway resistance with impairment of ventilation at high speeds. The symptoms are a snoring noise at in- and expiration during high-speed exercise. The disease is easily detected by endoscopy (Fig. 5.). The disease was previously attributed to a variety of viral infections, but according to Clarke et al. (1987) it must be considered as a multi factor disease. The disease is mostly self limiting within a variable time interval.

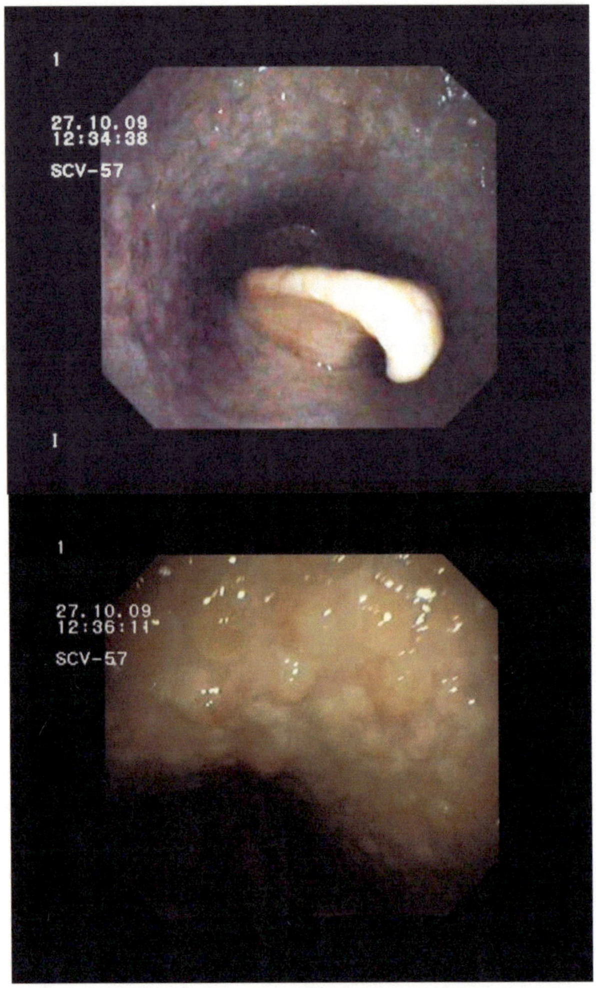

Fig. 5. Endoscopic view of follicular pharyngitis. Upper panel: swollen pharynx and an oblique view on the side and apex of the epiglottis. Lower panel: a closer view of the large lymphoid follicle of the dorsal pharayngeal wall

3.3.2 (Sub)chronic bronchitis

Cough and nasal discharge, caused by increased mucous production in the tracheo-bronchial tree, are common problems in equine medicine. It should be noticed that horses generally have a high threshold for coughing and thus cough is a strong indication for a respiratory disorder. In fact, coughing as clinical sign has an 80% sensitivity for diagnosing tracheo-bronchial disorder. Today, endoscopy is the common technique to diagnose respiratory diseases. For this purpose, 3 meter long human colonoscopes are inserted via the nasal passages and the rima glottis into the trachea. The scope is further advanced into the larger bronchi. Via the endoscope samples can be taken. Commonly, a tracheo-bronchial aspirate or a broncho-alveolar lavage (BAL) is performed. Occasionally cytobrush samples or small biopsies are collected . The endoscopic image in relation to the cytological and bacteriological findings of the samples mostly leads to the diagnosis. The use of lung function tests in horses is only limited to those techniques that require little cooperation. Most commonly the intrapleural pressure in relation to airflow parameters is measured (Fig 6.).

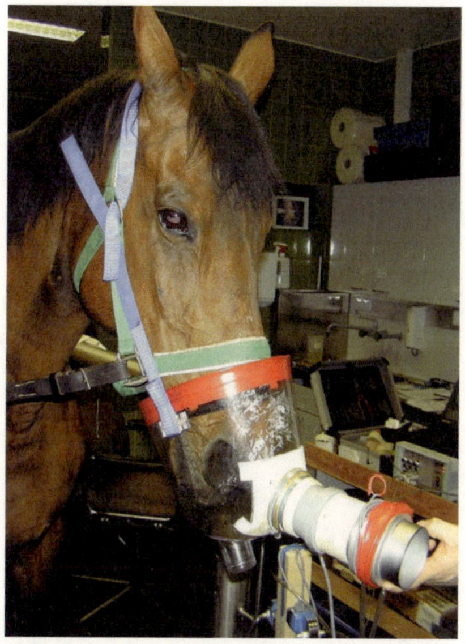

Fig. 6. Intrapleural pressure and air flow measurement in a horse. Intrapleural pressure is measured via an oesophageal balloon connected to a plastic tube. Flow is measured with a Fleisch type pneumotachograph connected to an airtight facemask

The two most important and frequent forms of bronchitis in the horse are Inflammatory Airway Disease (IAD) and Recurrent Airway Obstruction (RAO). In both conditions, a variable degree of airway hyperreactivity to inhaled dust particles plays a role (Ghio et al., 2006). In the case of RAO, next to bronchiolar pathology, secondary changes in the larger airways and in the alveoli will develop.

3.3.2.1 Inflammatory Airway Disease (IAD)

IAD is a respiratory syndrome, commonly observed in young performance horses (Burrell 1985; Sweeney et al., 1992; Burrell et al. 1996; Chapman et al. 2000; Wood, et al. 1999; Christley et al. 2001; MacNamara et al.1990; Rush Moore et al. 1995), but it is not exclusively a disease of the younger horse. Gerber et al. (2003a) showed that many asymptomatic well-performing show-jumpers and dressage horse have signs of IAD. These horses are generally 7-14 years, which is older than the age of affected flat race horses that mostly is between 2 to 5 years.

Although a universally accepted definition of IAD does not exist, a working definition was proposed by the International Workshop on Equine Chronic Airway Disease. IAD is defined as a non-septic airway disease in younger, athletic horses that does not have a clearly defined aetiology (Anonymous, 2003). This approach was reconfirmed in the ACVIM Consensus Statement (Couëtil, 2007).

The incidence of IAD in thoroughbred and standardbred racehorses is estimated between 11.3 and 50% (Burrell 1985; Sweeney et al., 1992; Burrell et al. 1996; Chapman et al. 2000; Wood, et al., 1999; MacNamara et al., 1990; Rush Moore et al., 1995).

The clinical symptoms are often so subtle, that they may go unnoticed. In that case, disappointing racing performance may be the only indication for the presence of IAD. Endoscopic examination is the major help in diagnosing IAD. Mucous accumulation in the airways is commonly observed. The result of cytology of collected BAL fluid (BALF) samples is an important parameter for diagnosing the disease. Various inflammatory cells can be seen in cytospins of BALF samples (Fig. 7.). In contrast to RAO, slightly increased numbers of eosinophil granulocytes may be observed.

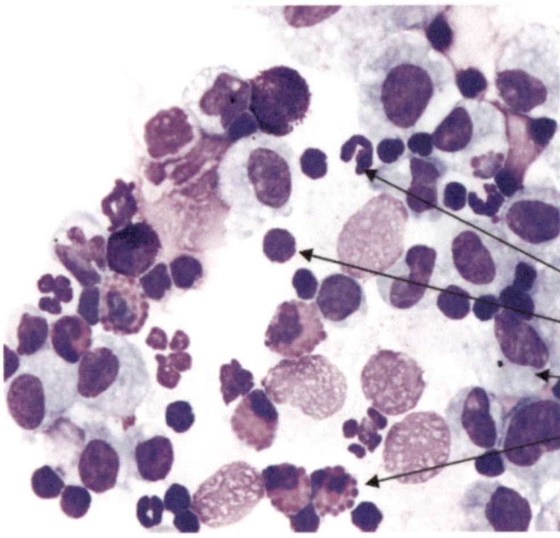

Fig. 7. Cytospin of BALF of a horse with IAD. Romanowsky stain (Foto kindly provided by Dr. C.Tumel, Veterinary School of Toulouse)

There is consensus that the clinical symptoms (Anonymous, 2003; Couëtil, 2007) should include airway inflammation and lung dysfunction. However clinical signs are rather obscure and lung function test may only show very mild changes in respiratory resistance. At endoscopy the horses may have accumulated secretions in the trachea without necessarily displaying cough. Therefore, in contrast to other respiratory disorders, cough is an insensitive indicator of IAD in racehorses. IAD in racehorses seems to diminish with the time being in a training environment (Christley et al., 2001).

Respiratory virus infections do not appear to play a direct role in the syndrome (Anonymous, 2003), but there is still no consensus on their indirect role in the development of IAD. Bacterial colonisation of the respiratory mucosa is regularly detected (Wood et al., 2005). This could be associated with decreased mucociliary clearance. Poor mucosal clearance on its term could be the result of ciliar damage by dust or toxic gases such as ammonia. Common isolates include *Streptococcus zooepidemicus*, *S. pneumoniae*, members of the Pasteurellaceae (including *Actinobacillus* spp), and *Bordatella bronchiseptica*. Some studies have demonstrated a role for infections with Mycoplasma, particularly with *M. felis* and *M. equirhinis* (Wood et al., 1997; Hoffman et al., 1992).

It is estimated, however, that 35% to 58% of IAD cases are not caused by infections at all. Fine dust particles are assumed to be the trigger of these cases (Ghio et al 2006). Once IAD has established, long-term stay in conventional stables does not seem to worsen the

IAD symptoms (Gerber et al., 2003a). Christley et al. (2001) reported that intense exercise, such as racing, may increase the risk of developing lower airway inflammation. Inhalation of dust particles from the track surface or of floating infectious agents may enter deep into the lower respiratory tract during hard exercise and cause impairment of pulmonary macrophage function together with altered peripheral lymphocyte function (Moore, 1996). In theory, intense exercise in cold weather may allow unconditioned air to gain access to the lower airways and cause airway damage (Davis & Foster, 2002), but studies in Scandinavia showed unequivocal results.

Many authors (Sweeney et al., 1992; Hoffman, 1995; Christley et al., 2001; Holcombe et al., 2001) consider the barn or stable environment the important risk factor for development of respiratory disease in young horses. Interestingly, a study in Australia by Christley et al. (2001) reported that the risk of development of IAD decreased with the length of time horses were in training and thus stabled. An explanation for this finding is the development of tolerance to airborne irritants, a phenomenon that has been demonstrated in employees working in environments with high grain dust levels (Schwartz et al., 1994). IAD of the horse partly fit within the clinical picture of the human organic dust toxic syndrome (ODTS). Some evidence for this idea was presented by van den Hoven et al. (2004) et al., who could show inflammation of airways caused by nebulisation of *Salmonella* endotoxin.

3.3.2.2 Recurrent airway obstruction

Recurrent airway obstruction (RAO) is a common disease in horses. In the past, it used to be known as COPD, but as the pathophysiological mechanisms are more similar to human asthma than to human COPD, the disease is called RAO since 2001 (Robinson, 2001). The disease is not always clinically present, but after environmental challenge, horses show moderate to severe expiratory dyspnoea, next to nasal discharge and cough (Robinson, 2001). Exacerbation of disease is caused by inhalation of environmental allergens, especially hay dust, that cause severe bronchospasm and in addition hypersecretion too. The mucosa becomes swollen while accumulated mucous secretions further contribute to airway narrowing (Robinson, 2001). During remission, clinical symptoms may subside completely, but a residual inflammation of the airways and a hyperreactivity of the bronchi to nebulized histamine still remain present. A low degree of alveolar emphysema may develop as well, caused by frequent episodes of air trapping. In the past, severe end-stage emphysema was often diagnosed, but today this is rather uncommon and only sporadically occurs in old horses after many years of illness. The commonly accepted allergens that cause or provoke an exacerbation of RAO are especially spores of *Aspergillus fumigatus* and *Fusarium* spp.

Although the RAO share many similarities with human asthma, an accumulation of eosinophils in the BALF at exacerbation has never been reported. An asthma attack in humans is characterized by an early-phase response of bronchoconstriction, occurring within minutes of exposure to inhaled allergens. This phase is followed by a late asthmatic response with the continuation of airway obstruction and the development of airway inflammation. Mastcells play an important role in this early asthmatic response (D'Amato et al., 2004; Van der Kleij et al., 2004). The activation of mast cells after inhaling allergen results in the release of mastcell mediators, including histamine, tryptase, chymase, cysteinyl-leukotrines, and prostaglandin D2. These mediators induce airway smooth muscle contraction, clinically referred to as early-phase asthmatic response. Mastcells also release proinflammatory cytokines that, together with other mastcell mediators, have the potential to induce the influx of neutrophil and eosinophil granulocytes and the bronchoconstriction that are involved in the late-phase asthmatic response. Activation of other type of mastcell receptors can also induce mastcell degranulation or amplify the Fc-RI mediated mastcell activation (Deaton et al., 2006).

In horses suffering RAO, such an early-phase response seems not to appear, whereas in healthy horses the early phase response does appear (Deaton et al., 2006). This early-phase response may be a protective mechanism to decrease the dose of organic dust reaching the peripheral airways (Deaton et al., 2006). Apparently in the horse with RAO, this protective mechanism has been lost and only the late-phase response will develop. The time of exposure to dust plays a determining role, as was shown by studies with exposure to hay and straw for 5 hours. This challenge caused an increase of histamine concentrations in BALF of RAO-affected horses, but not in control horses. In contrast, exposure of only 30 minutes to hay and straw did not result in a significant increase in BALF histamine concentration of RAO horses (McGorum et al., 1993b). A study of McPherson et al., 1979 showed that exposure to hay dust of at least 1 hour is needed to provoke signs. Also Giguère et al. (2002) and others (Schmallenbach et al., 1998) provided evidence that the duration of exposure to organic dust must be longer than 1 hour. They are the opinion that the necessary exposure to provoke clinical signs of airway obstruction varies from hours to days in RAO affected horses.

The role of IgE-mediated events in RAO is still puzzling. Serum IgE levels against fungal spores in RAO horses were significantly higher than in healthy horses, but counts of IgE receptor-bearing cell in BALF were not significantly different between healthy and RAO affected horses (Kunzle et al., 2007). Lavoie et al. (2001) and Kim et al. (2003) held a T-helper cell response of type 2 responsible for the clinical signs, similar to human allergic asthma. However, their results are in contradiction with results of other research groups who could not find differences in lymphocyte cytokine expression patterns in cases with exacerbation of RAO compared to a control group (Kleiber et al., 2005).

The diagnosis of RAO is made if at least 2 of the following criteria are met: expiratory dyspnoea resulting in a maximal intra pleural pressure difference ($\Delta pPlmax$) > 10 mm H_2O before provocation or > 15 mm H_2O after provocation with dust or by bad housing conditions. Any differential granulocyte count of > 10% in BALF is an indication for RAO. If symptoms can be ameliorated with bronchodilator treatment, the diagnosis is totally established (Robinson, 2001). In some severe cases the arterial PaO_2 may be below 82 mmHg. After provocation with hay dust, RAO patients may reach equally low arterial oxygen levels too. Keeping the animals for 24 hours on pasture will quickly reduce clinical symptoms to a subclinical level.

The visible morphological changes are primarily located in the small airways and spread reactively to the alveoli and major air passages (Kaup et al., 1990a,b). Lesions may be focally, but functional changes may manifest themselves well throughout the bronchial tree. Bronchial lumina may contain a variable amount of exudate and may be plugged with debris. The epithelium is infiltrated with inflammatory cells, mainly neutrophil granulocytes. Furthermore, epithelial desquamation, necrosis, hyperplasia and non purulent peribronchial infiltrates may be seen. Fibrosing peribronchitis spreading in neighbouring alveolar septa was reported in severely diseased animals (Kaup et al, 1990b). The extent of these changes in the bronchioles is related to decrease of lung function, but changes may be distinctly focal in nature (Kaup et al., 1990b). Especially the function of Clara cells is important for the integrity of the bronchioles. Mildly diseased animals show loss of Clara cell granules next to goblet cell metaplasia even before inflammatory changes occur in the bronchioles. This together with the ultrastructural alterations found by Kaup et al. (1990b) supports the idea of the damaging effects of dust and LPS. In severely affected horses Clara cells are replaced by highly vacuolated cells. Reactive lesions may be seen at the alveolar levels. These include necrosis of type I pneumocytes, alveolar fibrosis and variable degree of type II pneumocyte transformation. Furthermore, alveolar emphysema with an increase in Kohns`pores can be present. These structural changes may explain the loss of lung compliance in horses with severe RAO.

Whether there is any causal relation between RAO and IAD is not yet established (Robinson 2001; Anonymous 2003). In both disorders, however, a poor climate in the stables plays a role. It could be theoretisized that IAD eventually may result in RAO, but Gerber et al. (2003a) suggest there is no direct relation between IAD and RAO. In RAO the hyperreactivity induced by histamine nebulization or to air allergens is manifold more severe than in IAD, were only a mild bronchial hyperreactivity often can be shown.

Since long time, based on observations made on members of generations of horse families, it was believed that RAO has a hereditary component. Just recently Ramseyer et al. (2007) provided very strong evidence of an inherited predisposition to RAO on the basis of findings in two groups of horses. The same research group could demonstrate that mucin genes are likely to play a role too (Gerber et al., 2003b) and that the IL4RA gene located on chromosome 13 is a candidate for RAO predisposition (Jost et al., 2007). The results gathered so far suggest that RAO seems to be a polygenic disease. Using segregation analysis for the hereditary aspects of the pulmonary health status for two stallion families, Gerber at al. (2009) showed that a major gene plays a role in RAO. The mode of inheritance in one family was autosomal dominant, whereas in the other horse family RAO seems to be inherited in an autosomal recessive mode.

3.3.2.3 Silicosis

Pulmonary silicosis results from inhalation of silicon dioxide (SiO_2) particulates. It is uncommon in horses; only in California a case series has been published. Affected horse showed chronic weight loss, exercise intolerance, and dyspnoea (Berry et al., 1991).

4. Conclusion

It may be questioned whether our pets, especially dogs, cats and horses are to be considered as victims of or "Sentinels" for air pollution. They are actually victims of human activities, just like man himself. On the other hand, the dog, horse and cat breeds, as we know them today, were all bred by man during and after the process of domestication. If the horse (*equus caballi*) had not been domesticated by man, it would have become extinct long ago. The counter trade of this help is that horses have to adapt themselves to what they become offered by man. Feed, shelter, veterinary care, but also misuse and exposure to health compromising factors. Hence, horses like other companion animals and production animals are exposed to the same environmental factors as man and thus may serve as "Sentinels for environmental risks ". Due to their shorter life span, dogs and cats may express health problems by adverse environment during life or at post mortem at an earlier moment than man. Horses may display chronic effects of dust inhalation that are useful observations in comparative medicine. In the opinion of the authors, the combination of veterinary and human medical epidemiological data is a very powerful tool to identify environmental risk factors for man and its animal companions.

5. References

Anderson, L. J. Sandison, A. T. & Jarrett, W. F. H. (1969). A British abattoir survey of tumours in cattle, sheep and pigs. *Veteimary Record* , Vol.84, pp 547-551.

Anonymous (1972). Threshold limiting values for substances in work room air with intended changes for 1972. *American Con. Of Government Hygienists, Cinicinatti, USA.*

Anonymous (2003). Inflammatory airway disease: defining the syndrome. Conclusions of the Havemeyer Workshop October 2002, Michigan State University, Michigan, USA. *Equine veterinary. Education,* Vol., 15, pp. 61-63.

Bailey, C.J, Reid, S.W.J., Hodgson, D.R.,& Rose, R.J. (1999). Impact of injuries and disease on a cohort of two- and three-year-old Thoroughbreds in training. *Veteriary Record,* Vol. 145, pp.487-493.

Byrne, B.A., Prescott, J.F., Palmer, G.H., Takai, S., Nicholson, V.M., Alperin, D.C. & Hines, S.A.(2001). Virulence plasmid of Rhodococcus equi contains inducible gene family encoding secreted proteins. *Infection and immunity,* Vol. 69, 650-656.

Berry, C.R., O'Brien, T.R., Madigan, J.E. & Hager, D.A. (1991). Thoracic radiographic features of silicosis in 19 horses. *Journal of Veterinary Internal Medicine,* Vol. 5, pp. 248-256.

Bertone, E.R., Snyder, L. &. Moore, A.S. (2002). Environmental Tobacco Smoke and Risk of Malignant Lymphoma in Pet Cats . *American J ournal oJ Epidemioogyl,* Vol. 156, pp.268-273.

Brockmeier, S.L., Patrick G. Halbur, P.G. & Thacker, E.L. (2002). Chapter: 13 Porcine Respiratory Disease Complex. in: Brogden, K.A., Guthmiller, J.M. (editors): *Polymicrobial Diseases.* ASM Press, Washington (DC). pp. 231–258.

Buckley, T., Creighton, A., Fogarty,U. (2007). Analysis of Canadian and Irish forage, oats and commercially available equine concentrate feed for pathogenic fungi and mycotoxins. *Irish Veterinary Journal, Vol.* 60, pp. 231-236.

Bukowski, J.A. & Wartenberg, D (1997). An alternative Approach for Investigating the Carcinogenicity of Indoor Air Pollution: Pets as Sentinels of Environmental Cancer Risk. *Environmental health perspectives.,* Vol. 105, pp. 1312-1319.

Burrell, M.H. (1985). Endoscopic and virological observations on respiratory disease in a group of young Thoroughbred horses in training. *Equine Veterinary Journal, Vol.* 17, pp. 99-103.Burrell, M.H., Wood, J.L., Whitwell, K.E., Chanter, N., Mackintosh, M.E.,& Mumford, J.A.(1996). Respiratory disease in thoroughbred horses in training: the relationships between disease viruses, bacteria and environment. *Veterinary Record,* Vol.139, pp. 308-313.

Catcott, E.J. (1961). Effects of air pollution on animals. In: World Health Organization Monograph series, no.41, WHO, Geneva, pp.221-231.

Chapman, P.S., Green, C., Main, J.P.M., Taylor, P.M., Cunningham, F.M., Cook, A.J. & Marr, C.M. (2000). Retrospective study of the relationships between age, inflammation and the isolation of bacteria from the lower respiratory tract of thoroughbred horses. *Veterinary Record,* Vol. 146, pp. 91-95.

Christley, R.M., Hodgson, D.R., Rose, R.J., Wood, J.L., Reids, S.W., Whitear, K.G., Hodgson, J.L. (2001a). A case-control study of respiratory disease in Thoroughbred racehorses in Sydney, Australia. *Equine Veterinray Journal,* Vol. 33, pp. 256-264.

Christley, R.M., Hodgson, D.R., Rose, R.J., Hodgson, J.L., Wood, J.L.N. & Reid, S.W.J. (2001b). Coughing in thoroughbred racehorses: risk factors and tracheal endoscopic and cytological findings. *Veterinary Record,* Vol. 148, pp.99-104.

Clarke, A.F. (1987). A review of environmental and host factors in relation to equine respiratory disease. *Equine Veterinary Journal,* Vol. 19, pp. 435-441.

Clarke, A.F. (1993) Stable dust--threshold limiting values, exposures variables and host risk factors. *Equine Veterinary Journal,* Vol. 25, pp. 172-174.

Clarke, A.F., Madelin, T.M. & Allpress, R.G. (1987). The relationship of air hygiene in stables to lower airway disease and pharyngeal lymphoid hyperplasia in two groups of Thoroughbred horses. *Equine Veterinray Journal,* Vol. 19, pp. 524-530.

Church, S., Wyn-Jones, G.,Parks, A.H. & Ritchie, H.E. (1986). Treatment of gutteral pouch mycosis. *Equine Veterinary Journal,* Vol. 18, pp. 362-365.

D'Amato, G., Liccardi, G., Noschese, P., Salzillo, A., D'Amato, M. & Cazzola, M.(2004). Anti-IgE monoclonal antibody (omalizumab) in the treatment of atopic asthma and allergic respiratory diseases. *Current drug targets. Inflammation and allergy,* Vol. 3, pp. 227–229.

Couëtil, L.J., Hoffman, A.M., Hodgson, J., Buechner-Maxwell, V., Viel, L., Wood, J.L.N. & Lavoie, J-P. (2007). Inflammatory Airway Disease of Horses. *Journal of Veterinary Internal Medicine,* Vol. 21, pp. 356-361.

Davis, M.S. & Foster, W.M. (2002). Inhalation Toxicology in the Equine Respiratory Tract (Last Updated: 28-Feb-2002) In Lekeux P. (Ed.): *Equine Respiratory Diseases,* International Veterinary Information Service, Ithaca NY (www.ivis.org), 2002; B0319.0202

Deaton, C.M., Deaton, L., Jose-Cunilleras, E., Vincent, T.L., Baird, A.W., Dacre, K. & Marlin, D.J. (2007). Early onset airway obstruction in response to organic dust. *Journal of Applied Physiology,* Vol. 102, pp. 1071-1077.

Dodd, K.T., Gross, D.R.(1980). Ammonia inhalation toxicity in cats: a study of acute and chronic respiratory dysfunction. *Archives of environmental health* Vol. 35, pp. 6-14.

Donham, K. J. (1991). Association of environmental air contaminants with disease and productivity in swine. *American Journal of Veterinary Research,* Vol. 52, pp. 1723-30.

Enarson, D.A., Vedal, S. & Chan-Yeung, M. (1985). Rapid decline in FEV, in grain handlers: Relation to level of dust exposure. *American review of respiratory disease.,* Vol. 132, pp. 814-817.

Foster, W.M. (1999) Effects of Oxidants. In: Swift, D.L. & Foster, W.M., (eds). Air Pollutants and the Respiratory Tract. New York: Marcel Dekker, Inc., pp. 147-179.

Gerber, V., Robinson, N. E., Luethi, S., Marti, E., Wampfler, B. & Straub, R. (2003a). Airway inflammation and mucus in two age groups of asymptomatic well-performing sport horses. *Equine Veterinary Journal,* Vol.35, pp. 491-495.

Gerber,V., Robinson, N.E., Venta,P.J., Rawson,J., Jefcoat,A.M. & Hotchkiss,J.A. (2003b). Mucin genes in horse airways: MUC5AC, but not MUC2, may play a role in recurrent airway obstruction. *Equine Veterinary Jourrnal,* Vol. 35, pp. 252-257.

Gerber, V., Baleri, D., Klukowska-Rötzler, J., Swinburne, J.E. & Dolf, G. (2009). Mixed inheritance of equine recurrent airway obstruction. *Journal of Veterinary Intrenal Medine,* Vol. 23, pp. 626-630.

Ghio, A.J., Mazan, M.R., Hoffman, A.M. & Robinson, N.E. (2006). Correlates between human lung injury after paricle exposure and recurrent airway obstruction in the horse. *Equine Veterinary Journal*, Vol. 38, pp. 362-367.

Giguère, S., Viel, L., Lee, E., MacKay, R.J., Hernandez, J.& Franchini, M. (2002). Cytokine induction in pulmonary airways of horses with heaves and effect of therapy with inhaled fluticasone propionate. *Veterinary Immunology and Immunopathology*, Vol. 85, pp. 147–158.

Harry, E. G. (1978) 'Air pollution in farm buildings and methods of control: A review'. *Avian Pathology* Vol. 7, pp. 441-454.

Hayes, H.M. Jr, Hoover, R. & Tarone, R.E. (1981). Bladder cancer in pet dogs: a sentinel for environmental cancer? *American Journal of Epidemiology*, Vol. 114, pp. 229-233.

Harbison, M.L., & Godleski, J.J. (1983). Malignant mesothelioma in urban dogs. *Veterinary Pathology.*,Vol. 20, pp. 531-540.

Hillidge, C.J. (1986). Review of Corynebacterium (Rhodococcus) equi lung abscesses in foals: pathogenesis, diagnosis and treatment. *Veteinary Record*, Vol. 119, pp. 261-264.

Hirt, R.A., Vondrakova, K., de Arespacochaga, A.G., Gütl, A. & van den Hoven, R. (2007). Effects of cadmium chloride inhalation on airflow limitation to histamine, carbachol and adenosine 5'-monophosphate assessed by barometric whole body plethysmography in healthy dogs. *Veterinary J ournalJ*, Vol. 173, pp. 62-72

Hoffman, A.M., Baird, J.D., Kloeze, H.J., Rosendal, S. & Bell, M.(1992). Mycoplasma felis pleuritis in two show-jumper horses. *Cornell Veterinarian*, 82, pp. 155-162.

Hoffman, A.M. (1995). Small airway inflammatory disease in equids. In: *Proceedings of the American Colllege of Veterinary Internal Medicine*, Vol. 13, pp 754-757.

Holcombe, S.J., Jackson, C., Gerber, V., Jefcoat, A., Berney, C., Eberhardt, S. & Robinson, N.E. (2001). Stabling is associated with airway inflammation in young Arabian horses. *Equine Veterinary Journal*, Vol. 33, pp.244-249.

Jones, R.D., McGreevy, P.D., Robertson, a., Clake, a.F. & Wathes, C.M. (19987). Survey of the designs of racehorse stables in the south west of England. *Equine Veterinary Journal*, Vol. 19, pp. 454-457.

Jost, U., Klukowska-Rotzler, J., Dolf, G., Swinburne, J.E., Ramseyer, A., Bugno, M., Burger, D., Blott, S. & Gerber, V. (2007). A region on equine chromosome 13 is linked to recurrent airway obstruction in horses. *Equine Veterinary Journal*, Vol. 39, pp. 236-241.

Khan, A. J.. & Nanchal, R. (2007) Cotton dust lung diseases. *Current Opinion in Pulmonary Medicine*, Vol. 13, pp. 137-141.

Kaup, F.-J., Drommer, W. & Deegen, E. (1990a). Ultra structural findings in horses with chronioc obstructive pulmonorary diseases (COPD) I: alterations of thje larger conducting airways. *Equine Veterinary Journal*, Vol. 22, pp. 343-348.

Kaup, F.-J., Drommer, W., Damsch, S. & Deegen, E. (1990b). Ultra structural findings in horses with chronioc obstructive pulmonorary diseases (COPD) I: alterations of thje larger conducting airways. *Equine Veterinary Journal*, Vol. 22, pp. 349-355.

Kim, C.K., Kim, S.W., Park, C.S., Kim, B.I., Kang, H. & Koh, Y.Y.(2003). Bronchoalveolar lavage cytokine profiles in acute asthma and acute bronchiolitis. *Journal of Allergy and Clinical Immunology*, Vol. 112, pp. 64-71.

Kleiber, C., McGorum, B. C., Horohov, D.W., Pirie, R.S., Zurbriggen, A. & Straub, R. (2005). Cytokine profiles of peripheral blood and airway CD4 and CD8 T lymphocytes in horses with recurrent airway obstruction. *Veterinary Immunology and Immunopathology*, Vol. 104, pp. 91-97.

Kunzle, F., Gerber, V., van der Haegen, A., Wampfler, B., Straub, R. & Marti, E. (2007). IgE-bearing cells in bronchoalveolar lavage fluid and allergen-specific IgE levels in sera from RAO-affected horses. *Journal of Veterinray Medicine .A*, Vol. 54, pp. 40-47.

Lavoie, J.P., Maghni, K., Desnoyers, M., Taha, R., Martin, J.G. & Hamid, Q.A. (2001). Neutrophilic airway inflammation in horses with heaves is characterized by a Th2-type cytokine profile. *American Journal of Respiratory and Critical Care Medicine*, Vol. 164, pp. 1410-1413.

MacNamara, B., Bauer, S. & Lafe, J.(1990). Endoscopic evaluation of exercise-induced pulmonary hemorrhage and chronic obstructive pulmonary disease in association with poor performance in racing Standardbreds. *Journal American Veteterinary Medical Association*, Vol. 196, pp. 443-445.

McPherson, E.A., Lawson, G.H.K., Murphy, J.R., Nicholson, J., Breeze, R.G. & Pirie, H.M. (1979). Chronic obstructive pulmonary diseases (COPD) in horses: Aetiological studies.Responses to intradermal and inhalation antigenic challenge. *Equine Veterinary Journal*, Vol. 11, pp. 139-166.

Manzo, N.D., Slade, R. & Dye, J.A. (2010). Interrelationship of lung inflammation and air pollutant exposure on cellular oxidative stress and epithelial injury. *Proceedings 27th Symposium of the Veterinary Comparative Respiratory Society*, Plymouth, MA.

Marlin, D.J., Deaton, C.D., Smith, N.C., Roberts, C.A., Kelly, F., Harris, P. & Schroter, R.C. (2001) Development of a model of acute, resolving pulmonary oxidative stress in the horse by ozone exposure. In: *Proceedings World Equine Airways Symposium*, Edinburgh, pp 30.

Mc Donnell, P.E., Coggins, M.A., Hogan, V.J. & Fleming, G.T. (2008). Exposure assessment of airborne contaminants in the indoor environment of Irish swine farms. *Ann Agric Environ Med*, Vol. 15, pp. 323–326.

McGorum, B.C., Dixon, P.M. & Halliwell, R.E. (1993a) Responses of horses affected with chronic obstructive pulmonary disease to inhalation challenges with mould antigens. *Equine Veterinary Journal*, Vol. 25, pp. 261-267.

McGorum, B.C., Dixon, P.M. & Halliwell, R.E. (1993b). Quantification of histamine in plasma and pulmonary fluids from horses with chronic obstructive pulmonary disease, before and after "natural (hay and straw) challenges." . *Veterinary Immunology and Immunopathology*, Vol. 36, pp. 223–237.

McGorum, B.C., Ellison, J. & Cullen, RT. (1998). Total and respirable airborne dust endotoxin concentrations in three equine management systems. *Equine Veterinary Journal* , Vol. 30, pp. 430-434.

Mills, P.C., Roberts, C.A. & Smith, N.C. (1996). Effects of ozone and airway inflammation on glutathione status and iron homeostasis in the lungs of horses. *American Journal of Veterinary Researchs*, Vol. 57, pp. 1359-1363.

Moore, B. R. (1996). Lower respiratory tract disease. *Veterinary Clinics of North America: Equine Practice*, Vol. 12, pp. 457-472.

Muscatello, G., Anderson, G.A., Gilkerson, J.R. & Browning, G.F. (2006). Associations between the ecology of virulent *Rhodococcus equi* and the epidemiology of *R. equi* pneumonia on Australian thoroughbred farms. *Applied and Environmental Microbiology*, Vol. 72, pp. 6152-6160.

Neuberger, M., Rabczenkob, D. & Moshammer, H. (2007). Extended effects of air pollution on cardiopulmonary mortality in Vienna. *Atmospheric Environment,* Vol. 41, pp. 8549–8556.

Outram, A.K., Stear, N.A., Bendrey, R., Olsen, S., Kasparov, A., Zaibert, V., Thorpe, N. & Evershed, R.P. (2009). The Earliest Horse Harnessing and Milking. *Science* , Vol. 323, pp. 1332-1335.

Pickrell, J. (1991). Hazards in confinement housing - gases and dusts in confined animal houses for swine, poultry, horses and humans. *Veterinary and Human Toxicology,* Vol. 33: pp. 32-39.

Pirie, R.S., Dixon, P.M., Collie, D.D. & McGorum, B.C. (2001). Pulmonary and systemic effects of inhaled endotoxin in control and heaves horses. *Equine Veterinary Journal,* Vol. 33, pp. 311-318.

Pirie, R.S, Collie, D.D., Dixon, P.M. & McGorum, B.C. (2003). Inhaled endotoxin and organic dust particulates have synergistic proinflammatory effects in equine heaves (organic dust-induced asthma). *Clinical & Experimental Allergy,* Vol. 33, pp. 676-83.

Ragland, W.L. & Gorham, J.R. (1967). Tonsilar carcinoma in rural dogs. *Nature,* Vol. 214, pp. 925-926.

Ramseyer, A., Gaillard, C., Burger, D., Straub, R., Jost, U., Boog, C., Marti, E. & Gerber, V.(2007). Effects of genetic and environmental factors on chronic lower airway disease in horses. *Journal of Veterinary Internal Medicine,* Vol. 21, pp. 149-156.

Ranivand, L. & Otto, C. (2008). Feline asthma trends in Philadelphia, Pennsylvania 1896-2007. *26th Annual Symposium of the VCRS.*

Reif, J.S, Dunn, K., Ogilvie, G. K.& Harris, C.K. (1992). Passive smoking and canine lung cancer risk. *American Journal of Epidemiology,*Vol. 135, pp. 234-239.

Reif, J.S., Bruns, Chr. & Lower, K.S. (1998). Cancer of the Nasal Cavity and Paranasal Sinuses and Exposure to Environmental Tobacco Smoke in Pet Dogs. *American Journal of Epidemiology,* Vol. 147, pp. 488-492.

Reineroa, C.R., Amy E. DeCluea, A.E. & Rabinowitzb, P. (2009) Asthma in humans and cats: Is there a common sensitivity to aeroallegens in shared environments? *Environmental Research,* Vol. 109, pp. 634-640

Robinson, N.E. (Ed.) (2001): International workshop on equine chronic airway disease, Michigan State University 16-18, June 2000. *Equine Veterinary Journal,* Vol. 33, pp. 5-19.

Moore, B., Krakowka, S., Robertson, J.T. & Cummins, J.M. (1995). Cytologic evaluation of bronchoalveolar lavage fluid obtained from Standardbred racehorses with inflammatory airway disease. *American Journal of Veterinary Research,* Vol. 56, 562-567.

Scherer, M. & Unshelm, J. (1995). Untersuchungen zur Belastung von Mastschweinen sowie der Umgebung von Mastschweinstallen durch atembaren Feinstaub, stallspezifische Bakterien und Ammoniak. *Zentralblatt fur Hygiene und Umweltmedizin,* 196, pp. 399-415.

Schmallenbach, K.H., Rahman, I., Sasse, H.H., Dixon, P.M., Halliwell, R.E., McGorum, B.C., Crameri ,R. & Miller HR. (1998). Studies on pulmonary and systemic *Aspergillus fumigatus*-specific IgE and IgG antibodies in horses affected with chronic obstructive pulmonary disease (COPD). *Vet Immunol Immunopathol,* 66, 245-256.

Schwartz, D.A., Thorne, P.S., Jagielo, P.J., White, G.E., Bleuer, S.A. & Frees, K.L. (1994). Endotoxin responsiveness and grain dust-induced inflammation in the lower respiratory tract. *American Journal of Physiology*. Vol. 267, pp. L609-L617.

Sweeney, C. R., Humber, K.A. & Roby, K.A. (1992) Cytologic findings of tracheobronchial aspirates from 66 Thoroughbred racehorses. *American Journal of Veterinary Research*, Vol. 53, pp. 1172-1175.

Snyder, J.C., Reynolds, S.D., Hollingsworth, J.W., Li, Z., Kaminski, N. & Stripp, B.R.(2011). Clara cells attenuate the inflammatory response through regulation of macrophage behavior. *American Journal of Respiratory Cell and Molecular Biology*, Vol. 42, pp 161-171.

Tyler, W., Jones, J., Birks, E., Julian, D., Snyder, J., Pascoe, J., Steffey, E., Jarvis, K., Tarkington,B., Berry, J., Hinds, D. (1991). Effects of ozone on exercising horses. *Equine Exercise Physiology*, 3: pp. 490-502

van den Hoven, R., Duvigneau, J.C., Hartl, R.T., Riedelberger, K., Teinfalt, M. & Gemeiner, M. (2004) The Expression of Messenger RNA for Tumor Necrosis Factor-α; Interleukins 1ß, 6, 8, and 10;Tissue Growth Factor-ß; and Interferon-γ in bronchoalveolar lavage fluid cells from horses challenged with lipopolysaccharide and treated with clenbuterol. *Journal of Equine Veterinary Science*, Vol. 24, pp. 29-36.

van den Hoven, R. (2006) Equine exercise physiology - Transforming laboratory studies into practical concepts. *Pferdeheilkunde, Vol.* 22, pp. 525-530.

van den Hoven, R., Duvigneau, J.C., Hartl, R.T. & Gemeiner, M. (2006). Clenbuterol affects the expression of messenger RNA for interleukin 10 in peripheral leukocytes from horses challenged intrabronchially with lipopolysaccharides. *Veterinary Research Communications*, Vol. 30, pp. 921-928.

van den Hoven, R (2007). A jack-in-the-box of respiratory research: is the technique of barometric whole body plethysmography a disappointing surprise? *Veterinary Journal*, Vol. 173, pp. 250-251.

Wood, J.L.N., Newton, J.R., Chanter,N. & Mumford, J.A (2005). Association between respiratory disease and bacterial and viral infections in British racehorses. *Journal of Clinical Microbiologyl*,Vol. 43, pp. 120-126.

Vandenput, S., Istasse, L., Nicks, B.,& Lekeux, P. (1997). Airborne dust and aeroallergen concentrations in different sources of feed and bedding for horses. *Veterinary Quarterly*, Vol. 19, pp. 154-158.

Van der Kleij, H.P., Kraneveld, A.D., van Houwelingen, A.H., Kool, M., Weitenberg, A.C., Redegeld, F.A., Nijkamp, F.P.(2004). Murine model for non-IgE-mediated asthma. *Inflammation*, Vol. 28, pp. 115-125.

Wada, R., Kamada, M., Anzai, T., Nakanishi, A., Kanemaru, T., Takai, S. and Tsubaki, S., 1997. Pathogenicity and virulence of *Rhodococcus equi* in foals following intratracheal challenge. *Veterinary Microbiologie*, Vol. 56, pp. 301–312.

Wathes C.M., Phillips, V.R., Holden, M.R., Sneath, R.W., Short, J.L., White, R.P.P., Hartung J., Seedorf J., Schroder, M., Linkert, K.H., Pedersen, S., Takai, H., Johnsen, J.O., Groot Koerkamp, P.W.G., Uenk, G.H., Metz, J.H.M., Hinz, T., Caspary, V. & Linke S.(1998) Emissions of Aerial Pollutants in Livestock Buildings in Northern Europe: Overview of a Multinational Project. *Journal of Agricultural Engineering Research*, Vol. 70, pp. 3-9.

Wilson, M.R., Takov, R., Friendship, R.M., Martin, S.W., MacMillan, I., Hacker, R.R. & Swaminatan, S. (1986). Prevalence of respiratory diseases and their assocaition with growth rate and space in randomly selected swine herds. *Canadian Journal Veterinary Research,* Vol. 50, pp. 209-216.

Whittaker, A.G., Hughes ,K.J., Parkin, T.D.H. & Love, S. (2009). Concentrations of dust and endotoxin in equine stabling. *Veterinary Record,* Vol. 165, pp. 293-295.

Wood, J.L.N., Newton, J.R., Chanter, N., Mumford, J.A., Townsend, H.G.G., Lakhani, K.H., Gower, S.M., Burrell, M.H., Pilsworth, R.C., Shepherd, M., Hopes, R., Dugdale, D., Herinckx, B.M.B., Main, J.P.M., Windsor, H.M. & Windsor, G.D. (1999) A longitudinal epidemiological study of respiratory disease in racehorses: disease definitions, prevalence and incidence. In: *Equine Infectious Diseases VIII: Proceedings of the Eighth International Conference of Equine Infectious Diseases,* Eds: U. Wernery, J.F. Wade, J.A. Mumford and J.O. Kaaden, R&WPublications Ltd., Newmarket, pp. 64-70.

Wood, J.L., Chanter, N., Newton, J.R., Burrell, M.H., Dugdale, D., Windsor, H.M., Windsor, G.D., Rosendal, S. & Townsend, H.G. (1997). An outbreak of respiratory disease in horses associated with Mycoplasma felis infection. *Veterinary Record,* Vol. 140, pp. 388-391.

Woods, P.S., Robinson, N.E., Swanson, M.C., Reed, C.E., Broadstone, R.V. & Derksen, F.J. (1993). Airborne dust and aeroallergen concentration in a horse stable under two different management systems. *Equine Veterinary Journal,* Vol. 25, pp. 208-213.

Part 4

Air Pollution and the Impact on Vegetation

9

Comparative Analysis of Bioindicator and Genotoxicity Indicator Capacity of Lichens Exposed to Air Pollution

Sümer Aras, Demet Cansaran-Duman, Çiğdem Vardar and Esin Başaran
Ankara University, Science Faculty,
Department of Biology, Biotechnology Section, Ankara
Turkey

1. Introduction

Air pollution is very harmful on both the environment and on living organisms at high concentrations. Furthermore polycyclic aromatic compounds (PACs), heavy metals and halogenated aliphatic hydrocarbons have been shown to be genotoxic to the living organisms (Grant, 1998). Polycyclic aromatic hydrocarbons (PAHs) are capable of forming covalent interaction with nucleophilic centres of DNA (Piraino et al., 2006). They also cause base pair substitutions, frameshift mutations, deletions, S-phase arrest, strand breakage and a variety of chromosomal alterations. (Singer and Grunberger, 1983; Dipple, 1985; Baird et al., 2005). Further studies have also pointed out that long-term exposure to air pollution can cause cancer (Cohen and Pope, 1995; Sawicki, 1977; Jerrett et al., 2005).

Unfortunately, current physical and chemical methods for estimating air genotoxicity provide insufficient information to accurately quantify the risk to biota (Piraino et al., 2006).

In contrast to physical and chemical methods, biological methods allow the direct assessment of the genotoxic potential of air pollutants. Plants have been widely used as bioindicators in many studies up to now. Lichens can also be considered sensitive and efficient indicators of genotoxicity. They in particular have been widely used as trace element atmospheric biomonitors as they are widespread and capable of absorbing elements directly from the atmosphere and accumulating them in their tissues (Aras et al., 2010).

This chapter reports the results of biomonitoring experiments aimed at assessing the genotoxic potential of air pollutants. We tried to detect DNA damage using molecular marker technonolgy. Development of this technology has provided new tools for the detection of genetic alteration by looking directly at the level of DNA sequence and structure. Various types of molecular markers are available. In particular the PCR (Polymerase Chain Reaction) based molecular markers are useful for DNA analysis in complex genomes. With the PCR reaction almost any type of mutational event can be screened e.g.: point mutation, small insertion, deletion and rearrangement (Conte et al., 1998).

The experimental data generated in our studies could be used to develop a thematic map of air genotoxicity with the aim of defining the air quality due to the presence of genotoxic stressors.

2. Air pollution

According to the Environmental Protection Agency, air pollution is a mixture of solid particles and gases in the air. Car emissions, chemicals from factories, dust, pollen and mold spores may be suspended as particles in the air (Forman and Alexander, 1998). Some air pollutants are poisonous and inhaling them can increase the human health problems. Air pollution represents a serious threat to both the environment and living organisms. Millions tons of toxic pollutants are released into the air each year. Following activities are the major reasons of such kind of pollution; mobiles (cars, buses, trucks, etc.) and industrial sources (factories, refineries, power plants, etc.) (Wolterbeek, 2002).
Presumably heavy metal accumulation is the most important component of air pollution and also the major part of the environmental pollution. Heavy metals can easily mobilize, disperse and to some extent produce toxic effects, which in turn can lead to growth inhibition and decline in crop yield (Wolterbeek, 2002). Toxic heavy metals in air, soil, and water are global problems that are a growing threat to the environment. There are hundreds of sources of heavy metal pollution, including the coal, natural gas, paper, glassware and industries.
Environmental pollution is a serious problem in many parts of the world. The major impacts of air pollution can be stated as health problems with exposed human populations, forest decline, loss of agricultural productivity, contamination of ecosystems, etc. Those problems has been a cause of increasing public concern throughout the world. Concern about atmospheric pollutants and contaminants underlies the efforts to establish control programmes in many countries. The necesarry quantitative information on chemical element air pollution is generally obtained by modelling of the dispersion (source oriented approach, making use of a priori known information on emission sources) or by field measurements of the immission (receptor oriented approach) (Wolterbeek, 2002; Garty et. al., 2002). Monitoring air pollution is a complex process because of the high number of potentially dangerous substances, the difficulty of estimating their synergistic or antagonistic effects, the large spatial and temporal variation of pollution phenomena, the high cost of recording instruments, and hence the low sampling density of a purely instrumental approach. For these reasons it is hard to establish a region-wide monitoring system to reveal environmental risk assessment levels. Increasing awareness of the potential hazards of large scale contamination of ecosystems by pollutants has highlighted the need for continuous monitoring of the levels of contaminants in the environment.

3. Bioindication and biomonitoring

In a general sense, biomonitoring may be defined as the use of bioorganisms (biomonitors) to obtain quantitative information on certain characteristics of the biosphere (Wolterbeek, 2002; Garty et al., 2002). In general bioindicators are organisms that can be used for the identification and qualitative determination of human generated environmental factors, while biomonitors are organisms used for the quantitative determination of contaminants (Conti and Cecchetti, 2001). The use of cosmopolite organisms to assess pollution has developed notably during the last few decades. Such organisms assume environmental contaminants and may be used as indicators of the bioavailability of a given contaminant over time, allowing, in certain cases, comparison between contamination levels in geographically differenet areas (Conti and Cecchetti, 2001).

The use of living organisms as indicators for environmental stability has long been widely recognised. Higher plants, animals, alga, fungi, bacteria and lichen have been employed as bioindicators and biomonitors in air, soil and water pollution surveys over the past few decades (Garty, 1999; Galun et. al.,1987; Castello et al., 1990; Bargagli, 1998; Conti and Cecchetti, 2001; Wolterbeek, 2002). The fundamental aim of biomonitoring can be explained as supplying data for an effective ecological control system. In particular, biomonitoring should act as an early warning system by providing information about the sensitivity of living organisms.

4. Use of lichens as indicators and monitors of air pollutant effects

Lichens are prominent examples of symbiotic organisms, in which alga and fungi form an intimate biological union (Nash, 1996). They are slow-growing associations of fungi (mycobionts) and green alga or cyanobacteria (photobionts).

Lichens were recognised as potential indicators of air pollution as early as the 1860's Britain and Europe, since then lichens have played prominent roles in air pollution studies throughout the world because of their sensitivity to different gaseous pollutants, particularly sulphurdioxide (Marquez, 2008). They have also been found to act as acummulators of trace and radioactive elements.

Lichens are made up of a few distinct characters morphologically. The most obvious structure is the lichen thallus. The form of the thallus is a result of the fungal species involved in the simbiont. The thallus is the main body of the lichen. The top surface is called cortex which is normally a layer of tightly packed hyphae. There is an algal layer below this where the photobiont lives. Unlike higher plants, lichens have no roots or a well developed cuticle and they strongly depend on deposited material from the atmosphere to obtain their mineral nutrients. On the other hand the lichen surface, structure and roughness facilitate the interception and retention of particles (Marquez, 2008). These features of lichens, combined with their extraordinary capability to grow at a large geographical range and to accumulate mineral elements far above their need, rank them among the best bioindicators of air pollution (Cansaran-Duman et al., 2009).

Lichens in particular have been widely used as trace element atmospheric biomonitors as they are widespread and capable of absorbing elements directly from the atmosphere and accumulating them in their tissues. As a result of these proporties of lichens, several papers have been published on heavy metal monitoring of lichens in different geographic areas, even Antarctic regions (Garty et al., 1977; Ölmez et al., 1985; Bargagli, 1989; Bermudez et al., 2009; Villarini et al., 2009).

Lichen biomonitoring is often used as receptor based method in air quality studies. It can be useful in risk assessment for human health and it can be a powerful tool for administrators inolved in environmental planning.

5. Genotoxic effects of air pollution and genotoxicity assessment strategies

Mixed environmental toxicants are also known to affect the genetic structure of natural populations. In other words, in living organisms heterogenous air pollutants are considered as a major source of DNA damage. They act either through the direct action of the toxicant at the DNA level (direct mutagenic effect) or via toxicant-mediated mortality and/or curtailment of reproduction (population genetic effects). Furthermore polycyclic aromatic compounds (PACs), heavy metals and halogenated aliphatic hydrocarbons, have been shown to be genotoxic to the living organisms (Grant, 1998). Polycyclic aromatic hydrocarbons (PAHs) are capable of covalent interaction with nucleophilic centres of DNA (Piraino et al., 2006). They also cause base pair substitutions, frameshift mutations, deletions, S-phase arrest, strand breakage and a variety of chromosomal alterations. (Singer and Grunberger, 1983; Dipple, 1985; Baird et al., 2005)

5.1 Comparative analysis of air pollution genotoxicity by molecular markers (RAPD and AFLP) and bioindicator capacity in the exposed samples of *Pseudevernia furfuracea* province of Kayseri

We conducted studies in our laboratory in this context and the studies concluded that lichens could be utilized as sensitive and efficient indicators of genotoxicity in addition to their bioindicator capacity. Different studies which were applied to evaluate the genotoxic potential of the atmospheric environment by lichens were analysed comparatively. We aim to detect DNA damage using Randomly Amplified Polymorphic DNA (RAPD) and Amplified Fragment Length Polymorphism (AFLP), very sensitive molecular tools for the detection of DNA fragmentation and chromosomal mutations (Citterio et al., 2002).

One of our genotoxicity work reported the results of biomonitoring experiments, aimed to assess the genotoxic potential of air pollutants throughout the Kayseri Province (central Anatolia). For this investigation *Pseudevernia furfuracae* L. Zopf lichen species was chosen as a suitable bioindicator since its sensitivity to organic and inorganic compounds were well documented. Heavy metal accumulation was analysed by using Atomic Absorption Spectrometer (AAS) and effects of environmental pollution on DNA was investigated by RAPD and AFLP analysis in lichen species. *P. furfuracea* collected from Çat Forests located around the province of Sivas (central Anatolia), and then exposed at 12 polluted sites in the province of Kayseri. Lichen samples were transplanted to the different pollution sources in province of Kayseri for two time periods (dry and wet seasons) of the year (Fig 1) (Aras et al., 2010). Economic development in the province is facilitated by the close proximity of a major road network, with arterial roads extending from east to west and north to south. This network supports heavy traffic, a major source of air pollution to the region. Also land use in the province is diverse: apart from agriculture there are both large and small industrial districts located around the city. Interestingly, both industrial districts are characterised by different activities (mechanical, chemical, textile, food). There are definite boundaries which distinguishes urban and suburban sites in the city of Kayseri (central Anatolia). Pollution sites were urban roadsides, urban sites, urban park sites, industrial sites, rural areas and shanty areas (Table 1). Urban sites were chosen at least 10 m away from a main road, and the samples from urban roadsides were selected from sites close to the city centre, along busy main roads. Urban roadside samples were chosen between 0 and 5 m, usually not more then 2 m away from the busy road. Urban park sites were chosen from two large parks of Kayseri. Industrial sites were chosen from the industrial parts of the city. Shanty areas were selected from two shanty zones around the city and rural samples were chosen from south of the Kayseri which were more than 10 km away from any source of pollution (Aras et al., 2010).

The experimental data generated in this study provided a thematic map of air genotoxicity for the province of Kayseri with the aim of defining the air quality due to the presence of genotoxic stressors. The study area was characterized by the presence of numerous industrial activities, such as steel works, glassworks and metallurgical, mechanical, chemical and other industries. These were potential sources of heavy metals and other mixed pollutants in the environment (Aras et al., 2010).

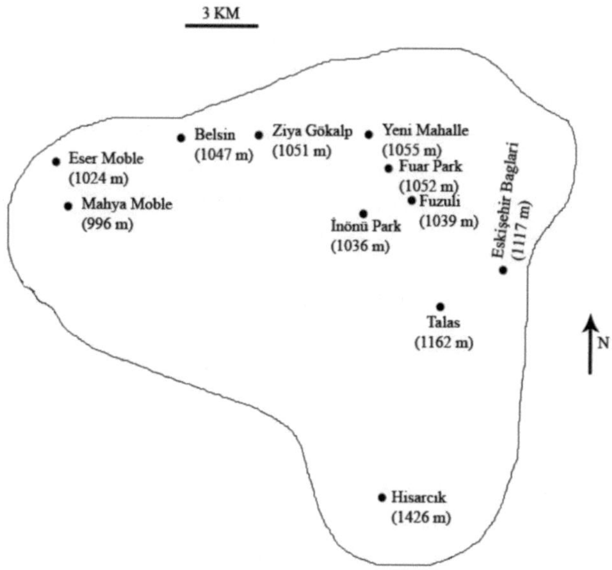

Fig. 1. Map of the study area (Aras et al., 2010)

Control
Hisarcık
Industry
Eser Moble
Mahya Moble
Urban Park
Fuar Park
İnonu Park
Shanty
Ziya Gökalp
Eşkişehir Bağları
Urban
Yenimahalle
Belsin
Urban roadsite
Fuzili
Talas street

Table 1. Study area of samples

The AAS data were evaluated via pollution scales, and specimens were analysed using RAPD and AFLP-PCR to detect the probable pollution effects and changes on DNA molecules. Polymorphism was calculated in relation to the appearance of new bands and disappearance of normal bands considering the control's band patterns.

Seven primers used in the study yielded new band appearance and disappearance in the exposed samples compared with the control. For example, the primer TubeA01 (5′CAGGCCCTTC3′) yielded eight clear, reproducible bands in the control which only one of them was disappeared in the exposed samples, Urban Park II, Urban I, II and Urban roadside II in wet season. Same primer yielded seven bands in control for the dry season and more band appearances and disappearences were noticed. Three primers TubeA02 (5′TGCCGAGCTG3′), TubeA03 (5′AGTCAGCCAC3′), TubeA04 (5′AATCGGGCTG3′) yielded higher polymorhism values compared to the other primers for the wet season in which the TubeA04 was displayed the highest polymorphism. TubeA02 (5′TGCCGAGCTG3′) displayed only one band disappearance in all the samples in dry season. TubeA12 (5′TCGGCGATAG3′) also showed maximum three band disappearance out of 14 in dry season (Aras et al., 2010). The results of RAPD analysis displayed an interesting distinction in band patterns of wet and dry seasons. According to our results, the differences among the band patterns between dry and wet seasons were remarkable. Many studies have documented the differences in cell wall permeability due to factors like precipitation and subsequent better performance of lichens as a bioaccumulator during wet period. In the RAPD analysis, different band patterns were obtained for the samples collected during dry and wet season but the polymorphism ratios of the primers were not informative to make any suggestion about the harmful effect of the genotoxic agents in a certain season. But apparently DNA polymerization during PCR reactions were affected in a way and different band patterns were obtained (Aras et al., 2010).

With the aim of verifying the effect of environmental pollutants on the genetic material of the lichen samples, AFLP analysis was performed from the same *P. furfuracea* lichen samples exposed to polluted sites in the province of Kayseri. The primers used in the are the combination of E22-M3, E32-M7, E32-M3, E22-M6 and E32-M6. The AFLP profiles showed substantial differences between unexposed and exposed lichen samples, with apparent band changes in the number of amplified DNA fragments at different locations.

The nineteen primers used displayed significant differences between the control and polluted samples collected from various parts of Kayseri. The highest number of band appearance and disappearence was determined at the samples collected from Urban road site-I for wet season and Shanty-II for the dry season with all of the five primer pairs used. But the other areas also displayed high band appearance and disappearance compared to the original control samples. Polymorphisms obtained, were due to loss and/or gain of amplified bands observed in exposed samples compared with the control (Atienzar et. al., 1999). Meanwhile, 5 pairs of primer combination gave a total of 231 bands in wet period, 147 bands in dry period. Different polymorphic bands were detected at each location and periods for different primer combinations. Average value of polymorphism (P) obtained from the amplification of the primers used in the research was P (%) = 45.02 for wet period and 64.62 for dry period. In all cases, polymorphism was due to the loss and/or the gain of amplified bands in the exposed samples compared with the control.

In addition, genomic template stability ratios (GTS) were calculated. GTS implies qualitative measure reflecting changes in RAPD and AFLP profiles. Changes in RAPD and AFLP profiles were expressed as reductions in GTS (a qualitative measure reflecting the obvious changes of the number and intensity of DNA bands in DNA patterns generated by toxicant exposed) in relation to profiles obtained from control samples.

In RAPD analysis, we obtained 96 polymorphic bands in wet season (Aras et al., 2010). In AFLP analysis we obtained 104 polymorphic bands in the same season. 48 polymorphic bands were observed by RAPD analysis in dry season. But 95 polymorphic bands were observed by AFLP analysis in the same season. Similarly GTS results were compared in two different seasons per method. In RAPD analysis the highest GTS value (75%) was obtained from Park-I samples, in AFLP analysis the highest GTS value was obtained from Rural-II (85.8%) samples in wet season. In dry season, Urban-II (87.5%) and Urban roadsite-I (87.5%) samples showed the highest GTS values by RAPD analysis and Rural-II (88.8%) samples showed the highest GTS values by AFLP analysis.

According to results, AFLP profile changes provide all sensitive markers to detect genotoxicity in lichens. Thus, the AFLP method has been successfully used as a sensitive means of detecting DNA damage and shows potential as a reliable and reproducible assay for genotoxicity. Furthermore, DNA affects in conjunction with other biomarkers from higher levels of biological organization would to be a powerful ecotoxicological tool. Our studies revealed that results of AFLP analyes were paralled with atomic absorbtion results.

We also studied on some methods to establish a correlation between RAPD, AFLP and AAS results. We tried to construct a special mathematical model between the RAPD results and AAS values. Excel correlation function was used in order to design this model but we have not been able to construct a model showing the direct correlation between the band patterns and AAS data yet.

5.2 Comparative analysis of air pollution genotoxicity by molecular markers (RAPD) and bioindicator capacity of the exposed samples of *Pseudevernia furfuracea* province of Ankara

In this study, we aimed to describe the heavy metal contents of exposed *P. furfuracea* lichen samples to various polluted areas in the province of Ankara (central Anatolia). Lichen samples collected from the Yenice Forest (province of Karabük) and transplanted in bags to different sites in Ankara and exposed to pollution (Fig 2). Samples were exposed for three and six months and then the concentrations of six trace elements (Cd, Cu, Mn, Ni, Pb and Zn) in *P. furfuracea* were determined. Heavy metal concentrations of *P. furfuracea* were determined by inductively coupled plasma-mass spectrometry (ICP-MS) technique.

Ankara, the capital of Turkey, is located in central Anatolia, and it is the second most crowded city in the country. Ankara has gained a bad reputation for the black smog that hung over the city during the winter. The primary source of the pollution was the usage of coal, as a main fuel for residential and industrial heating during winter times. Since the 1980s, however, Turkey has made investments for the development of an extensive network of natural gas pipelines that serve all the major cities and most towns. Consequently, natural gas has replaced coal as the primary source of fuel in most the populated centers, and urban air has become cleaner than the past.

The districts chosen for the study are given in Fig 2. Kızılay is the downtown of the city. Sıhhıye is the closest district to Kızılay. Ulus is the downtown of the old town of the city and Dışkapı is close to Ulus. All these districts are very crowded and polluted areas especially in terms of vehicular traffic. Tandogan, Emek, Yenimahalle and Etlik are the major residential areas where natural gas is the main energy source used for heating. Mamak is a shanty area where coal burning is still used for heating purposes. Eryaman is a suburban site which is 30 km far from downtown and natural gas is the only energy source used.

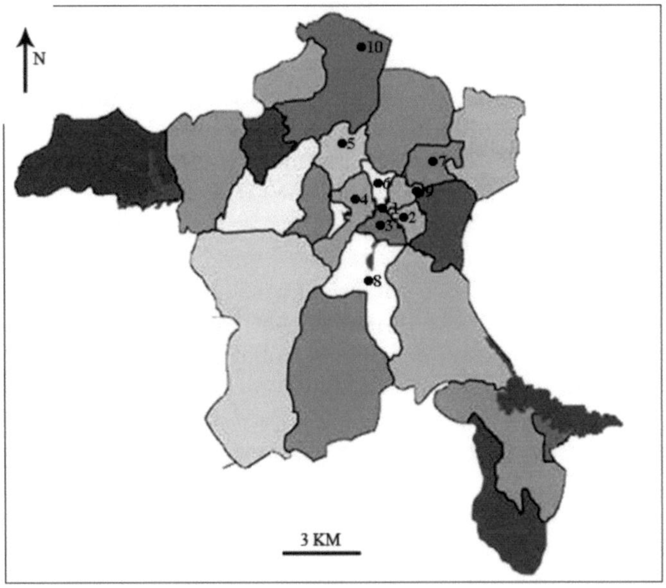

Fig. 2. Ankara
Site 1= Kızılay, Site 2= Sıhhıye, Site 3= Ulus, Site 4= Tandogan, Site 5= Emek, Site 6= Yenimahalle, Site 7= Etlik, Site 8= Mamak, Site 9= Dışkapı, Site 10= Eryaman

Control	Karabuk, Yenice, North of Yalnızca plateau, GPS:41°15′ N, 32°34′ E, 1200m
Site 1	Eryaman-5th street, in front of Ugur taxi
Site 2	Dışkapı- in front of Yıldırım Beyazıt Hospital
Site 3	Mamak- in front of Muhabere Okul Komutanlığı
Site 4	Kızılay- in front of Turkish Air Force
Site 5	Sıhhıye- in front of Refik Saydam National Public Health Agency
Site 6	Ulus- TBMM Garden
Site 7	Etlik Kasalar-in front of The Ministry of Healt of Turkey Etlik İhtisas Hospital
Site 8	Yenimahalle- İvedik street, in front of Yayla taxi station
Site 9	Emek-8. street, opposite of Karacaoğlan restraurant
Site 10	Tandogan-in front of Ankara University

Table 2. The localities of the lichen samples used in Ankara

Comparisons of the Pb concentrations of the samples from exposed sites with the control yielded very significant variations. Kızılay and Sıhhıye with the highest human activities, together with high vehicular density congestion, showed the highest Pb with the values of, 47.00 $\mu g g^{-1}$ in three months and 52.80 $\mu g g^{-1}$ in six months which were significantly higher than the control site, 22.70 $\mu g g^{-1}$. It can be concluded that Pb concentration is the highest in Kızılay district because it is the central part of the city where human activities and density of traffic are very intense.

The mean Cd concentration in Eryaman, Emek and Tandogan are slightly higher than the control site. The other districts; Dışkapı, Mamak, Kızılay and Sıhhıye showed significantly higher Cd concentrations than control site in six months. The concentrations of Cd in three and six months were significantly higher in Dışkapı, Mamak, Kızılay and Sıhhıye than the control site, probably indicating the contaminants from motor vehicles, dust raised by metal industry and other human activities. The most important sources of Cd pollution were regarded as fossil fuels used by the vehicles, metal business, plastics, house tools construction and sewer. Increases in trace element concentration confirm that vehicular traffic plays a prominent role on air pollution of Ankara.

In addition, DNA alterations in the exposed lichen samples were aimed to be described by RAPD analysis. Out of 17 decamer oligonucleotide primers tested, six of them showed clear and reproducible bands. In RAPD analyses some of the primers displayed significant differences between the control and polluted samples exposed to various parts of the city of Ankara.

The number of band appearance and disappearance was the highest in the samples collected from Mamak district for three months and Dışkapı for the six months for all the six primers used. The size of the disappearing bands ranged from 220 bp to 1600 bp. But the other areas also displayed high band appearance and disappearance compared with the original control sample.

For example, the primer OPOO3 (5′CTGTTGCTAC3′) yielded 11 clear, reproducible bands in the control which only one of them was disappeared in the exposed samples in Emek, Mamak, Dışkapı, Kızılay and Sıhhıye in three months. Same primer yielded more band appearances and disappearances in six months (Mamak, Dışkapı, Kızılay and Sıhhıye). One of the other primers used in the study OPB16 (5′TTTGCCCGGA3′), also showed maximum three band disappearance (Mamak) out of 12 in six months. As could be seen from results the sample exposed in Mamak yielded 15 polymorphic bands and the sample from Dışkapı showed 14 polymorphic bands. Sıhhıye and Kızılay districts followed them with 13 and 11 polymorphic bands, respectively in three months. In six months Kızılay district samples showed the highest polymorphism with 15 band variation. The results prove that these districts are the most populated and polluted areas in Ankara (Cansaran-Duman et al., under review).

The highest DNA band variation was recorded in the samples exposed for six months in Dışkapı. *P. furfuracea* samples transplanted to Mamak and Dışkapı showed higher DNA variation than the other sites after three and six months. This result might be correlated with the elevated Pb accumulation as a result of very dense vehicular traffic.

Results indicated that both accumulating trace elements and DNA variations by RAPD analysis were the highest in samples exposed for six months.

5.2.1 Comparison of chemical content and RAPD profiles in the exposed samples of *P. furfuracea* province of Ankara

The results for Zn, Cd and Pb elements from Dışkapı and Mamak samples were found high, as expected. In these districts coal burning in stoves is common which stimulates Zn and Cd accumulation in that region and also the areas are hollow where circulation and reverse inversion do not exist. All these factors are the main reasons for air pollution in these districts and explain the elevated levels of heavy metals recorded for areas. Likewise, the results of RAPD analysis yielded the highest band variations in the samples from these districts.

According to the results of chemical analysis, the sample exposed in Eryaman which is 30 km away from downtown (Kızılay) yielded the lowest metal values which are close to the values obtained from the control sample. In Eryaman accumulation in the sample is not high

as the central natural gas system is used commonly for heating. Likewise, RAPD profiles of Eryaman sample displayed the most similar band pattern with the control among all the samples. According to our observations obtained from chemical analysis, Mamak and Dışkapı were the most polluted areas because of usage of stoves for heating, Kızılay and Ulus were polluted because of the traffic.
We also recorded a slight increase in metal contents of the samples from three months to six months. Likewise DNA band variation intensities were increased and in accord the GTS values were decreased in the samples exposed for 6 months. As a result, the present study confirms lichens as efficient metal accumulators and their appropriate use in biomonitoring studies. The concentrations of six elements detected in *P. furfuracea,* after exposure in bags in the urban area of Ankara, compared with the element content in control site gave a clear indication of urban air contamination by trace elements. The correlation between Cd, Cu, Mn, Ni, Pb and Zn confirm that vehicular traffic plays a prominent role in Ankara's air pollution.

5.3 Comparison of the results obtained from two different provinces (Ankara-Kayseri, Turkey)

The experiments were carried out in Ankara and Kayseri using the bag technique with the same type of lichens. Both cities were in central Anatolia region and their climate had similar aspects. Same types of lichens were transplanted using the bag technique in especially shanty areas and the parts with traffic of both cities.
When the RAPD results were compared, it was observed that there were numerous band changes for the lichen species especially near the highways like Kızılay, Sıhhiye, Ulus in Ankara and Kayseri.
Band changes were observed clearly for the lichen species in regions with high consumption of coal as a fuel; shanty areas in Kayseri and Mamak, Dışkapı provinces of Ankara. RAPD results were compared with the atomic absorption spectrometer data. It was identified that results for Mg, Zn and Mn showed high values in Sıhhiye, Mamak, shanty areas and near the highways in Kayseri but the results were not correlated with mathematical expressions and models.

5.4 Comparative analysis of heavy metal accumulation in the samples of *Pseudevernia furfuracea, Evernia prunastri, Usnea hirta, Hypogymnia physodes* province of Iron-steel Factory in Karabük and examples to genotoxicity

5.4.1 Accumulation of heavy metals in *Pseudevernia furfuracea*

The study was initiated with the objective to provide baseline information on metal accumulation in lichen species growing in and around the Iron-steel Factory in Karabük, Turkey. *P. furfuracea* lichen specimens were collected from every 5 km starting from around the Iron-steel Factory located in the central area of Karabük province (Anatolia), up to Yenice Forest (Table 3). The locations of the districts were also given in the map (Fig 3). Zn, Cu, Mn, Fe, Pb, Ni, Cd, Cr concentration were analyzed in the samples collected from polluted and unpolluted areas. In the study *P. furfuracea* sample from Yenice Forest was used as a control. The reason for the choise of Yenice Forest was the abundance of species diversity, and therefore sample collection might cause a very low impact on natural population density. The forest is among the 100 forested areas that must be urgently taken under protection according to World Wildlife Fund (WWF) researches. The present investigation has involved the collection of ten *P. furfuracea* samples growing on *Pinus* sp. from 10 sites in and around Karabük Iron-steel Factory area, Karabük, Turkey. The metal concentration of *P.*

furfuracea samples collected from Yenice Research Forest (Yenice-Karabük) and Karabük Iron-steel Factory were analyzed by atomic absorbsion spectrometry. The thalli of *P. furfuracea* were used to determine the levels of eight metals (Zn, Cu, Mn, Fe, Pb, Ni, Cd, Cr).

Locality No	**Date of collection**	**GPS co-ordinates**	**Locality name**	**Altitude** (m)
1	15.11.2005	44°62′ N, 45°73′ E	Karabük, Yenice, Kuzdağ district	1125
2	15.11.2005	41°15′ N, 32°35′ E	Karabük, Yenice, Kabaklı kaya	1140
3	15.11.2005	41°13′ N, 32°28′ E	Karabük, Yenice, vicinity of Hamzakıran district	1140
4	15.11.2005	41°14′ N, 32°35′ E	Karabük, Yenice, Dikilitaş	1125
5	15.11.2005	41°12′ N, 32°25′ E	Karabük, Yenice, vicinity of Kuzdere, Hamdioğlu district	1400
6	15.11.2005	41°15′ N, 32°34′ E	Karabük, Yenice, North of Yalnızca plateau	1200
7	15.11.2005	41°11′ N, 32°27′ E	Karabük, Yenice, Acısu Center	1375
8	15.11.2005	41°14′ N, 32°33′ E	Karabük, Yenice, Kazancıoğlu district	1750
9	15.11.2005	41°12′ N, 32°29′ E	Karabük, Yenice, Hacıömerler district	1380
10	15.11.2005	41°12′ N, 32°29′ E	Karabük, Yenice, Kızılgöz kayası	1385
11*	15.11.2005	41°10′ N, 32°24′ E	Karabük, Yenice, vicinity of Cami district	1100

* 11.= control sample

Table 3. The localities of the lichen samples used in Karabük

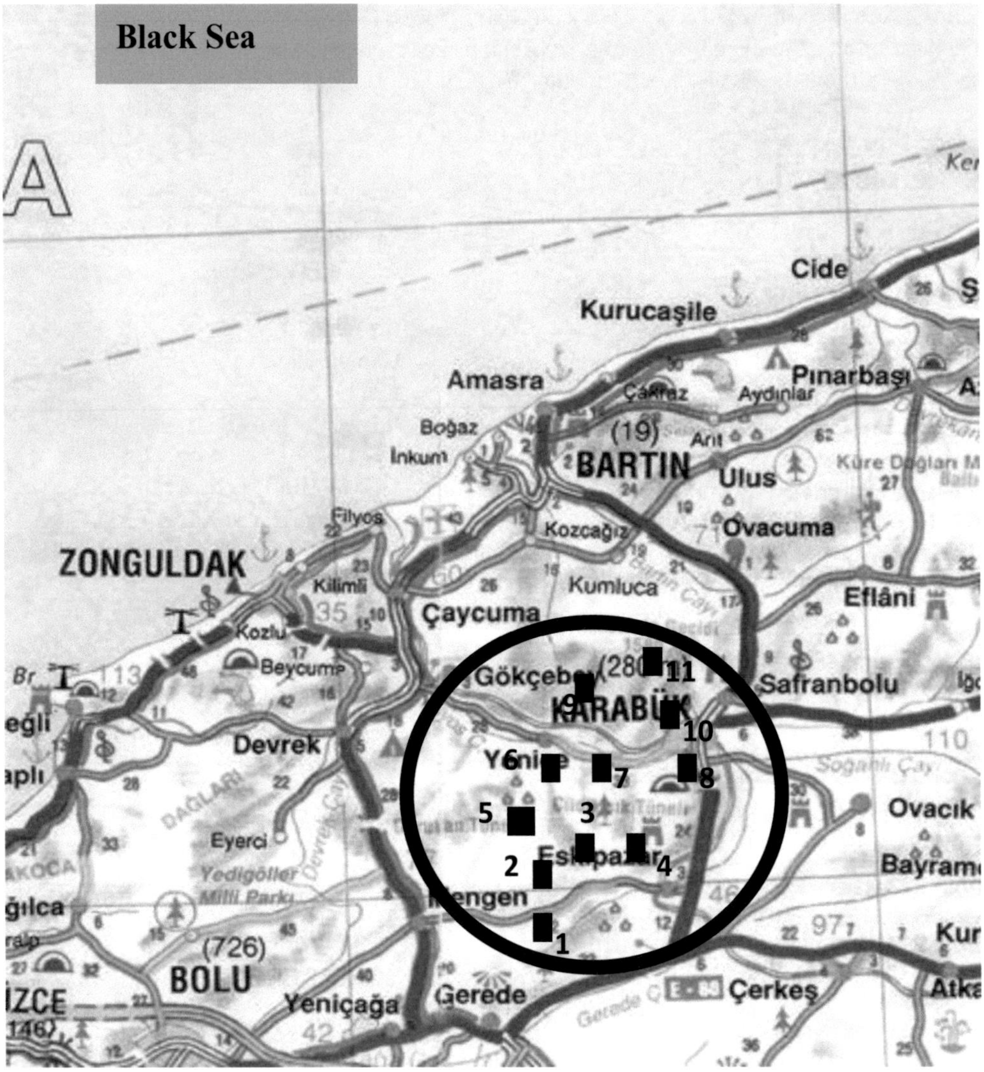

Fig. 3. Map of the Karabük

The 5th and 8th stations which were located close to Iron-steel Factories and major motor vehicle traffic, manifested the highest level of Mn (Table 3). At site 8, combustion of coal and other kind of fuels rather than natural gas seem to be the reason of air pollution. Site 10 with the highest human activities, together with high vehicular density congestion, showed the highest Pb levels with the value of 9.750 mg/kg which is significantly higher than that of the control site (4.00 mg/kg) (Table 4). The most important sources of Cr pollution are indicated as industrial activities like refining works and Iron-steel Factories (Cansaran-Duman et al., 2009).

Results of the study displayed significant variations among the concentration of these elements between stations. As expected, the pollution sources such as Iron-steel Factory, roads and railroads, industry, heavy traffic and waste treatment plants have major impact on the heavy metal accumulation in *P. furfuracea* and in accordance to their location samples

8 and 10 displayed high element accumulation. Surprisingly, although Yenice Forest is under protection, results of our study showed that the region is becoming polluted by the influence of many pollution sources in the area. The present study also confirms the efficient metal accumulation capacity of lichens.

Site	Zn		Cu		Mn		Fe	
	N	Mean±SD	n	Mean±SD	n	Mean±SD	n	Mean±SD
1	5	26.280±0.158	5	3.090±0.111	5	45.496±3.197	5	2379.000±44.193
2	5	25.680±0.316	5	2.050±0.100	5	32.500±0.174	5	1273.030±17.408
3	5	24.410±0.095	5	1.940±0.032	5	41.730±2.103	5	356.460±7.906
4	5	30.162±3.455	5	2.770±0.047	5	56.290±0.142	5	965.250±15.811
5	5	25.610±0.063	5	3.350±0.190	5	119.860±0.380	5	766.350±15.812
6	5	28.078±0.040	5	3.300±0.016	5	92.950±0.016	5	199.030±3.162
7	5	40.628±0.561	5	2.990±0.551	5	71.890±1.059	5	1558.960±46.833
8	5	32.360±0.158	5	4.450±0.047	5	112.970±1.091	5	3016.000±13.835
9	5	31.362±0.119	5	3.550±0.016	5	31.720±2.087	5	1560.132±6.665
10	5	53.802±0.340	5	4.560±0.047	5	54.080±0.190	5	1185.600±39.845
11*	5	28.640±0.174	5	1.810±0.010	5	42.250±0.965	5	918.452±7.471
ANOVA								
F ratio		0.065		6.750		0.051		0.081

Site	Pb		Ni		Cr		Cd	
	n	Mean±SD	n	Mean±SD	n	Mean±SD	n	Mean±SD
1	5	7.200±0.158	5	5.170±0.079	5	4.540±0.047	5	0.725±0.002
2	5	4.900±0.158	5	2.090±0.063	5	2.950±0.032	5	0.690±0.008
3	5	5.100±0.152	5	1.490±0.063	5	2.730±0.031	5	0.490±0.019
4	5	6.000±0.073	5	2.350±0.158	5	3.242±0.024	5	0.706±0.007
5	5	4.130±0.095	5	1.974±0.052	5	2.940±0.016	5	0.618±0.005
6	5	3.000±0.079	5	1.450±0.079	5	2.730±0.007	5	0.632±0.004
7	5	3.150±0.080	5	1.240±0.159	5	2.620±0.032	5	0.668±0.002
8	5	4.700±0.160	5	2.490±0.063	5	4.100±0.063	5	0.671±0.003
9	5	4.600±0.158	5	1.880±0.063	5	3.440±0.063	5	0.720±0.006
10	5	9.750±0.128	5	4.190±0.079	5	3.390±0.079	5	0.770±0.007
11*	5	4.000±0.035	5	2.100±0.100	5	2.280±0.007	5	0.630±0.007
ANOVA								
F ratio		2.717		3.199		5.712		1.651

Table 4. Average Zn, Cu, Mn, Fe, Pb, Ni, Cr and Cd concentrations in *P. furfuracea* ($\mu g g^{-1}$) in the Karabük city, with Standart deviations (SD)

5.4.2 Accumulation of heavy metals in *E. prunastri*

Heavy metal concentrations of *E. prunastri* samples taken from polluted sites and control group are summarized in Table 5. All stations were statistically analyzed to determine their relationships with respect to each heavy metal. SPSS 11.5 analysis was used to show the relationships of the stations and some results were shown with tables (Table 5) (Cansaran-Duman et al., 2011).

The highest levels of Manganase (Mn) in the *E. prunastri* were found in sites 8 (82.7 µg/g), 5 (77.0 µg/g) and 6 (73.7 µg/g). In order to compare the ability of the lichen species to accumulate some heavy metals, they were compared with the element concentrations in the baseline material. For example, the highest levels of Mn in the *E. prunastri* were found in the site 8 (82.7 µg/g) (control is 28.8 µg/g) (Table 5) (Cansaran-Duman et al., 2011).

Site	Zn		Cu		Mn		Fe	
	n	Mean±SD	n	Mean±SD	n	Mean±SD	n	Mean±SD
1	5	46.698±0.058	5	3.673±0.007	5	34.425±5.665	5	943.032±5.238
2	5	43.359±0.772	5	3.002±0.025	5	44.185±0.641	5	443.061±0.417
3	5	24.139±0.826	5	1.903±0.190	5	32.201±0.706	5	1023.900±3.069
4	5	16.532±0.058	5	1.549±0.007	5	30.428±2.646	5	540.792±6.833
5	5	21.895±0.261	5	1.724±0.082	5	77.026±1.209	5	419.541±11.863
6	5	17.891±0.029	5	1.558±0.021	5	73.773±2.687	5	775.832±8.405
7	5	16.215±0.161	5	2.022±0.019	5	57.955±1.699	5	1289.000±11.765
8	5	19.055±0.008	5	2.693±0.014	5	82.773±0.685	5	2187.200±71.983
9	5	17.860±0.555	5	1.601±0.009	5	26.333±1.409	5	786.969±0.638
10	5	15.526±0.316	5	1.551±0.015	5	54.663±2.812	5	827.505±32.554
11*	5	12.543±0.332	5	1.469±0.032	5	28.830±0.172	5	460.228±0.302
ANOVA								
F ratio		0.026		0.156		0.254		0.046

Site	Pb		Ni		Cr		Cd	
	n	Mean±SD	n	Mean±SD	n	Mean±SD	n	Mean±SD
1	5	5.171±0.236	5	7.819±0.201	5	2.719±0.017	5	0.620±0.001
2	5	0.316±0.005	5	2.665±0.010	5	2.718±0.029	5	0.644±0.002
3	5	1.037±0.033	5	4.627±0.082	5	3.364±0.011	5	0.604±0.004
4	5	0.958±0.092	5	1.862±0.073	5	1.801±0.007	5	0.682±0.001
5	5	1.011±0.097	5	1.970±0.071	5	1.821±0.017	5	0.624±0.006
6	5	0.995±0.143	5	0.950±0.137	5	2.329±0.013	5	0.505±0.003
7	5	3.087±0.886	5	1.830±0.485	5	5.752±0.012	5	0.630±0.003
8	5	1.606±0.473	5	2.290±0.445	5	4.650±0.091	5	0.696±0.003
9	5	2.863±0.578	5	2.426±0.375	5	2.851±0.096	5	0.560±0.018
10	5	1.030±0.264	5	2.312±0.209	5	2.395±0.011	5	0.609±0.003
11*	5	1.315±0.292	5	0.594±0.037	5	1.694±0.029	5	0.306±0.006
ANOVA								
F ratio		1.466		0.318		0.023		0.081

Table 5. Average Zn, Cu, Mn, Fe, Pb, Ni, Cr and Cd concentrations in *E. prunastri* (µgg^{-1}) in the Karabük city, with Standart deviations (SD)

Comparisons of the Pb concentrations of *E. prunastri* specimens from polluted sites with the control yielded very significant variations. Sites 1, 7 and especially sites 9 with the highest human activities, together with high vehicular density congestion, showed the highest Pb with the values of, 5.17 μg/g which were significantly higher than the control site (1.31 μg/g). It could be concluded that Pb concentration was highest in site 1 as it is the central part of the city where human activities and density of traffic are very intense. Similar kinds of observations were made by Cansaran-Duman *et al.* (2009) while studying *Pseudevernia furfuracea* thalli as an indicator of air pollution in the same province (Karabük) in Turkey. All sites, especially site 8 (0.69 μg/g), showed significantly higher Cd concentrations than the sample from control site (0.30 μg/g) (Cansaran-Duman et al., 2011).

In the same study the samples were also evaluated to detect DNA damage in thallus, caused by environmental pollutants. The region surveyed in the study suffers from substantial historical and current air contamination principally due to the presence of the steel and iron industry, which have been active since 1925 until now. The use of biological responses to contaminant exposure by lichen species has become a useful tool in environmental quality evaluation and risk assesment (Cansaran-Duman et al., 2011).

5.4.3 Chemical content in *Hypogymnia physodes* and *Usnea hirta*

Lichen *Hypogymnia physodes* and *Usnea hirta* samples were collected in 2005 from 10 stations around Iron-steel Factory in Karabük, Turkey (Fig. 3). *H. physodes* and *U. hirta* samples from Yenice Forest were used as a control. The aim was to evaluate the bioaccumulation ability and to determine the environmental impact of an Iron-steel Factory in Karabük (Cansaran-Duman, 2011). The analytical results were compared statistically by using Statistical Package for the Social Sciences (SPSS). As expected, the study area (Yenice Forest, Karabük) was chosen as control site (site no 11) (Table 2) showed significantly lower impact in comparison to other site (site no 1-10). Compared with the two lichen species, *H. physodes* was the species with the highest accumulation capacity while *U. hirta* had the lowest one. These criteria attested the best suitability for *H. physodes,* followed by *U. hirta* (Cansaran-Duman, 2011).

Around the Karabük Iron-steel Factory, the highest levels of Zinc (Zn) in the *H. physodes* were found in site 4 (33.1 $\mu g g^{-1}$), site 8 (30.2 $\mu g g^{-1}$) and site 5 (30.1 $\mu g g^{-1}$), respectively. Sites 6, 7, 10 and 1 were determined close to each other value in the *H. physodes* species (Fig4). Also, the highest levels of Zinc (Zn) in the *U. hirta* were found in sites (1, 5 and 7) (21.1, 21.4 and 21.2 $\mu g g^{-1}$, respectively). In addition to these sites, Zn concentration in sites 3(19.0 $\mu g g^{-1}$) and 9 (20.6 $\mu g g^{-1}$) was high value in *U. hirta.* Zn concentration in the lichen samples was linearly related to the vehicle traffic, railway and activity of industrial units. The highest levels of Manganase (Mn) in the *H. physodes* were found in sites 2 (195.8 $\mu g g^{-1}$), and 4 (202.7 $\mu g g^{-1}$), respectively. The highest levels of Mn in the *U. hirta* were found sites 3 (195.9 $\mu g g^{-1}$) and 4 (150.3 $\mu g g^{-1}$) with a control value of 19.3 $\mu g g^{-1}$. We considered that both of samples were higher for Mn concentration in site no 4. The reason for this, motor vehicles are known to be a source of Mn in urban areas (Monaci *et al.*, 2000) and could explain the reason of elevated Mn concentrations in site 4 (Cansaran-Duman, 2011).

Comparisons of the Pb concentrations of the *H. physodes* and *U. hirta* species from polluted sites with the control yielded very significant variations, especially *U. hirta* species. Sites 3, 5, 8 and especially sites 1, 10 (Fig 3) with the highest human activities, together with high vehicular density congestion, showed the highest Pb with the values of, 8.78 $\mu g g^{-1}$ in *U. hirta* species which were significantly higher than the control site, 1.32 $\mu g g^{-1}$ (Fig. 4). It could be concluded that Pb concentration was highest in sites 1 and 10 because they are the central

part of the city where human activities and density of traffic are very intense (Cansaran-Duman, 2011).
Although the highest levels of chromium (Cr) in the *H. physodes* was found in site 7 (3.86 $\mu g g^{-1}$), and 8 (4.56 $\mu g g^{-1}$), respectively, these sites slightly higher than the controls. The chromium (Cr) concent in sites 5 (6.75 $\mu g g^{-1}$), 6 (4.18 $\mu g g^{-1}$), and 7 (3.15 $\mu g g^{-1}$) were significantly higher than control site (1.96 $\mu g g^{-1}$) in *U. hirta*. The most important sources of Cr pollution are indicated as industrial activities like refining works and Iron-steel Factories (Cansaran-Duman, 2011).
Copper (Cu) contents in the *H. physodes* samples ranged from 2.44 to 3.94 $\mu g g^{-1}$. Cu content in site 7 (3.11 $\mu g g^{-1}$) in *U. hirta* was significantly higher than control site (1.59 $\mu g g^{-1}$). These two species showed high Cu concentration. Nickel (Ni) concentrations in site 3 was found as 10.81 $\mu g g^{-1}$ *H. physodes* and 8.66 $\mu g g^{-1}$ in site 9 of *U. hirta* (Cansaran-Duman, 2011).
The mean Cd concentration in sites 1 (0.85 $\mu g g^{-1}$), 8 (0.84 $\mu g g^{-1}$) and 10 (0.87 $\mu g g^{-1}$) are slightly higher than the control site (0.73 $\mu g g^{-1}$) in *H. physodes*. All sites, especially site no 4 (0.61 $\mu g g^{-1}$), showed significantly higher Cd concentrations than control site (0.17 $\mu g g^{-1}$) in *U. hirta*. The concentrations of Cd in *U. hirta* were significantly higher in all sites than from the control site, probably indicating the contaminants from motor vehicles, dust raised by metal business and other human activities. The most important sources of Cd pollution were regarded as fossil fuels used by the vehicles, metal business, plastics, house tools construction and sewer (Markert, 1992). Markert (1992) was recorded the Cd levels in between 0.01 and 0.3 $\mu g g^{-1}$ for unpolluted natural environments and also reported that all of the study sites have been polluted except rural sites.

5.4.4 Comparison *H. physodes, U. hirta, P. furfuraceae* heavy metal accumulation

The presence of heavy metals in *P. furfuracea* was already reported in Cansaran-Duman *et al.* 2009 and to allow the comparison with *U. hirta* and *H. physodes* species in the present review the results are expressed. It was compared *H. physodes, U. hirta* with *P. furfuracea,* there was no different in Zn concentration. Especially, sites 7 and 10 in *P. furfuracea* were significantly higher than *U. hirta* and *H. physodes* (Fig. 4).
In the site 10, Pb concentration of *P. furfuracea* was highest than *U. hirta* and *H. physodes*. Thus, *H. physodes* and *U. hirta* the highest concentration of Pb which can be related to a selective cation uptake as was informed previously by Cansaran-Duman et al. 2009. The authors attributed this finding to a greater affinity between Pb cations and the lichen cell wall exchange sites that are probably strongly attached to binding sites (Fig 4).
H. physodes in the levels of Fe metal accumulated in site 8 were similar to those obtained by Cansaran-Duman et al. 2009 in *P. furfuracea* species from same site of Iron-steel Factory in Karabük. Highest Fe concentration was found in *P. furfuracea* to site no 1, while Fe concentration was lower in *U. hirta* and *H. physodes* (Fig 4).
The order of magnitude of accumulation on Mn at 2, 4 and 10 sites were *H. physodes* > *U. hirta* > *P. furfuracea*. Although *U. hirta* showed the highest levels of Mn in site 3, *H. physodes* was higher accumulated than *P. furfuracea* and *U. hirta* sites 5 and 9 (Fig 4). Mn could be tracer of both eolic dust particles as well as vehicular traffic, since this element has recently been used as a substitute for Pb in additives (Ardeleanu *et al.*, 1999).
Our findings are consistent with data reported by Cansaran-Duman et al. (2009), who have monitored *P. furfuracea* species in the same district. In concordance with findings made by Cansaran-Duman et al. (2009) on *P. furfuracea,* this study demonstrated the importance of

heavy metals accumulation on the choose of lichen species. Heavy metals were the highest concentration around the Iron-steel Factory in Karabük, which is in accordance with previous data (Cansaran-Duman et al., 2009).

Generally, previous studies that used *P. furfuracea* as a passive biomonitor in a Iron-steel Factory in the province of Karabük showed concentrations higher than *U. hirta* species found in the present study. *P. furfuracea* and *H. physodes* were close quarters to heavy metal accumulation in Iron-steel Factory. According to our observations in result of this study, *H. physodes* higher than the values to *P. furfuracea* (Cansaran-Duman et al., 2009).

The presence of heavy metals in *P. furfuracea* was already reported in Cansaran-Duman et al. 2009. Moreover, results of this study are comparison with *U. hirta* and *H. physodes* species. It was observed that the accumulation of metals in *P. furfuracea* was similar to the one observed in *H. physodes*, with significantly higher values of all elements in samples exposed in the Karabük. Moreover, in previous study *P. furfuracea* and in the current study *H. physodes* were proved to be almost similar bioaccumulator than *U. hirta* species.

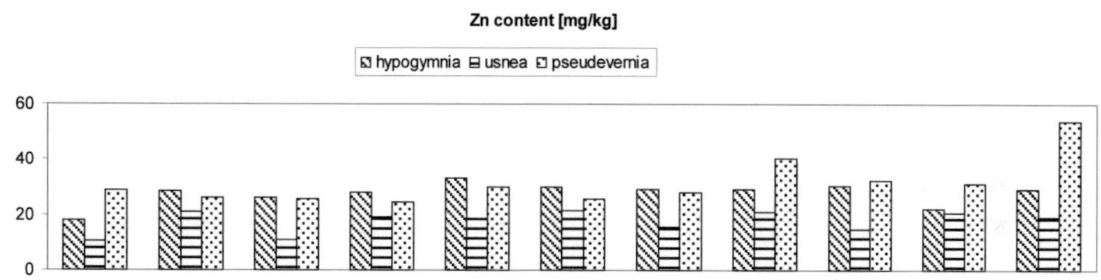

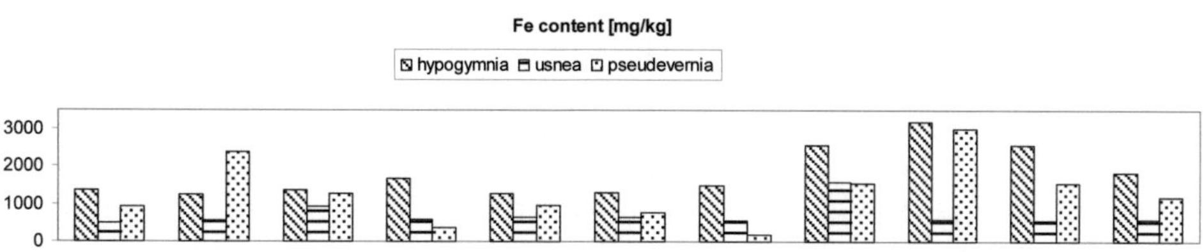

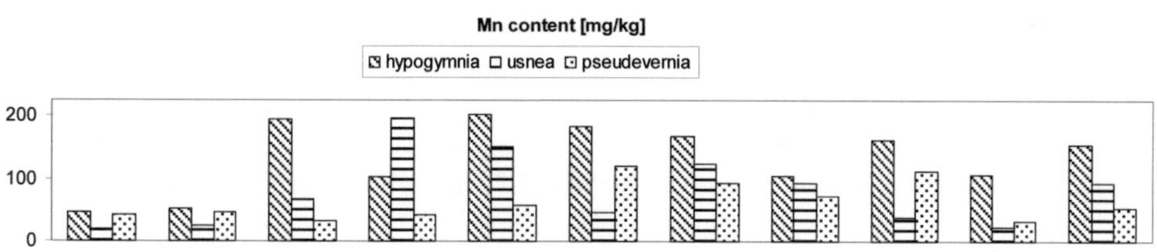

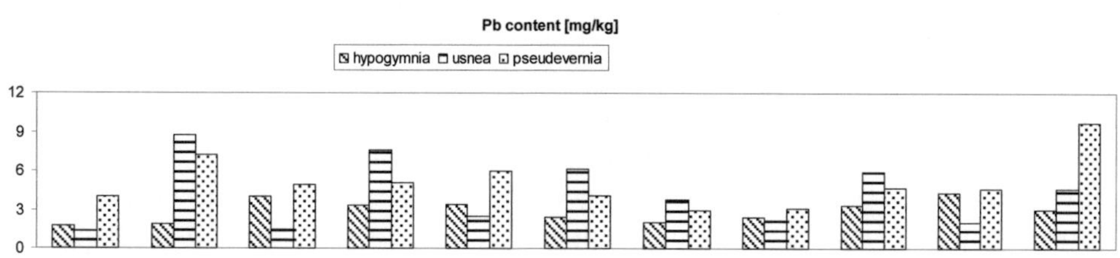

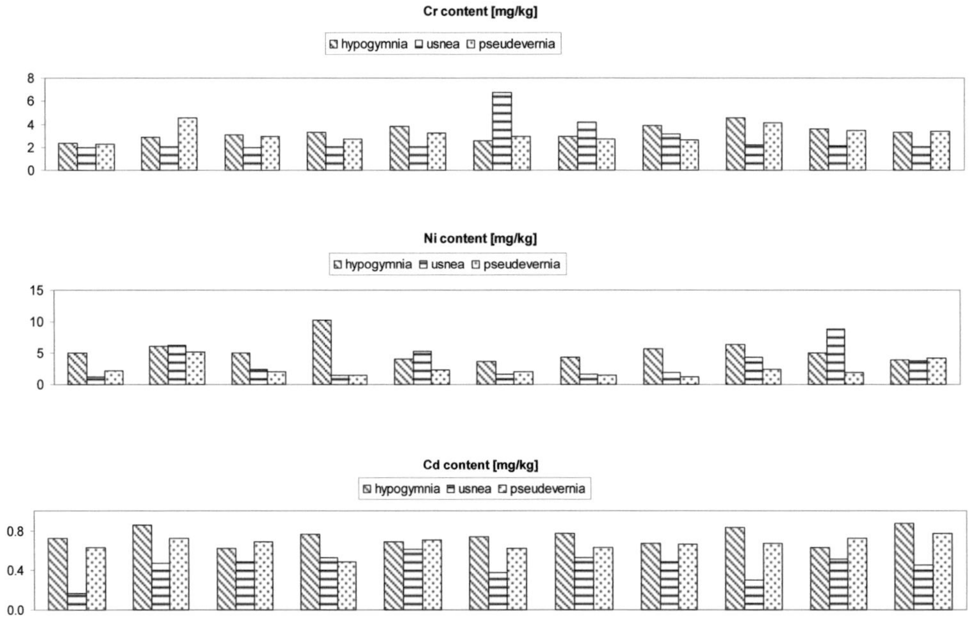

Fig. 4. Comparison *H. physodes, U. hirta, P. furfuraceae* heavy metal accumulation

5.4.5 RAPD profiles of *P. furfuraceae*

In this study the genotoxic effects of various environmental pollutants were tested with the samples collected from their natural habitats from the Karabük region. 10 of the primers yielded clear and reproducible bands in RAPD analysis. Among the primers used, TubeA03 (AGTCAGCCAC) showed the highest polymorphism while TubeA02 (TGCCGAGCTG) and TubeA04 (AATCGGGCTG) showed monomorphic band patterns.

Additionally, genomic template stability ratios (GTS) were calculated. GTS related to the level of DNA damage, the efficiency of DNA repair and replication, Atienzar et al., (1999) could explain the appearance and disappearance of bands. The lowest values were obtained in sample numbers 10, 9 and 8. Generally in samples 8, 9 and 10, the lowest GTS values were obtained, which might imply the sensitivity of lichens to genotoxic stressors near the Iron-steel Factory. Previous studies have also indicated that mutations, chromosomal rearrangements and other DNA lesions could be the reason for the variation in RAPD band patterns. It was demonstrated that changes in RAPD profiles induced by pollution could be regarded as modifications in genomic template stability. On the other hand mutations could be displayed as an appearance of new bands in RAPD assay if the same locus hosts mutations in a sufficient number of cells which might require at least 10% of the mutations (Atienzar et al., 2000).

In a another study, a controlled experiment with one type of heavy metal treatment was also conducted in our laboratory in order to show the effects of a genotoxic agent under *in vitro* environment (Aras et al., 2010). In the study a clean sample of *P. furfuracea* collected from Yenice Forest was exposed to various doses of Pb in different time intervals. Although the polymorphism percentages were not calculated in the study obvious band changes were visualized especially after 24 and 48h. of Pb treatments. Results of the study displayed that even only one kind of stressor (Pb) might induce DNA changes in *P. furfuracea* samples under *in vitro* conditions (Aras et al., 2010). Thus, it might become easier to explain the level of polymorphisms recorded, in the studies in samples exposed to polluted environment.

5.4.6 RAPD profiles of *E. prunastri*

Genotoxic contamination was monitored by RAPD analyses with *Evernia prunastri* lichen samples exposed to differently polluted sites in Karabük (Table 3). A clear genotoxic influence is demonstrated in *E. prunastri* exposed naturally to railways, motorways and the Iron-steel Factory in Karabük. DNA damage assessed by RAPD analyses in the *E. prunastri* province of the Iron-steel Factory in Karabük, Turkey. Results of this study was exhibited similar results *P. furfuraceae* lichen species in Karabük (Cansaran-Duman et al., 2011).

5.4.7 Comparison of *P. furfuraceae* and *E. prunastri* RAPD fingerprint results

The high values of DNA damage were obtained by both lichen species (*E. prunastri* and *P. furfuracea)* exposed naturally to various polluted sites around the Iron-steel Factory in Karabük. Cansaran-Duman et al. (2011) have recorded high deterioration of DNA integrity in *E. prunastri* exposed at the same site in 2005, while data presented here showed moderate DNA damage in *P. furfuracea* exposed at the same location. But direct comparison of results may be difficult due to differences in the biology of the employed species, different season and duration of exposure and because of constant alternations in quality and quantity of air pollution. However, the level of DNA damage measured in *P. furfuracea* was lower in *E. prunastri* in the same region. However, these measurements cannot provide detailed data on a variety of specific chemicals or their interactions responsible for genotoxic impact during the investigated period, and are therefore used here only as an indication of polluted status (Stambuk et al., 2009). This may be due to adaptive mechanisms developed in lichen species continuously inhabiting polluted environments. Nevertheless, further studies which focuse especially on the influence of other ecological factors on DNA damage are needed.

6. Conclusion

The results of all these studies pointed the advantages of biomonitoring with lichens over instrumental monitoring. Lichens accumulate most of the elements of the periodic table, are usable at low expense, do not depend on electricity for their operation, do not need treatment and are easy to hide, thus discouraging vandalism. In contrast to physical and chemical methods, biological methods allow the direct assessment of the genotoxic potential of air stressors. Thus biological data can be used to estimate environmental and the potential impact on other organisms, including humans.

Ultimately, usage of lichens may allow the ecotoxicological examination of the link between molecular alternations and measurable adverse effects at higher levels of biological organizations. The techniques also might provide an early warning system with a higher sensitivity than the conventional techniques.

7. Acknowledgements

Our studies were supported by TUBİTAK (The Scientific and Technological Research Council of Turkey, Projects with no. 109T046) and Ankara University, Management of Scientific Research Projects with No. 2003-0705080.

8. References

Atienzar F.A., Conradi M., Evenden A.J., Jha A.N., Depledge M.H. Qualitative assesment of genotoxicity using random amplified polymorphic DNA: comparison of genomic template stability with key fitness parameters in Daphnia magna exposed to benzo[a]pyrene, *Environ. Toxicol. Chem.* 18 (1999) 2275-2282.

Atienzar F.A, Cordi B., Donkin M.E. Comparison of ultraviolet-induced genotoxicity detected by random amplified polymorphic DNA with chlorophyll fluorescence and growth in a marine macroalgae, Palnaria palnata. *Aquat Toxicol.* 50 (2000) 1-12.

Aras S., Kanlıtepe Ç., Cansaran-Duman D., Halıcı M.G., Beyaztaş T. Assessment of air pollution genotoxicity by molecular markers in the exposed samples of Pseudevernia furfuracea (L.) Zopf in the Province of Kayseri (Central Anatolia*). J. Environ. Monit.* 12 (2010) 536-543.

Ardeleanu A., Loranger S., Kennedy G., Gareau L., Zayed J. Emission Rates and Physicochemical Characteristics of Mn Particules Emitted by Vehicles Using Metylcyclopentadienyl Manganese Tricarbonyl (MMT) As an Octan Improver. *Water, Air and Soil Pollut.* 115 (1999) 411-427.

Baird W.M., Hooven L.A., Mahadevan B. Carcinogenic polycyclic aromatic hydrocarbon-DNA adducts and mechanism of action. Environ. Mol. Mutagen. 45 (2005) 106–114.

Bargagli R. Trace elements in terrestrial plants. In: *An Ecophysiological Approach to Biomonitoring and Biorecovery.* Berlin, Springer, (1998), pp. 179.

Bargagli R. Determination of metal deposition patterns by epiphytic lichens. *Toxicol. Environ. Chem.* 18 (1989) 249-256.

Bermudez G.M.A., Rodriguez J.H., Pignata M.L. Comparison of the air pollution biomonitoring ability of three Tillandsia species and the lichen Ramalina celastri in Argentina. *Environ. Res.* 109 (2009) 6-14.

Cansaran-Duman D., Atakol O., Atasoy I., Kahya D., Aras S., Beyaztaş T. Heavy metal accumulation in Pseudevernia furfuracea (L.) Zopf from the Karabük Iron-steel Factory in Karabük, Turkey. *Z Naturforsch* C. 9/10-64c (2009) 717-723.

Cansaran-Duman D., Beyaztaş T., Atakol O., Aras S. Assesment of the air pollution genotoxicity by RAPD in Evernia prunastri L. Ach. province of Iron-steel Factory in Karabük, Turkey. *J Environ. Sci. China.* 23/5 (2011). DOI: 10.1016/S1001-0742(10)60505-0.

Cansaran-Duman D. Study on accumulation ability of two lichen species (Hypogymnia physodes (L.) Nyl and Usnea hirta (L.) Weber ex F.H. Wigg) at Iron-steel Factory site, Turkey. *J Environ Biol* 32 (2011) in press.

Cansaran-Duman D., Aras S., Atasoy İ. Accumulation of trace elements in Pseudevernia furfuracea (L.) Zopf in Ankara and assessing the genotoxicity. *Zeitschrift fur Naturforschung* Section C. 2011. under reviwer.

Castello J.D., Leopold D.J., Smallidge P.J. Pathogens, patterns, and processes in forest ecosystems. *Biosci* 45 (1990) 1-2.

Citterio S., Aina R., Labra M., Ghiani A., Fumagalli P., Sgorbati S., Santagostino A. Soil genotoxicity: a new strategy based on biomolecular tools and plants bioindicators. *Environ. Sci. Tech.* 36 (2002) 2748-2753.

Cohen A.J, Pope I. Lung cancer and air pollution. *Environ. Health Perspect.* 103-8 (1995) pp. 219–224.

Conte C., Mutti I., Puglisi P., Ferrarini A., Regina G.R.G., Maestri E., Marmiroli N. DNA fingerprint analysis by PCR based method for monitoring the genotoxic effects of heavy metals pollution. *Chemosph.* 37 (1998) 2739-2749.

Conti M.E., Cecchetti G. Biological monitoring: lichens as bioindicators of air pollution assessment: a review. *Environ. Pollut.* 114 (2001) 471-492.

Dipple A. Polycyclic aromatic hydrocarbon carcinogens. In: R.G. Harvey, Editor, Polycyclic *Aromatic Hydrocarbons and Carcinogenesis*, American Chemical Society Press, Washington, DC, pp. 1-17, (1985).

Forman-Richard T.T., Alexander L.E. Roads and Their Major Ecological Effects. *Annual Review of Ecology and Systematics*, 29 (1998) 207-231.

Galun M., Siegal S.M., Cannon M.L., Siegel B.Z., Galun E. Ultrastructural localization of uranium biosorption in Penicillium digitatum by stem X-ray microanalysis. *Environ. Pollut.* 43 (1987) 209-218.

Garty J., Fuchs C., Zisapel N., Galun M. Heavy metals in the lichen Caloplaca aurantia from urban, suburban and rural regions in Israel (a comparative study). *Water Air and Soil Pollut.* 8 (1977) 171-188.

Garty J. Lithobionts in the Eastern Mediterranean. In: Seckbach, J. (ed.) (invited chapter). *Enigmatic Microorganisms and Life in Extreme Environments*. Dordrecht, The Netherlands, Kluwer Academic Publishers, pp. 255-276, (1999).

Garty J., Levin T., Cohen Y., Lehr H. Biomonitoring air pollution with desert lichen Ramalina maciformis. *Physiol Plant.* 115 (2002) 267-275.

Grant W.F. Higher plant assay for the detection of genotoxicity in air polluted environments. *Ecosyst Health.* 4 (1998) 210-229.

Jerrett M., Arain A., Kanaroglou P., Beckerman B., Potoglou D., Sahsuvaroglu T. A review and evaluation of intraurban air pollution exposure models. *J Expo Anal Environ Epidemiol* 15(2) (2005) 185–204.

Marques A.P.V. Positional responses in lichen transplant biomonitoring of trace element air pollution (2008)-IOS Press ISBN: 978-1-58603-928-8.

Markert B., Presence and significance of naturally occuring chemical elements of the periodic system in the plant organism and consequences for future investigations on inorganic environmental chemistry in ecosystem. *Veget.* 103 (1992) 1-30.

Nash T. Lichen Biology.In: Lichens as indicators of air pollution (C. Gries), pp 240-253. Cambridge Univ Pres, United Kingdom, 1996.

Nimis P.L., Bargagli R. Linee guida per l'utilizzo di licheni epifitic some bioaccumulatori di metalli in traccia. In: Piccini C, Salvati S. eds.) *ANPA - Atti del Workshop Biomonitoraggio della qualità dell'aria sul territorio nazionale.* pp. 279—289, 1999, Rome, Italy.

Olmez I., Gulovali M.C., Gordon G.E. Trace metal element concentration in Lichens near a coal fired power plant. *Atmos. Environ.* 19 (1985) 1663-1669.

Piraino F., Aina R., Palin L., Prato N., Sgorbati S., Santagostino A., Citterio S. Air quality biomonitoring: Assessment of air pollution genotoxicity in the Province of Novara (North Italy) by using Trifolium repens L. and molecular markers. *Sci. Total Environ.* 372 (2006) 350-359.

Sawicki, E. Air Pollution and Cancer in Man.. Lyon, France: International Agency for Research on Cancer. IARC No. 16, 127-57, (1977).

Singer, B., and Grunberger, D. Molecular Biology of Mutagens in pres and Carcinogens, Plenum Publishing Corp., New York, (1983).

Stambuk A., Pavlica M., Vignjevic´ G., Bolaric´ B., Klobucˇar G.I.V. Assessment of genotoxicity in polluted freshwaters using caged painter's mussel, *Unio pictorum*. *Ecotoxicol.* 18 (2009) 430- 439.

Villarini M., Fatigoni C., Dominici L., Maestri S., Ederli L., Pasqualini S., Monarca S., M. Moretti. Assessing the genotoxicity of urban air pollutants using two in situ plant bioassays. *Environ Pollut.* 157 (2009) 3354-3356.

Wolterbeek B. Biomonitoring of trace element air pollution: principles, possibilities and perspectives. *Environ Pollut* 120 (2002) 11-21.

10

Monitoring Epiphytic Lichen Biodiversity to Detect Environmental Quality and Air Pollution: the Case Study of Roccamonfina Park (Campania Region - Italy)

Aprile G. G., Catalano I., Migliozzi A. and Mingo A.
Department of Arboricoltura, Botanica e Patologia Vegetale,
University of Napoli "Federico II", Portici (NA)
Italy

1. Introduction

Biomonitoring of air pollution, i.e. monitoring environmental pollution through the use of living organisms (Nimis & Skert, 1999), may be based either on the tendency of some organism to accumulate pollutants in their tissues (bioaccumulation) or on the changes that occur in the composition of animal and plant communities after exposure to pollutants (bioindication). Compared to instrumental monitoring, the use of biomonitors allows to measure as a whole the global effect that abiotic and biotic factors exert on biota, what is not possible by just analyzing the concentrations of single selected pollutants in the environment.
Biomonitoring provides useful information about the global conditions affecting the environment over a given area. Of course it should not be considered as a substitute of instrumental monitoring, but rather, a necessary complement of it. It may also be suited to screen areas subjected to any risk of contamination, so helping to plan landscape policies and to set land nets of air quality (ANPA, 2001). Biomonitoring of air pollutants can be passive or active. Passive methods observe organisms growing naturally within the area of interest. Active methods detect the presence of air pollutants by placing test organisms of known response and genotype into the study area (Szczepaniak & Biziuk, 2003).
Biomonitoring may be obtained by using organisms either as bioindicators or as bioaccumulators. Bioindicators are defined as organisms that allow to identify human-generated environmental pollutants and to determine their level on a scale of qualitative determination (Conti & Cecchetti, 2001). A good bioindicator should present:

- high sensitivity to environmental pollutants;
- low mobility in space;
- long living cycle;
- wide distribution over the studied area;
- high genetic evenness.

Bioaccumulators are defined as organisms that reflect the chemical content of atmosphere and can so be used for the quantitative determination of contaminants (Conti & Cecchetti, 2001). A good bioaccumulator should present:

- high tolerance to environmental pollutants;

- wide distribution over the studied area;
- low mobility in space;
- long living cycle;
- high capability to accumulate pollutants.

Due to their morphological and physiological characters, lichens are doubtless among the most suited organisms available for the studies on atmospheric quality, either as bioindicators or as bioaccumulators. Lichens in effect are long-lived, slow-growing organisms, that show a good constancy of morphology over time and do not shed parts during growth. In addition, since these no-rooted organisms are not provided with protective structures in their tissues, such as cuticle or stomata, they absorb passively during their entire life cycle any element present in the atmosphere, either by rain or by particulate deposition (Costa et al., 2002). Moreover, the lichen surface, structure, and roughness facilitate the interception and retention of particles (Sloof & Wolterbeek, 1993). Thus, metal absorption by lichens depends not only on intercellular absorption (exchange process and/or intercellular accumulation) but also on entrapment of particles that contain metals.
Lichens are particularly sensitive to environmental stresses, especially with regard to pollution, eutrophication and climate change (ANPA, 2001), since the metabolism of these organisms is directly dependent on gas exchange. These organisms respond rapidly to atmospheric changes, particularly if determined by anthropogenic factors. Thus, biodiversity of epiphytic lichens may be kept as a good indicator of air pollution (Nimis et al., 1989; Piervittori, 1999; Giordani et al., 2002; Loppi et al., 2002).

1.1 The Lichens

1.1.1 What are lichens?

More than two thousands years ago, Theophrastus, the father of botany, coined the term "lichen" to denote the product of the actions of some unknown organism on tree barks (Ozenda & Clauzade, 1970). Up to nineteenth century, lichens were considered as individually recognizable organisms (Ozenda & Clauzade, 1970). Lichens were recognized to be composed by two different organisms in 1869 (Schwender, 1869), but it was not clear what was the kind of association between the two biological partners. Today, the common view is that they represent one of the most interesting case of symbiosis, since they are constituted by the association of a fungus, called *mycobiont* (Ascomycetes, rarely Basidiomycetes), and an unicellular alga (Chlorophyceae or Cyanobacteriae), called *photobiont*. Yet the question of whether it should be considered as a symbiosis or an extremely evolved parasitism is controversial (Ahmadjian, 1993).
The coexistence between the two partners attains a very high degree of morphological constancy and physiological equilibrium, so that the product of this association, the so called "lichenic *thallus*" (Fig. 1), may assume the rank of an unitary organism. This is also due to its capability to produce metabolites that neither fungus nor alga would be able to produce when living alone. This kind of association is commonly treated as a symbiosis. The benefit to the fungus is in that it obtains organic matter produced by algae through photosynthesis; whereas the algal partner, thanks to the protection offered by fungal mycelium, acquires the ability to survive in dry substrates where normally, it could not live alone (Nimis & Skert, 1999). However, the Lichens, do not reproduce sexually as a whole organism: the fungus maintains its gamic reproduction, whereas the alga just propagates by scission and loses its ability to produce zoospores (Hawksworth, 1988; Ahmadjian 1993). Thus, some author consider quite problematic to define these associations as a valid

taxonomic category, but rather tend to consider them as a kind of "lichenized fungi" (Tehler, 1996). In fact, the typical asexual strategy of lichens is that of *fragmentation*: in the most simple cases, a single fragment may grow into a new lichen; as well as lichens may produce specialized microscopic particles composed of algal cells enveloped by fungal *ifae* (*soredia*) that produce a new individual. In other cases, small peaces of thallus including the photosynthetic partner (*isidia*) may reproduce the whole organism. Lichen fragments, soredia and isidia can be transported at great distances by wind and water.

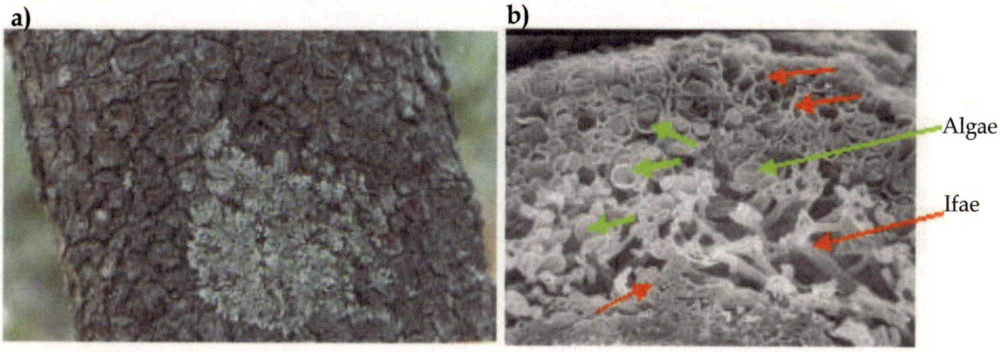

Fig. 1. Lichen thallus and its section.
a) Thallus *in situ* of *Physcia biziana* (A.Massal.) Zahlbr. var. *leptophylla* Vezda. b) Cross section SEM (1700X) of the thallus

1.1.2 Lichen character

Anatomically, fungal *ifae* constitute the most conspicuous part of a thallus. In the most simple early-evolved ones, the so called *homeomerous* (Fig. 2a) lichens, fungal ifae and algal cells are just assembled in a homogeneous and undifferentiated interlacement. A more complex structure is found in *heteromerous* thalli (Fig. 2b), the most widespread lichens, where different layers are recognizable: an *upper cortex*, a *medulla* layer and a *lower cortex*, constituted by fungal pseudo-tissues; and a photobiont layer housed between the upper cortex and the medulla.

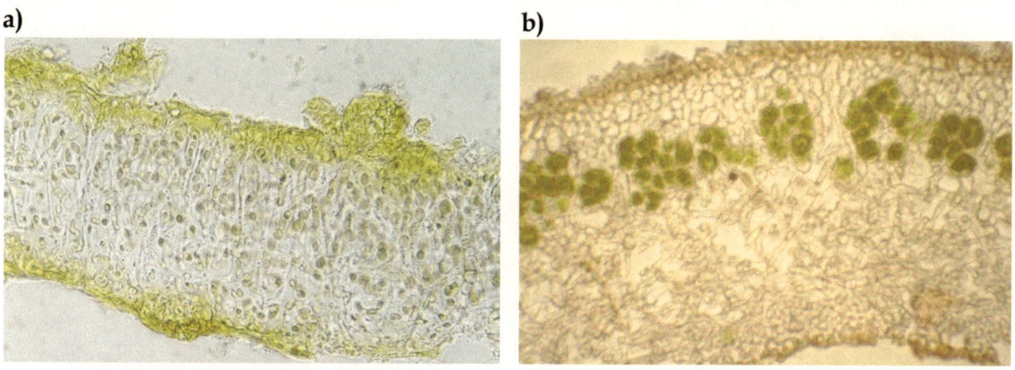

Fig. 2. Thallus structure.
TEM sections of lichen thalli. a) homeomerous; b) heteromerous

Most of the lichens belong to one of the three main morphological categories: the *crustose* type (Fig. 3a), strictly adhering to their substratum and not provided of a lower cortex; *foliose* type (Fig. 3b), with a kind of leaf appearance, provided with both upper and lower cortex, more or less attached to its substrate; and *fruticose* type (Fig. 3c), generally branched with variable shapes (sections from circular to flat) and structure (pendulous strands or hollow upright stalks).

a) b) c)

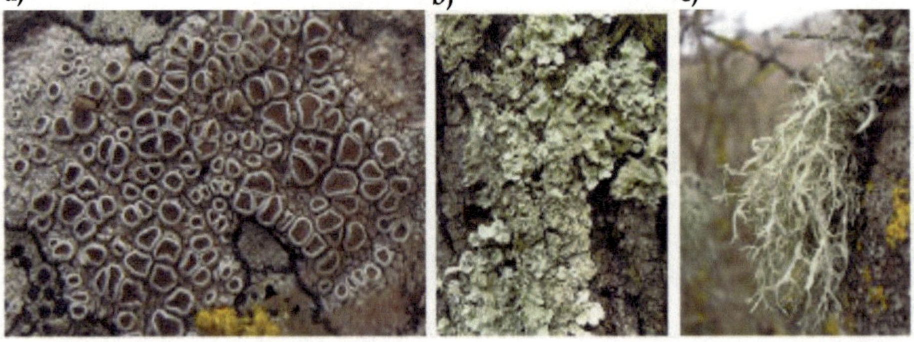

Fig. 3. Main lichen morphological types.
Lichen growth types. a) crustose (*Lecanora chlarotera* Nyl.); b) foliose (*Flavoparmelia caperata* (L.) Hale); c) fruticose (*Ramalina farinacea* (L.) Ach.)

Depending on their particular morpho-physiological attributes, lichens may colonize the most variable substrates, such as soil (*terricolous* types Fig. 4a), rocks (*saxicolous* types Fig. 4b), tree barks (*epiphytes* types Fig. 4c). The vast majority of lichens are adapted to tolerate a wide range of changing environments; but there are also some species that are strictly confined to particular habitats. Homeomerous lichens are generally less tolerant to changing environments, and are the first to disappear if even few variable are subject to change. Heteromerous on the contrary, and particularly crustose lichens, tend to show a wider adaptation and are found in a greater range of different environments, going from Antarctic to equatorial deserts.

The ability of lichens to tolerate such hard environmental conditions is probably due to their capacity to switch quickly from an active to a latent living state, through the rapid dehydration of the thallus. For this reason, lichens are considered to be *poikilohydric* organisms, that is they may survive to extreme low levels of water content (Nash, 1996).

a) b) c)

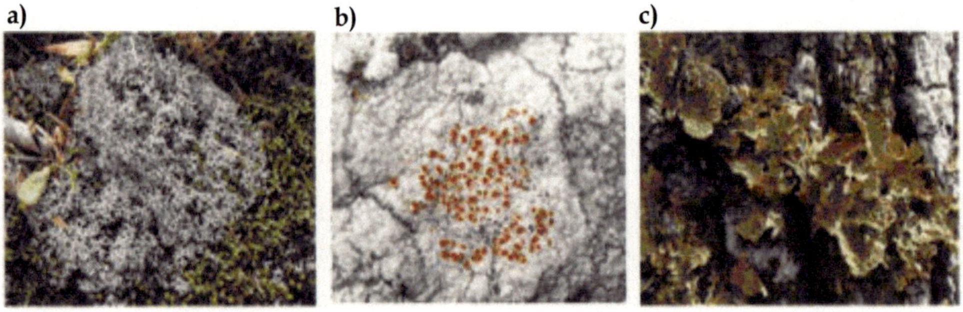

Fig. 4. Main lichen substrates.
Lichens growing on their natural substrates. a) soil (*Cladonia rangiformis* Hoffm., *terricolous* types); b) rocks (*Caloplaca erythrocarpa* (Pers.) Zwackh, *saxicolous* types); c) tree barks (*Lobaria pulmonaria* (L.) Hoffm., *epiphytes* types)

1.2 Techniques of biomonitoring

Several techniques were proposed to detect air quality by sampling the communities of epiphytic lichens (Barkman, 1963; De Sloover & Leblanc, 1968; Hawksworth & Rose, 1970, Ferry et al., 1973, Nash & Wirth, 1988; Richardson, 1992; Cislaghi & Nimis, 1997; Purvis, 2000, Van Dobben et al., 2001). A project aimed to develop an objective and reproducible model of bioindication, suited to different air pollutants, was launched during the 80s in Switzerland (Liebendöerfer et al., 1988; Herzig et al., 1987; Herzig & Uregh, 1991), giving rise to the Index of Air Purity (IAP), adopted with just small differences by several countries. In Italy, an index of lichen biodiversity (IBL) was proposed by Nimis (ANPA, 2001) as a way to provide an indirect evaluation of air quality. According to this method, the frequency of occurrence of epiphytic lichen species within a sampling grid provides information on the long-term effects of air pollutants, eutrophication and anthropogenic factors on sensitive organism (Asta et al., 2002).

1.3 Aims of the study

In this study, lichen distribution was examined in a district of Campania region to monitor the evolution of air quality at landscape scale. A new methodological approach was tried by overlapping the results of biomonitoring samplings to land cover maps, in order to highlight the relations between air pollution and land use patterns (Pinho et al., 2004; Paoli et al., 2006; Pinho et al., 2008).

The main objectives of this work were the following: i) to evaluate air quality on the studied area with the aid of IBL index, and reporting geo-referenced data on a thematic map; ii) to relate lichen distribution and biodiversity to land use spatial patterns; iii) to put the basis for a comparative analysis focussed on changes induced by the present socio-economic evolution of the plain, from agriculture to industrial and tertiary; iv) to provide a reproducible protocol for monitoring air quality, identifying clusters of lichen species linked to particular land use models and formulating previsions about environmental quality on areas characterized by similar dynamics.

2. Materials and methods

2.1 Study area

This study was conducted in the large area (84 km^2) of Roccamonfina volcano (Campania Region, South Italy, Long. 13°58'18''; Lat. 41°17'43'') (Fig.5). This area was interested by active volcanism from 630,000 to 50,000 years ago. This complex is a kind of *stratovolcano* that was subjected to the collapse of the crater area, generating a caldera. Its volcanic apparatus is somehow similar to that of Vesuvius, particularly as for constitution, insulation and morphology (De Rita & Giordano, 1996). Moreover, the contiguous area of the regional park are, at present, experiencing a rapid process of industrial reconversion.

Annual rains over the area range from 916 to 1046 mm (Fig.6), with a predominant autumn-winter distribution and a dry summer period ranging from 1 to 3 months. Mean temperatures get their maximum value during the months of July-August and their minimum in January. According to the phytoclimatic classification of Italy (Nimis & Martellos, 2008), the area belongs to the humid sub-Mediterranean belt.

The survey area can be divided in two main districts: (a) Roccamonfina's regional park characterized by woodland of *Castanea sativa* Mill., with small scattered villages; (b) the southern part of the study area, with small urban agglomerations, industrial and commercial zones localized on the lower part of the volcanic hill and on the plain.

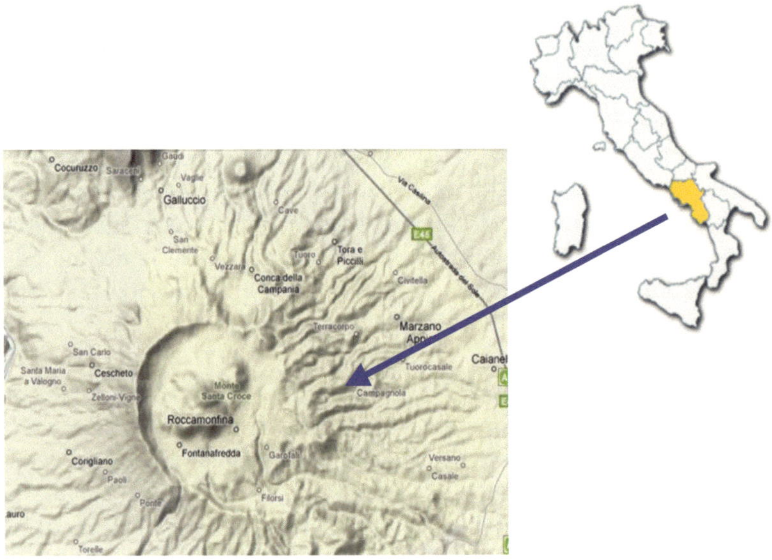

Fig. 5. Location of the study area.
The study was conducted in a volcanic area of Campania region (Italy)

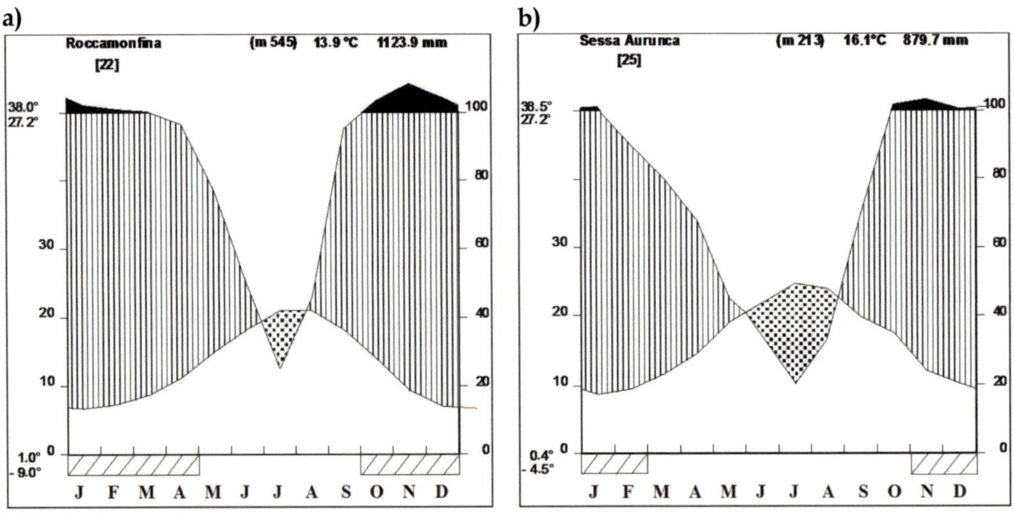

Fig. 6. Thermopluviometric diagrams.
Two stations differing for the relative importance of the dry summer period are represented in the figure. a) Roccamonfina; b) Sessa Aurunca. Data are averages of about 50 years (1951-99)

This area was selected since it includes both regions characterized by a high index of naturalness and districts more disturbed by anthropic influence, with a gradient of environmental pollution decreasing from urban settlements to the agro-forestry areas.

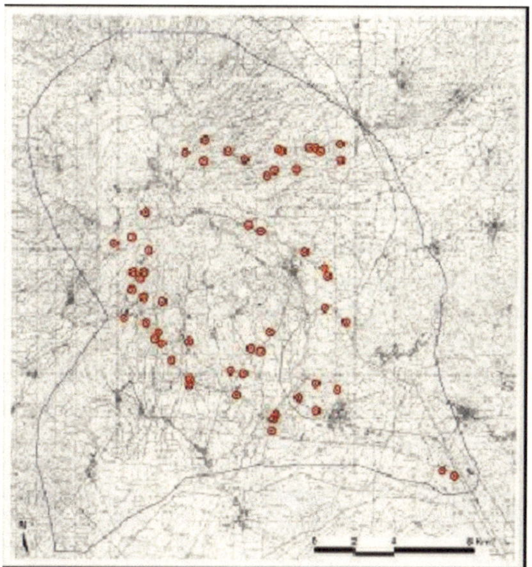

Fig. 7. Topographic map of the study area.
The red circles represent the location of the 56 sampling stations

2.2 Lichen sampling

Location of sampling stations and trees follows the guidelines by ANPA (2001). Fifty-six random sampling stations (1 km^2 each) were selected (Fig. 7), each divided in 4 square areas of 250 x 250 m. Relevés of epiphytic lichen vegetation were carried out on chestnut (*Castanea sativa* Miller.) and oak trees (*Quercus pubescens* Willd.). The selected trees were provided with similar bark properties (subacid, mesotrophic to oligotrophic barks with similar water storing capacity). The trees had to satisfy the following requirements (Nimis, 1999b; Asta et al., 2002): (a) free-standing well-lit tree; (b) inclination of the trunk not exceeding 10°; (c) circumference larger than 60 cm; (d) absence of evident factors of disturbance or pathologies. Damaged or decorticated parts of the trunk, parts with knots, parts corresponding to seepage tracks after rain, parts with more than 25% cover of bryophytes, were excluded from the samplings.

The sampling grid consisted of four vertical ladders of 10x50 cm, each divided in five 10x10 cm unit areas; each of the four ladders was applied to one of the four cardinal points, with the base at 100 cm from the ground (Fig. 8). To exclude from sampling any unsuited part of the trunk, a shift from verticality up to 20° clockwise was allowed when positioning individual ladders (Castello & Skert, 2005).

In each station, about 3-12 trees were sampled, for a total of 89 chestnut and 119 oak trees. The species represented in each unit of the grids were listed, and a frequency value ranging from 0 to 5 was obtained for each species in each ladder, depending on the number of unit in which the species was found. This values were summed for each of the four cardinal points and averaged for the number of trees in the station, obtaining four indices of lichen biodiversity each referred to one of the four cardinal points. The sum of these four indices was kept as the lichen biodiversity indicator for the station (Castello & Skert, 2005).

Thalli were collected during field sampling and brought to laboratory for any dubious identification. Taxonomic determination was done referring to Nimis (1987; 1992; 1993a;

1993b), Ozenda & Clauzade (1970), Clauzade & Roux (1985) and Poelt (1969.). We also took in account lichen floras of the nearby areas of Sannio e Daunia, Vesuvius, Matese and Partenio mounts (Garofalo et al. 1998-1999; Aprile et al. 2001; 2002-2003a; 2002-2003b) . Nomenclature follows Nimis & Martellos (2008). Ecological indices, expressing the level of biological tolerance of the species regarding environmental features, were calculated referring to the criteria of ITALIC system (http://dibiobs.univ.trieste.it). The ecological attributes included response to pH, light levels, water availability, geographical distribution and evenness. In addition, we determined two indices related to the degree of human impact on the atmosphere, the indices of *eutrophycation* and *poleophoby*. The former is related to the frequency of lichens tolerating (or escaping) nitrogen compounds dispersed as dust in the atmosphere. The second is related to the frequency at which lichens tolerating (or escaping) urban environment are found, so accounting for the general degree of human disturbance over an area. Finally, indices related to lichen morphology, photobiont association and reproductive type were calculated.

Fig. 8. Example of the sampling grid on a chestnut tree.
The sampling grids were positioned on each tree at 1 m from ground level in the four cardinal directions. See text for details

2.3 Map analysis

The surveys were geo-referenced using GPS stations (GPS MAP 60 CS) in three-dimensional mode with the static method and calculation of the average position. Thus, in addition to the floristic and ecological data, all lichen species positions were also stored in a Data Base Management System and processed with a pattern analysis to obtain information about their arrangement in space. Subsequently, the continuous grid "Air Quality map", was obtained in Ilwis 3.3/3.5 environment, by data point interpolation, constrained by barriers, using the Moving Average Algorithm with the weight function "Inverse Distance" (Fig. 9):

$$W = \frac{1}{d^n} - 1 \quad (1)$$

where W is the weight, . n is the weight exponential and d represents the relative distance of point to output pixel, given by:

$$d = \frac{D}{D_0} \quad (2)$$

in which D = Euclidean distance of point to output pixel and D_0 = limiting distance.

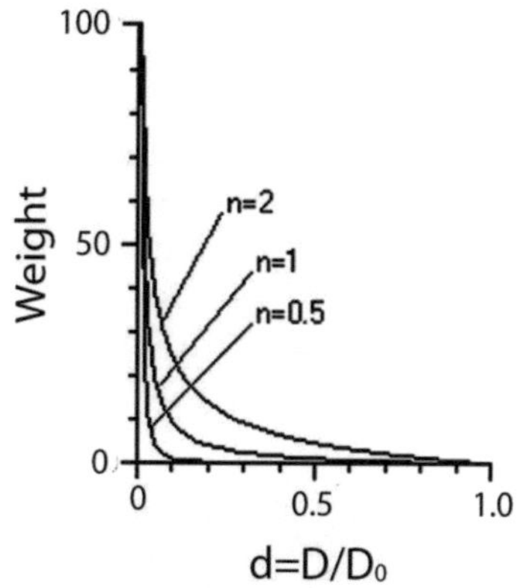

Fig. 9. Curves of the Inverse Distance Weight (IDW).

The selected function for the interpolation is quite sensible to sudden changes of the point data value, across the space. So, for any unsampled location, it is possible to predict a value based on the assumption that things that are close to one another are more alike than those that are farther apart. Using a barrier during the interpolation operation, the estimated cell value is calculated inside the space limited by the barrier

The air quality map was then overlapped to a land-use map, elaborated from Corine 2000 Land Cover map and field data. Dem, slope and aspect maps from morphological data were used in the cross-tabulation raster operations to produce a multi-thematic geo-database.

2.4 Statistical analysis

Lichen data were analyzed with Principal Component Analysis (PCA) and with cluster analysis with group average method (UPGMA), based on continuous values of the frequency of individual species and using the Euclidean distance. Two dendrograms were obtained that identified the main clusters of lichen species and locations. The clusters of species were used for identifying lichen indicator assemblages of species, whereas the clusters of locations were interfaced in GIS with land use patterns. Once obtained the main clusters of the stations, for each group the average values assumed by the main indicator indices were calculated and the corresponding graphic spectra were produced.

The results of cluster analysis were also overlapped with Corine Land Cover data, in order to define the features of land use in the main groups of station. An index of "naturalness" was calculated for each cluster as the ratio of the total surface falling into class level 3 (natural vegetation) against the sum of classes 1 and 2 (settlements and agricultural areas). Pearson correlation analysis was then applied to detect any common trend among the index of naturalness obtained by land use analysis, the IBL index obtained by lichen species frequencies and all the other indices obtained by lichen indicator attributes such as eutrophication, poleophoby, etc.

3. Results

3.1 Lichen flora

A total of 48 epiphytic lichen species (Fig. 10) were identified on the 208 trees examined in the study area, 59 % of which were foliose, 31 % crustose and 10 % fruticose. Most of the lichens found were typical of the vegetation unit of *Xanthorion parietinae* and *Parmelion caperatae*, characterized by the presence of relatively common species and so not expressing a high biodiversity value. No endangered species were found over the area.
The ecological indicators of this lichen flora were mainly falling in the middle of the ranges for most of the selected criteria of bioindication. Species richness, on the contrary, was found to change considerably among the locations, providing a relatively higher indicator value.
Most of the species were typical of humid-sub humid Mediterranean climate, though a considerable number of species resulted to be relatively unusual for this climatic belt. Among these, *Caloplaca herbidella, Anaptychia ciliaris, Ochrolechia balcanica.* All these species are usually found in cooler environments, such as temperate and boreal arctic areas.

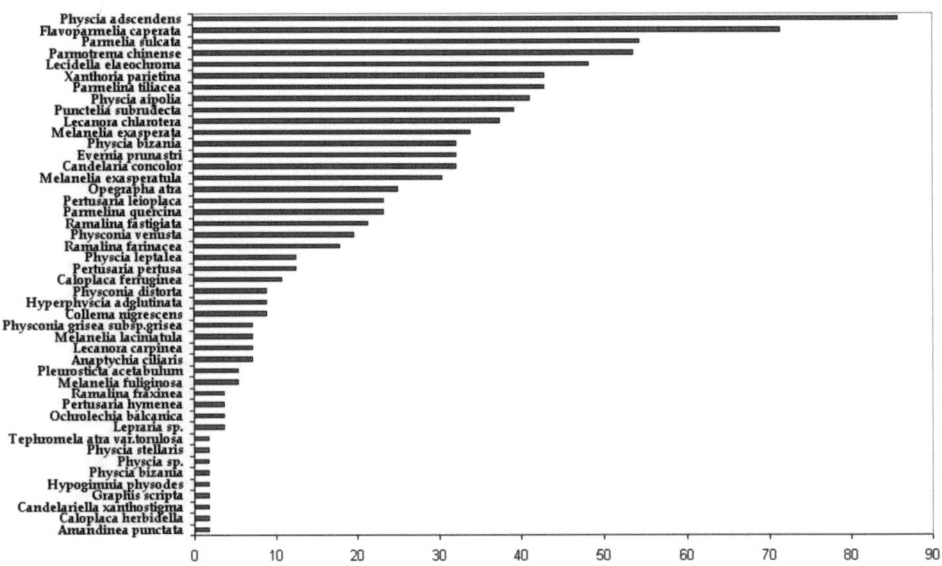

Fig. 10. List of the lichen species found on the study area.
The red bars represent % frequencies of the lichen species

Concerning poleophoby, most of the species resulted to be distributed in natural- semi natural habitats, or also in slightly disturbed environments. Concerning eutrophication, the species indicated wide ranges of adaptation, though indicating in the average values not very high levels of contamination.
The index of biodiversity value (IBL) ranged between 3 and 103 on the studied station. Lower IBL values were found in the proximities of the main roads and urban settlements, whereas the higher values were found in woodland and natural areas.

3.2 GIS analysis

A quantitative analysis of Corine land use classes revealed that deciduous forests and agro-mosaics occupy most of the study area (Fig. 11). Broadleaved forests are characterized mainly by chestnut groves that dominate on the slopes of the volcano. Oak woods are found close to the plain areas either as forest stands or along the borders of agricultural fields. Thus, this landscape appears highly fragmented, with land properties separated by vast forest corridors, which represent an important factor of ecological continuity. The area occupied by settlements is mainly distributed at the lower altitudes, with a scattered distribution, but ecologically relevant due to the dense net of road and highways that interconnect them.

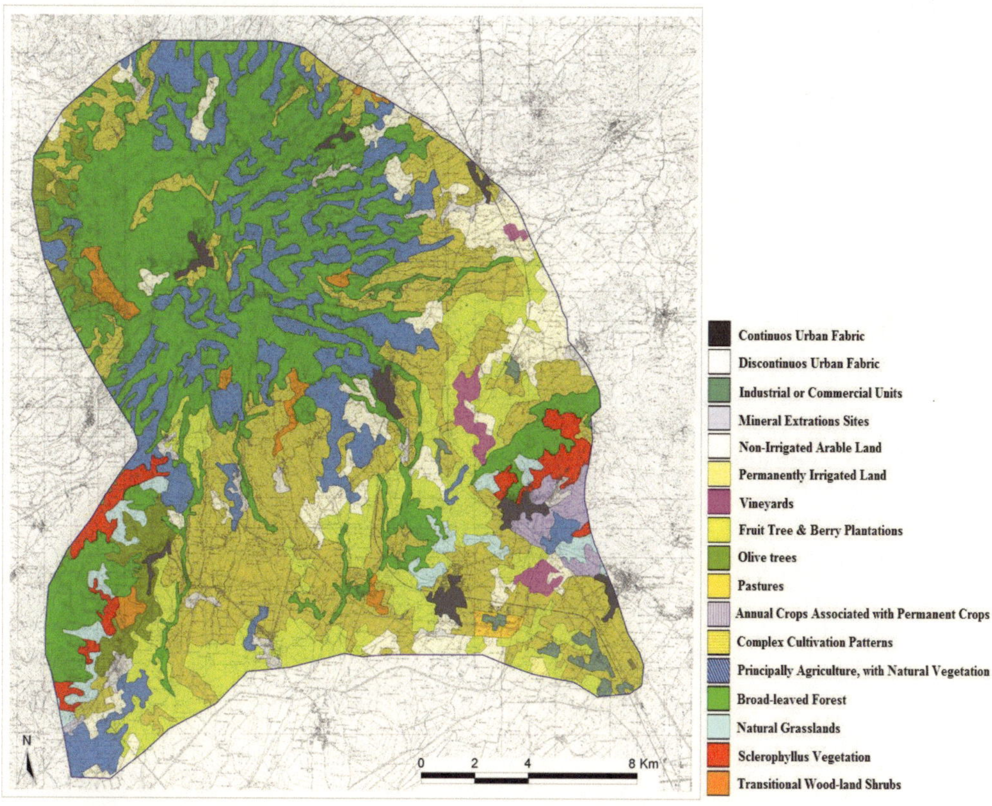

Fig. 11. Corine land cover map

Using point interpolation tools of ILWIS 3.5, all the floristic data, based on the values assumed by IBL index over the sampled area, were spatialized in GIS environment and after a "slicing" operation, a map of air quality was produced (Fig. 12). Six classes of IBL were identified by quantitative analysis. Following, the map obtained was overlapped with the DEM of the area, the slope map and the Land Cover map. Low IBL values were found in the plains near the highway exits, where vehicular transit is quite intense and cars and camions are often subjected to stop for variable time. Low-intermediate values of IBL were generally found at low altitude in all the areas subjected to industrialization, whereas higher IBL values were found in all the areas covered by vegetation, and particularly at higher altitudes, where deciduous forests substitute the agricultural fields.

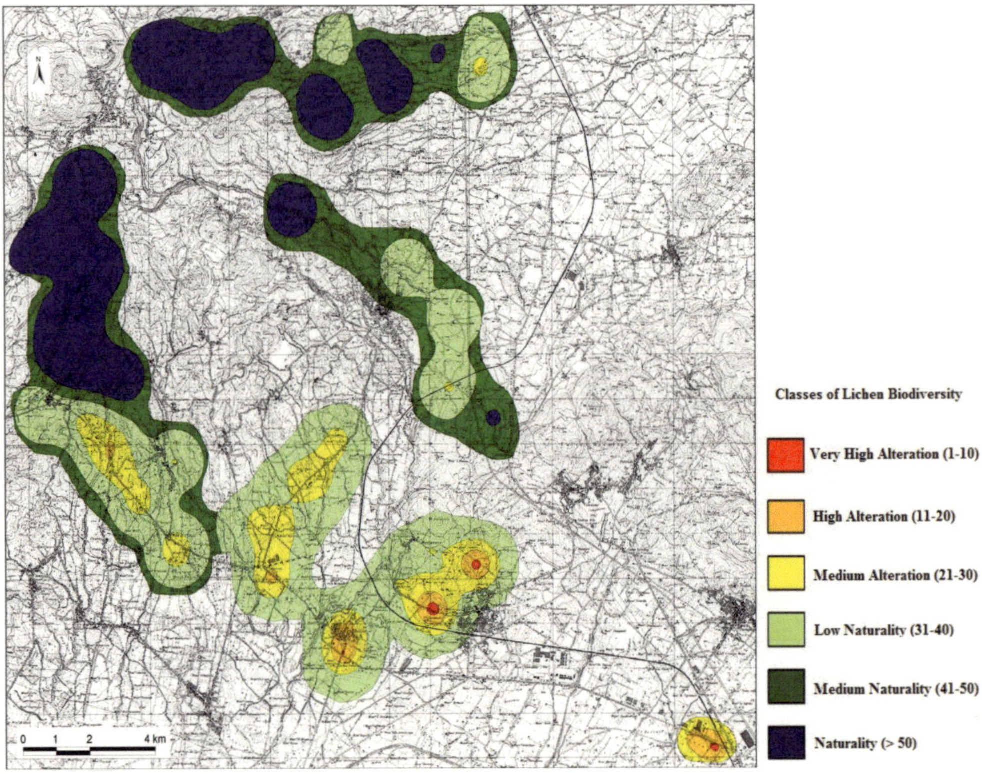

Fig. 12. Map of air quality

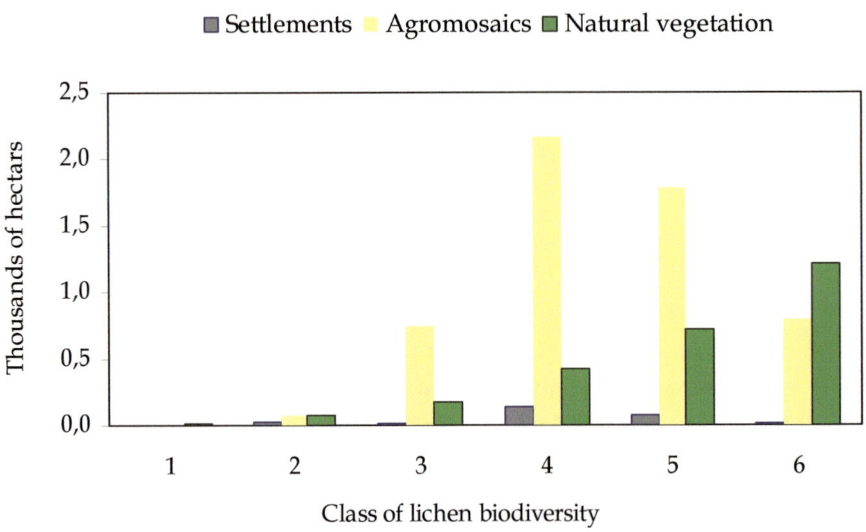

Fig. 13. Overlay of IBL belts with the map of land use cover.
1: Very high alteration; 2: High alteration; 3: Medium alteration; 4: Low naturality; 5: Medium naturality; 6: Naturality

The overlay between the six IBL belts and the map of land use confirms that maximum lichen biodiversity values are found in environments characterized by natural vegetation and agromosaics (Fig. 13).

3.3 Classification of the stations

Cluster analysis applied to sampling station revealed the presence of 4 main groups (Fig. 14), characterized by an increasing level of lichen biodiversity index. Cluster A included 28 stations mainly distributed in the most anthropized environment, mainly agricultural areas quite close to urban settlements and highways, at an average altitude of 130 m s.l.m.; IBL value in this cluster was 20.9, whereas the index of naturalness assumed the value of 0.22. Cluster B was composed by 6 stations distributed in environments quite similar to the previous ones, at a mean altitude of 206 m s.l.m., with IBL value of 44.3 and naturalness index of 0.17. Cluster C included 10 stations distributed in more natural environments, at a mean altitude of 395 m s.l.m., with IBL value of 61.4 and index of naturalness of 3.71. Cluster D included 12 stations in natural and agricultural areas, at a mean altitude of 411 m s.l.m., with IBL value of 76.3 and naturalness index of 1.94.

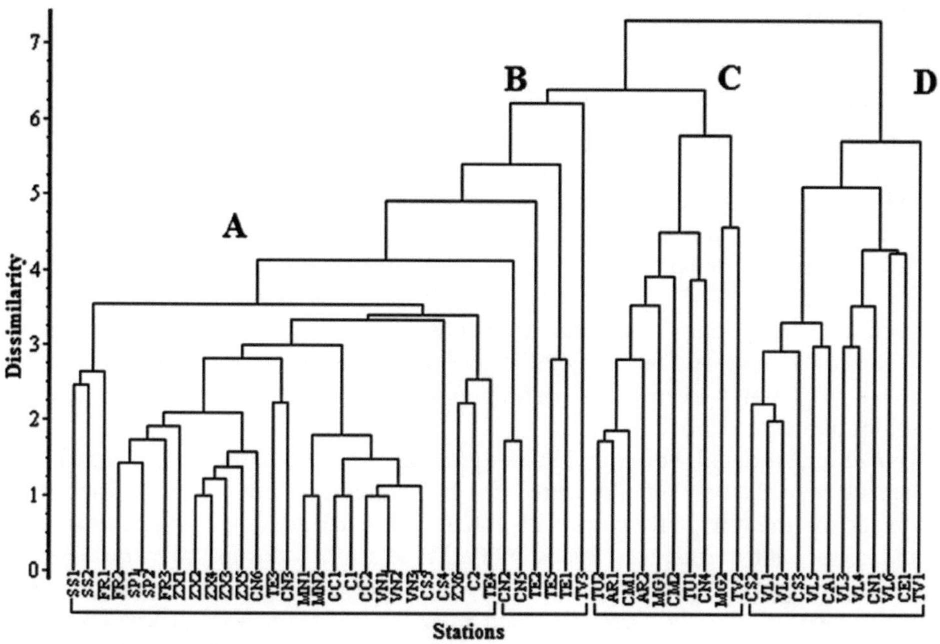

Fig. 14. Dendrogram of the sampling stations.
Four clusters (A, B, C, D) resulting by multivaried analysis applied to the sampling stations. See text for details

3.4 Correlation analysis and ecological indices

The index of lichen biodiversity was directly correlated with altitude (R=0.96, P<0.05) and with the frequency of *common* species (R=0.97, P<0.05). It was also directly correlated with *crustose* (R=0.96, P<0.05) and *fruticose* (R=0.97, P<0.05) species, and inversely correlated with *foliose* ones (R=-0.97, P<0.05). The index of naturalness was inversely correlated with the frequency of the species adapted to highly eutrophized environments ($R=-0.98$, $P<0.05$), and directly correlated with the presence of *extremely rare* species ($R=0.96$, $P<0.05$). The correlation between the IBL and the index of naturalness was relatively high ($R=0.68$) but not significant.

Among the ecological indicators, the index of eutrophication and the index of poleophoby showed consistent trends with naturalness indicators expressed by land use patterns. Cluster C, the one provided with the highest naturalness value, presented the highest frequency of species relatively unadapted to eutrophication and to urban environment, and the lowest value of the species characterized by the opposite adaptation.

4. Discussion

A good consistency was found between the values assumed by IBL index across space and the environmental quality presumed by the analysis of the Corine land cover maps. IBL resulted to be higher in natural and semi-natural areas, whereas it was very low around urban areas and settlements. Low values of IBL were also found in areas relatively close to natural environments but subjected during recent times to rapid changes toward industrialization. It was particularly interesting the data relative to agricultural areas. IBL assumed intermediate values in these areas, detecting a relatively lower value of air quality. In fact, air quality in agricultural areas is expected to decrease if compared to natural environments, due to the extensive use of chemicals for plant defence, particularly fungicides. It is noteworthy that agriculture in this area usually takes place in close interconnection with natural environments. Much of the areas classified as agromosaics belonged to the subcategories of complex systems, that often are bordered or intermixed with natural woods. So, this study also proved that IBL index is a sufficiently sensible tool to detect environmental patterns, even when these patterns are not strongly evident if analyzed at a large scale detail. In many cases, the patterns of IBL values fitted with the field-based land use pattern recognition in a considerably higher extent if compared with Corine land cover analysis. Thus, a bioindicator based recognition of environmental quality was proved to guaranty even higher definition and predictivity than common landscape geographic approach.

On a scale of higher detail, however, in some cases the analysis based on just the IBL value did not result to be the best suited to detect the global quality of environment. The trends of IBL index across the four main groups identified by cluster analysis, did not assume very high and significant correlation values with the selected indicator of naturalness. IBL resulted to be highly correlated with the massive presence of common lichen species, whereas it was not strongly correlated with presence of rare species. These last species, however, were strongly correlated with naturalness. The highest values of IBL were not found in cluster C, the one with highest naturalness value, but in cluster D, were chestnuts and mixed woods were strongly intermixed with agricultural lands. Moreover, the species negatively correlated to eutrophication and adaptation to urban environment were found in a higher proportion in cluster C compared to cluster D. We may hypothesize that in cluster D, the impact of mechanical interventions linked to agriculture and also to the mechanization of chestnut harvesting, could have negatively affected the most sensitive elements of lichen flora, but nevertheless the IBL was high because the most common lichen species grew in a higher number due to the best availability of growth substrates and for the best climate given the highest altitude. Thus, the quantitative approach of IBL analysis

should always be integrated with the study of the qualitative indicators expressed by the lichen flora in order to get a fine tuning of the previsions about air quality.
In conclusion, measuring lichen biodiversity through the IBL index allowed us to characterize the state of environment on the studied area. This method put the basis for a wider study on the dynamic evolution of a district interested by a process of industrial reconversion very close to a natural reserve. A good responsiveness of statistical clustering towards geo-statistical features of lands was found, confirming the reliability of IBL index as an indicator of environmental quality. The assemblages of lichen species were found to reflect not only environmental quality on a local scale, but also the more complex set of variables related to land use patterns at landscape level.

5. Acknowledgments

We kindly thank Prof. Annamaria Carafa for helping us in the microscopic analysis of the lichen samples and for cooperating to the interpretation of image analysis.

6. References

Ahmadjian, V. (1993). *The Lichen symbiosis* (John Wiley and Sons Eds.), ISBN 0-471-57885-1, New York

ANPA (2001). *I.B.L. Indice di biodiversita` lichenica*. ANPA, Manuali e linee guida 2/2001, pp. 85, ISBN 88-448-0256-2

Aprile, G.G.; Garofalo, R.; Cocca, M.A. & Ricciardi, M. (2001). La flora lichenica del complesso Somma-Vesuvio (Napoli). *Allionia,* Vol.38, pp. 195-205, ISSN 0065-6929

Aprile, G.G.; Garofalo, R.; Cocca, M.A. & Ricciardi, M. (2002-03a). I licheni dei monti del Partenio (Appennino Campano). *Allionia*, Vol.39, pp. 77-86, ISSN 0065-6929

Aprile, G.G.; Garofalo, R.; Cocca, M.A. & Ricciardi, M. (2002-03b). I licheni del Matese (Appennino molisano - campano). *Allionia*, Vol.39, pp. 87-109, ISSN 0065-6929

Asta, J.; Erhardt, W.; Ferretti, M.; Fornasier, F.; Kirschbaum, U.; Nimis, P.L.; Purvis, O.W.; Pirintsos, S.; Scheidegger, C.; Van Haluwyn, C. & Wirth, V. (2002). Mapping lichen diversity as an indicator of environmental quality, In: *Monitoring with Lichens – Monitoring Lichens*, P.L. Nimis, C. Scheidegger and P. A. Wolseley, pp. 273-279, Kluwer Academic Publishing, ISBN 1-4020-0430-3, The Netherlands

Barkman, J.J., (1963). *De epifythen-flora en -vegetatie van Midden-Limburg (BelgiJ).* Verhandeling der Koninklijke Nederlandse Akademie van Wetenschappen, Afdeling Natuurkunde. Tweede Reeks, vol. 54, 1963

Castello, M. & Skert, N. (2005). Evaluation of lichen diversity as an indicator of environmental quality in the North Adriatic submediterranean region. *Science of the Total Environment,* Vol.336, No.1-3 (January 2005), pp. 201- 214, ISSN 0048-9697

Cislaghi C. & Nimis, P.L. (1997). Lichens, air pollution and lung cancer. *Nature,* Vol. 387, No. 6632 (May 1997), pp. 437-532, ISSN: 0028-0836

Clauzade, G. & Roux, C. (1985). Likenoj de Okcidenta Europo. *Bull. de la Soc Bot. du Centre-ouest,* Nouvelle Sèrie, Numèro Spècial: 7, ISSN 0154-9898

Conti, M.E. & Cecchetti, G. (2001). Biological monitoring: lichens as bioindicators of air pollution assessment - a review. *Environmental Pollution,* Vol.114, No.3 (October 2001), pp. 471-492, ISSN 0269-7491

Costa, C.J.; Marques, A.P.; Freitas, M.C.; Reis, M.A & Oliveira, O.R. (2002). A comparative study for results obtained using biomonitors and PM10 collectors in Sado Estuary. *Environmental Pollution*, Vol.120, No.1 (November 2002), pp. 97-106, ISSN 0269-7491

De Rita, D. & Giordano, G. (1996). Volcanological evolution of Roccamonfina volcano (Italy): origin of the summit caldera. *In: Volcano instability on the Earth and other planets*, Mc Guire et al. editors, Geol. Soc. Sp. Pub. 110: 209-224.

De Sloover, J. & Leblanc, F. (1968). Mapping of atmospheric pollution on the basis of lichen sensitivity. In: *Proceedings of the Symposium in Recent Advances in Tropical Ecology, International Society for Tropical Ecology*. Misra R, Gopal B, editors., pp. 42- 56, Varanasi Banaras Hindu University, India

Ferry, B.W.; Baddeley, M.S. & Hawksworth, D.L. (1973). *Air pollution and lichens* (eds). Toronto: University of Toronto Press, pp. 2-389

Garofalo, R.; Cocca, M.A; Aprile, G.G. & Ricciardi, M. (1998-1999). Licheni dei monti del Sannio e della Daunia (Appennino Campano). *Allionia*, Vol.36, pp. 53-65, ISSN 0065-6929

Giordani, P.; Brunialti, G. & Alleteo D. (2002). Effects of atmospheric pollution on lichen biodiversity (LB) in a Mediterranean region (Liguria, NW-Italy*). Environmental Pollution*, Vol.118, No.1 (June 2002), pp. 53-64, ISSN 0269-7491

Hawksworth, D.L & Rose, L. (1970). Qualitative scale for estimating sulphur dioxide air pollution in England and Wales using epiphytic lichens. *Nature*, Vol.227, pp. 145-148, ISSN: 0028-0836

Hawksworth, D.L. (1988). The variety of fungal-algal symbioses, their evolutionary significance and the nature of lichens. *Botanical Journal of the Linnean Society*, Vol.96, No.1 (January 1988), pp. 3-20, ISSN 1096-8339

Herzig, R. & Urech, M. (1991). Flechten als Bioindikatoren Integriertes biologisches Messsystem der Luftverschmutzung fqr das Schweizer Mittelland. *Bibliotheca Lichenologica*, Vol. 43, pp. 283, ISSN 1436-1698

Herzig, R.; Liebendfrfer, L. & Urech, M. (1987). Flechten als Bioindikatoren der Luftverschmutzung in der Schweiz: Methoden-Evaluation und. *VDI-Ber*, Vol.609, pp. 619-639

Liebendöerfer, L.; Herzig, R.; Urech, M. & Amman, K. (1988). *Staub-Reinhaltung der Luft*, Vol. 48, pp. 233-238

Loppi, S.; Giordani, P.; Brunialti, G.; Isocrono, D. & Piervittori, R. (2002). Identifying deviations from naturality of lichen diversity for bioindication purposes. In: *Monitoring with Lichens – Monitoring Lichens*, P.L. Nimis, C. Scheidegger and P. A. Wolseley, pp. 281-284, Kluwer Academic Publishing, ISBN 1-4020-0430-3, The Netherlands.

Nash, III, T.H. (1996). *Lichen biology* (Thomas H. Nash III Eds.), ISBN 0-521-45974-5, Cambridge.

Nash, T. & Wirth, V. (1988). *Lichens, bryophytes, and air quality*. (J. Cramer), ISBN 3443580092, Berlin.

Nimis, P. L. (1987). I macrolicheni d'Italia. chiavi analitiche per la determinazionme. *Gortania* , Vol. 8, pp. 101-220, ISSN 0391-5859

Nimis, P.L. (1990). Air Quality Indicators and Indices. The use of plants as bioindicators and biomonitors of air pollution. In: *Proceedings International Workshop on Indicators and Indices*, JRC Ispra, EUR 13060 EN: 93-126. Colombo A. & Premazzi G. (eds.)

Nimis, P. L., (1992), Chiavi analitiche del genere Caloplaca Th. Fr. in Italia. *Notiziario della Società Lichenologica Italiana*, Vol. 5: 9-28, ISSN 1121-9165

Nimis P. L., (1993a), *The Lichens of Italy. An annotated catalogue. Monografie XII.* (Museo Regionale di Scienze Naturali), ISBN 88-86041-02-0, Torino

Nimis, P. L., (1993b), Chiavi analitiche del genere Lecanora Ach. in Italia. *Notiziario della Società Lichenologica Italiana*, Vol. 6: 29-46, ISSN 1121-9165

Nimis, P.L., (1999b). Linee-guida per la bioindicazione degli effetti dell'inquinamento tramite la biodiversità dei licheni epifiti. *Proceedings of Workshop Biomonitoraggio della qualità dell'aria sul territorio nazionale.* Serie Atti/2 ANPA, pp. 267-277, ISBN 88-448-0021-7, Roma, 26-27 November 1998

Nimis, P.L. & Martellos, S. (2008). In: *ITALIC - The Information System on Italian Lichens. Version 4.0.* University of Trieste, Dept. of Biology, IN4.0/1. Avaiable from: <http://dbiodbs.univ.trieste.it/>

Nimis, P.L. & Skert, N. (1999). *Introduzione al biomonitoraggio con licheni epifiti* Amministrazione provinciale di Vicenza Dipartimento Ambiente. Corso di formazione. Bassano del Grappa, 16-18 Settembre 1999, Trieste

Nimis, P.L.; Ciccarelli, A.; Lazzarin, G.; Bargagli, R.; Benedet, A.; Castello, M.; Gasparo, D.; Lausi, D.; Olivieri, S. & Tretiach, M. (1989). I licheni come bioindicatori di inquinamento atmosferico nell'area di Schio-Thiene-Breganze. *Boll. Mus. civ. St. nat.Verona*, Vol. 16, pp. 1-154, ISSN 1590-8399

Ozenda, P. & Clauzade, G. (1970). *Les Lichenes*, Masson & C., Paris

Paoli, P.; Guttová, A. & Loppi, S. (2006). Assessment of environmental quality by the diversity of epiphytic lichens. *Biologia, Bratislava,* Vol.61, No.4, pp. 353-359, ISSN 0006-3088

Piervittori, R. (1999). Licheni come bioindicatori della qualità dell'aria : stato dell'arte in Italia. *Proceedings of Workshop Biomonitoraggio della qualità dell'aria sul territorio nazionale.* Serie Atti/2 ANPA, pp. 97-122, ISBN 88-448-0021-7, Roma, 26-27 November 1998

Pinho, P.; Augusto, S.; Branquinho, C; Bio, A.; Pereira, M. J.; Soares, A. & Catarino, F. (2004). *Chemistry, Vol.*49, No.1-3, pp. 377–389, ISSN 0167-7764

Pinho, P.; Augusto, S.; Ma´guas, C.; Pereira, M.J.; Soares, A. & Branquinho, C. (2008). Impact of neighborhood land-cover in epiphytic lichen diversity: Analysis of multiple factors working at different spatial scales.*Environmental Pollution,* Vol.152, No. 2 (March 2008), pp. 414-422, ISSN 0269-7491

Poelt, J. (1969). *Bestimmungschlüssel Europaischer Flechten.* Verlag von J. Cramer, Berlin.

Purvis W., (2000). *Lichens.* Natural History Museum, London/Smithsonian Institution, London, Washington D.C.

Richardson D. H. S. (1992). *Pollution Monitoring with Lichens.* Richmond Pub Co ISBN-10: 0855462892, Berkshire

Schwender, S. (1869). *Die Algentypen der Flechtengonidien. Programm für die Rectorsfeier der Universitä Basel* Vol.4, pp. 1-42

Sloof, J.E.; Wolterbeek, H.Th. (1993). Interspecies comparasion of lichens as biomonitors of trace-element air pollution. *Environmental Monitoring and Assesment,* Vol.25, No.2 (April 1993), pp. 149-157, ISSN 0167-6369

Szczepaniak, K. & Biziuk, M. (2003). Aspects of the biomonitoring studies using mosses and lichens as indicators of metal pollution. *Environmental Research*, Vol.93, No.3, (November 2003), pp. 221–230, ISSN 0013-9351

Tehler, A. (1996). Systematics, phylogeny and classification. In: *Lichen Biology*, Thomas H. Nash III ed. , pp. 217-239, Cambridge University Press, UK ISBN: 0521453682

van Dobben, H.F.; Wolterbeek, H.Th.; Wamelink, G.W.W. & Ter Braak, C.J.F. (2001). Relationship between epiphytic lichens, trace elements and gaseous atmospheric pollutants - *Environmental Pollution*, Vol.112, No.2 (April 2001), pp. 163–169, ISSN 0269-7491, Cambridge

11

Use of the Micronucleus Test on *Tradescantia* (Trad-MCN) to Evaluate the Genotoxic Effects of Air Pollution

José Roberto Cardoso Meireles and Eneida de Moraes Marcílio Cerqueira
Feira de Santana State University
Brazil

1. Introduction

Good air quality is the fundamental condition for maintaining the equilibrium of human health and ecosystems. However, starting at the time of the industrial revolution, human action has progressively introduced into the atmosphere compounds and particles that compromise the harmony of life on our planet: a process that is known as atmospheric pollution.

Thus, atmospheric pollution is defined as the result from excessive introduction of compounds and particles into the air layers that surround Earth (Marcondes, 1993). There are now many sources of atmospheric pollution, and prominent among these are industrial plant, factories and automotive vehicles powered by gasoline and diesel. Among the many pollutants that compromise air quality, particularly in large urban centers, are sulfur dioxide (SO_2), carbon monoxide (CO), nitrogen oxides, particles and aromatic polycyclic hydrocarbons, which can combine to form other compounds such as ozone and peroxyacetyl nitrate, thereby giving rise to a complex mixture of pollutants in the atmosphere.

The presence of these compounds in the air causes a variety of human health problems, such as abnormalities of the respiratory and cardiovascular systems, allergic reactions and development of lung and tracheal cancer (Brunekreef et al., 2002; Perera et al., 2002; Traversi et al., 2008).

Cancer is a disease that results from abnormalities in genes involved in controlling cell proliferation and differentiation and/or genes involved in DNA repair mechanisms and in inducing apoptosis (Hanahan & Weinberg, 2000). The association between atmospheric pollution and cancer therefore comes from the fact that many of the contaminants are mutagens, i.e. they are agents capable of inducing both point mutations (abnormalities involving only one or very few DNA bases) and chromosome aberrations (abnormalities that compromise the structure or the number of chromosomes).

2. Genotoxic effects of atmospheric pollution

Mutagenic effects from atmospheric contaminants have been demonstrated by many authors using different methodologies, both *in vivo* and *in vitro*, such as DNA adduct detection (Peluso et al., 2008), comet test (Ianistcki et al., 2009; Kawanishi et al., 2009), sister chromatids exchanges (SCE) and micronuclei (Kawanishi et al., 2009), identification of gene mutations in mice (Yauk et al., 2008) and Ames test (Coronas et al., 2008; Traversi et al., 2009).

Peluso *et at.* (2008) evaluated air genotoxicity among 79 workers at the MIE industrial complex (one of the largest steel and oil refinery industrial complexes in Southeast Asia) and among 72 people living nearby. Fifty individuals living in an area without industrial pollution were evaluated as controls. Peripheral lymphocytes were analyzed by means of the ^{32}P-postlabeling technique in order to identify DNA adducts, i.e. structures formed through covalent bonding of molecules to the DNA. The DNA adduct levels were significantly greater among the workers and individuals living close to the industrial complex than among the controls, thus demonstrating the potential that atmospheric pollution has for inducing mutations.

Greater occurrence of damaged DNA consequent to exposure to polycyclic aromatic hydrocarbons (PAHs) present in the atmosphere was observed by Ianistcki *et at.* (2009) through using the Comet test on *Helix aspersa* Müller, 1774. This study was conducted in areas of the metropolitan region of Porto Alegre (Brazil) where the atmosphere is contaminated with PAHs. Between the two fractions of particulate matter that were isolated (particles measuring between 2.5 μm and 10 μm in diameter and particles of diameter less than 2.5 μm), greater genotoxicity was observed consequent to exposure to the particles of smaller diameter, which may be related to greater absorption of these particles. Kawanaka *et at.* (2011) observed that ultrafine particles are deposited more efficiently in the alveolar region of human lungs than are larger particles. Thus, ultrafine particles play an important role as carriers of mutagenic agents.

Kawanishi *et at.* (2009) evaluated the genotoxicity of 3,6-dinitrobenzo[e]pyrene, a novel mutagen in ambient air and surface soil, *in vitro* and *in vivo*. Human cells of the HepG2 human lineage were analyzed regarding the frequencies of exchanges between sister chromatids and occurrences of *hprt* mutations and micronuclei, and the Comet test was used to evaluate occurrences of damaged DNA in IRC mice. A genotoxic effect caused by 3,6-dinitrobenzo[e]pyrene was observed in all the trials.

Yauk *et at.* (2008) also described damage to DNA, as seen through greater occurrences of gene mutations, in germinative cells from mice exposed to atmospheric pollution without protection. They observed that spermatozoid mutations were 1.6 times more frequent in these exposed animals than in animals whose exposure was diminished through using high-efficiency particulate-air (HEPA) protection. Furthermore, they identified high levels of DNA hypermethylation in spermatozoids from the exposed animals, which in their view, might have wide-ranging repercussions on chromatin structure, gene expression and genome stability. Genotoxic effects from atmospheric contaminants have also been shown using the Ames test (Coronas et al., 2008; Traversi et al., 2009).

The proven mutagenic action of atmospheric pollutants is the basis for the biological plausibility of epidemiological studies that have revealed associations between some types of cancer, especially lung cancer, and atmospheric pollution (Beelen et al., 2008; Eitan et al., 2010; Kapka et al., 2009; Trédaniel et al., 2009; Vineis et al., 2007).

Air quality monitoring therefore becomes essential and, within this context, the use of biological tests capable of detecting gene mutations or chromosome damage is considered to be an important preventive medical measure. Such tests are also effective for detecting the

additive and synergistic effects of air pollutants, unlike chemical analyses, which evaluate the action of a given genotoxic agent in isolation (Cohen et al., 2002; Wada et al., 2001).
Bioassays developed using plants and clones of the genus *Tradescantia* are considered to be valuable tools for assessing the mutagenic effects of environmental contaminants. For this reason, since the beginnings of genetic toxicology, they have been widely used in studies and biomonitoring programs.

3. Evaluation of the genotoxicity of atmospheric pollution using plants and clones of the genus *Tradescantia*

Carvalho (2005) highlighted several characteristics of higher plants that qualify them for use in environmental biomonitoring. These include the following:

- Just like human cells, higher plant cells are eukaryotic and present organization of the genetic material that resembles that of humans;
- Also in common with humans, higher plants present germinative and somatic cell lines that go through meiotic and mitotic cycles, respectively;
- Many higher plants present short reproductive cycles;
- Higher plants can be propagated vegetatively, thus ensuring genetic purity, which also favors the perpetuation of hybrids;
- Studies on the use of higher plants can be conducted both *in situ* and in the laboratory.

According to Alves et al. (2001), plants are generally more sensitive to pollution than animals are, which makes them particularly suitable for biomonitoring studies on atmospheric pollution.

3.1 The genus *Tradescantia*

Tradescantia is a genus of perennial herbaceous plants in the family Commelinaceae. It originates from the New World and comprises around 500 species with distribution going from southern Canada to northern Argentina (Watson & Dallwitz, 1992). In the 17th century, they were introduced into Europe, where they are used for ornamentation, just as they are used in most other countries.
The height of the plants varies in most species between 30 cm and 60 cm, and its oval-shaped leaves are distributed along the stem. The inflorescence, which may be axillary or terminal, is protected by bracts that may or may not be differentiated from the leaves. The flowers present radial symmetry and are hermaphroditic. The pedicles vary in size between the species, and the calyx and corolla are formed, respectively, by three sepals and three petals. The stamens are free and there are six of them, and the anthers have two thecae. The gynoecium has a trilocular sessile ovary, and each locule holds two ovules. The style is simple and the stigma is capitate. In most of the species, the fruits present a capsule with a locus, and each locus holds two seeds.
Plants and clones of the genus *Tradescantia* have been used for environmental monitoring since the time of the first studies investigating the mutagenic activity of chemical and physical compounds, in relation to components of the atmosphere, soil or water. In addition to its high sensitivity to mutagen action, its ease of cultivation, short life cycle and, especially, its large chromosomes that are few in number are characteristics that make plants and clones of the genus *Tradescantia* valuable instruments for environmental monitoring (Ma & Grant, 1982).

3.2 Species and clones of the genus *Tradescantia* used in genotoxicity bioassays

Most of the studies using *Tradescantia* plants, and particularly those developed in Europe, have been conducted using clone 4430, which is a diploid hybrid between *T. hirsutiflora* Bush (2461C), with a blue flower, and *T. subacaulis* Bush (2441), with a pink flower (Isidori et al., 2003; Klumpp et al., 2006; Ma et al., 1996). This clone, which was developed by Sparrow et al. in 1960, is very versatile and has, since then, been widely cultivated indoors, although it requires special conditions for it to grow and flower. On the other hand, it presents great sensitivity to the action of chemical and physical mutagens in the environment. Because it is sterile, it has the advantage that its genetic uniformity is maintained. It can be used both for the micronucleus test (Trad-MCN) and for the stamen hair mutation test (Trad-SHM).

The *Tradescantia* species most frequently used for carrying out Trad-MCN is *T. pallida* (Rose) Hunt. cv. *purpurea* Boom (Figure 1). This is a small-sized herbaceous plant (reaching a maximum height of 25 cm) with spear-shaped succulent leaves that is native to North America and Central America (Mexico and Honduras) (Lorenzi & Souza, 2008). The epidermis of the leaves presents large quantities of anthocyanin, which gives them a purple color, particularly in very bright light (Joly, 1998). Two large canoe-shaped bracts protect the inflorescence, which presents pink flowers.

Fig. 1. Tradescantia pallida

Although this species has sexual reproduction, its genetic uniformity can be ensured by means of vegetative propagation from a single stalk. It has shown sensitivity to the action of environmental mutagens resembling that of 4430 and #03 (Andrade Jr et al., 2008; Batalha et al., 1999; Guimarães et al., 2000; Meireles et al., 2009; Suyama et al., 2002). This species has been greatly used in studies conducted in South America because of the great adaptability of these plants to the climatic conditions of South American countries, unlike clone 4430, which is difficult to cultivate under these conditions. In addition, using plants that develop under natural conditions reduces the costs of the study and minimizes any problems of pest attacks, given that these plants are already biologically adapted to the environment (Rodrigues et al., 1997).

3.3 Genotoxicity bioassays using *Tradescantia*

Tests carried out using plants or clones of the genus *Tradescantia*, like most plant bioassays, are considered to be type II mutagenesis tests (Ennever et al., 1988), since they present high sensitivity and low specificity.

Four tests for assessing environmental genotoxicity have been developed using *Tradescantia*. One of these, the stamen hair mutation test (Trad-SHM) detects occurrences of point mutations. This test is carried out using clone 4430 and is based on the occurrence of recessive mutation that results in changing the blue color of stamen hairs to pink (Emmerling-Thompson & Nawrocky, 1982). It was first used to assess the genotoxic effects from ionizing radiation and, at the start of the 1970s, it was proposed that the test should be applied in studies on chemical mutagenesis (Underbrink et al., 1973). Additional studies using Trad-SHM to assess the genotoxic effects of chemical substances reconfirmed the high sensitivity that had previously be observed with this test in studies on radiation (Ma et al., 1994; Shima & Ichikawa, 1994). Trad-SHM has now been used for assessing mutagenicity in relation to both atmospheric and water pollution (Arutyunyan et al., 1999; Grant et al., 1992).

The other tests using *Tradescantia* that have been developed for environmental biomonitoring are cytogenetic studies based on occurrences of changes to the number and/or structure of the chromosomes: root-tip mitosis and pollen tube tests and the micronucleus test on tetrads of *Tradescantia* (Trad-MCN). In assessments on the genotoxic action of atmospheric pollutants, the main test used is Trad-MCN, which is the focus of the present chapter.

4. The micronucleus test on tetrads of *Tradescantia* (Trad-MCN)

4.1 Fundamentals and development

The micronucleus test on tetrads of *Tradescantia* (Trad-MCN) is currently the most widely used bioassay on plants for detecting genotoxins in the environment. According to a recent paper, approximately 160 chemicals have so far been tested and 100 articles on complex environmental mixtures have been published (Misík et al., 2011).

Trad-MCN consists of a set of procedures for exposing plants of the genus *Tradescantia* to contaminating agents, culminating in estimates of the micronucleus frequencies in mother cells of pollen grains at the tetrad phase. Micronuclei are structures that result from whole chromosomes or fragments of chromosomes that, because they do not bind to the spindle fibers, are not included in the nuclei of the daughter cells. Instead, they remain in the cytoplasm of the interphase cells, where they are observed as corpuscles resembling the nucleus, measuring 1/3 to 1/5 of the size of the nucleus. Since micronuclei result both from chromosomal fragments and whole chromosomes lagging behind in anaphase, they reveal the clastogenic and/or aneugenic action of a given mutagen (Holland et al., 2008) (Figure 2).

Estimation of the numbers of micronuclei in tetrads of *Tradescantia* was first proposed at the end of the 1970s, in a pioneering study conducted by Ma and coworkers (Ma et al., 1978). In that study, they used clone 4430 to compare the formation of micronuclei in tetrads (Trad-MCN) with stamen hair mutations in cells (Trad-SHM), after exposing the plants to 1,2-dibromoethane (DBE). The results obtained revealed that the sensitivity of Trad-MCN was approximately 30 times greater than that of Trad-SHM. The greater sensitivity of Trad-MCN in relation to Trad-SHM has been observed in several other studies, such as Gichner & Velemínnský (1999); Minouflet et al. (2005).

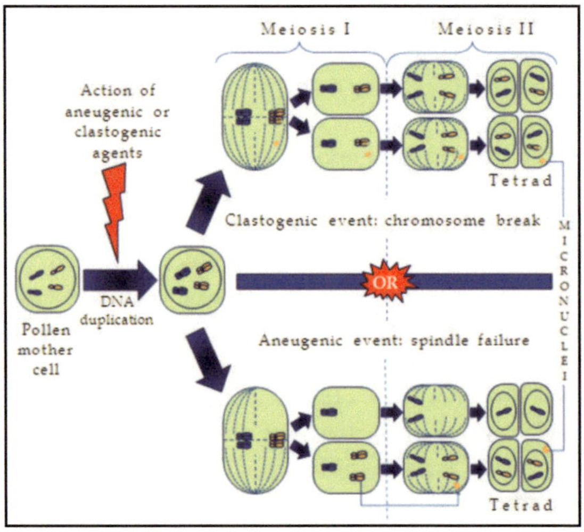

Fig. 2. Diagram of micronucleus formation

4.2 Methodology

4.2.1 Cultivation and maintenance of *Tradescantia* plants

The cultivation and maintenance conditions for *Tradescantia* plants depend primarily on the climatic conditions under which the study will be developed. In countries with a temperate climate, the plants can be cultivated in greenhouses, growth chambers or outdoors. In countries with a very rigorous winter, the cultivation should be undertaken in greenhouses or growth chambers. It is important to control the ventilation, temperature and air quality conditions, since these are factors that may induce occurrences of micronuclei. According to Klumpp et al. (2004), at a temperature of 11°C the frequency of spontaneous occurrence of micronuclei is 100 times greater, while exposure at a temperature of 42°C is incompatible with development of the mother cells of pollen grains. The ideal cultivation conditions consist of day/night temperatures of 21/19 ± 0.2°C, day/night relative humidity of 65/70% and a daylight cycle of 17h (Rodrigues, 1999). The substrate composition should consist of fertile soil with good drainage, which can be achieved by adding coarse sand and standardized organic matter to the soil. Commercially marketed soils such as ED73 can also be used (Klumpp et al., 2006). In addition to watering the soil, which should be done on alternate days, monthly use of fertilizers is recommended (NPK).

Following these procedures is the essential prerequisite for obtaining plants in numbers and quality that are sufficient to ensure that the study can be conducted and that the results will be reliable.

4.2.2 Exposure of the plants

When the genotoxicity of atmospheric pollutants is evaluated *in situ*, this involves exposing plants that have been cultivated under the conditions described above and been transported to the location under investigation (active monitoring), or it is done on plants in gardens, public squares and streets that were already planted out in these locations for ornamental purposes (passive monitoring).

In active monitoring, at least 15 plants per monitoring point are exposed for time periods that vary according to the protocol, depending on the degree of contamination at the location investigated. Fomin & Hafner (1998) considered that exposure for a six-hour period was insufficient for detecting genotoxic effects and suggested that the exposure duration should be between 10 and 24h. Ma et al. (1996) also considered that a five to seven-hour period was insufficient for detecting genotoxic effects and recommended that the exposure duration should be between 24 and 30h.

In biomonitoring the atmosphere, it is important that the overall duration of the study should not be short, given the variations in pollution levels that may occur according to how the weather varies across the different seasons. From our experience, we consider that 12 months is a period that more faithfully reflects the real effects of atmospheric pollution. To avoid contamination by soil pollutants, the vases containing the plants should be kept on stands at a height of between 1m and 2m. Figure 3 shows the dimensions of the stands that have been used in studies developed by researchers at our laboratory.

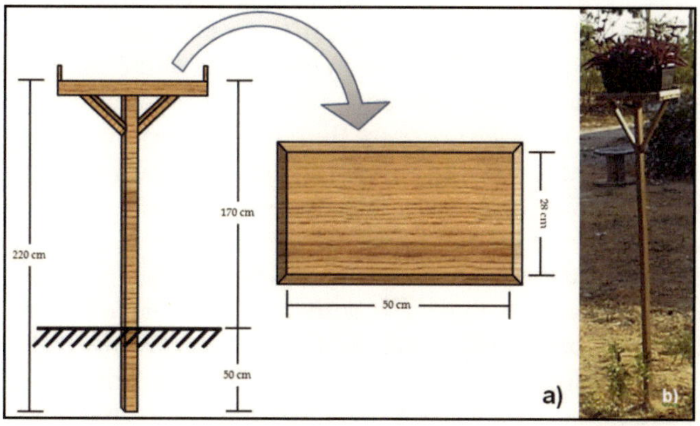

Fig. 3. Dimensions of the stands (a) and photography exposure mode (b)

The records in the literature reveal that the way in which plants have been exposed has varied greatly between the protocols adopted, according to the objectives and characteristics of each study, although this does not compromise the validity of the test.

Evaluations on the genotoxicity of atmospheric particles using Trad-MCH can be carried out in a laboratory, starting from air samples collected in glass fiber filters using a high-volume PM-10 sampler. The filter mesh used can be of different sizes in order to allow particle separation according to size. Exposure of the inflorescences (between 15 and 20 of them) can be done by immersing them for 24h in distilled water containing particle extract dissolved in dimethyl sulfoxide (DMSO).

Negative control plants should always be included in all protocols, i.e. plants grown in pollution-free areas and/or treated with distilled water or another compound that is known to be non-mutagenic, such as DMSO. A positive control, consisting of exposing plants to a known genotoxin, may also be included.

In assessing the genotoxicity of atmospheric pollutants, whether done through active monitoring or done through exposure in a laboratory, it is important to take the recovery time into consideration, i.e. the length of time after the exposure that is needed for the meiocytes to go from prophase I to the tetrad phase. This is the time taken for the meiotic cycle to run its course, and it will ensure that the damage induced from prophase I onwards will be observed in a sufficiently large number of cells. Therefore, the duration of meiosis in the *Tradescantia* plant or clone used in the test needs to be known: the closer the length of the recovery period is to the time taken for meiosis to occur, the greater the chance is that the damage induced through the exposure will be detected. Falistocco et al. (2000) observed that the duration of meiosis for clone 4430 was 84-86h and they therefore suggested that if the recovery time of 24h used in many protocols was insufficient to observe any damage, the experiment should be repeated with a longer recovery time.

4.2.3 Selection and preparation of inflorescences

The protocol for selecting and preparing inflorescences that is used in many Trad-MCN studies is the one proposed by Ma (1981). This protocol prescribes that young inflorescences should be collected and, after a recovery period of 24-30h, they should immediately be fixed in an acetic-alcohol solution (1:3). After 24h of fixing, the inflorescences are transferred to 70% ethanol and kept refrigerated (± 6°C) until the time of slide preparation.

To prepare the slides, the inflorescences are dissected with the aid of a surgical probe under a stereo microscope, in order to isolate the flower buds. Once the buds have been isolated, they are arranged in order of size and the one in the middle of the size range is selected first. This bud is transferred onto a glass slide and is dissected with the aid of a surgical probe, under a stereo microscope, in order to expose the anthers. After discarding the fragments of the bud, the anthers are macerated using a glass rod and drops of acetocarmine (2%) are added. After the maceration has been completed, a coverslip is placed on top of the preparation, which is then quickly heated up using a glass alcohol lamp (80°C). The coverslip is pressed down using finger pressure and then the preparation is placed under a microscope to identify whether there are sufficient numbers of tetrads. The steps for preparing the slides are presented in Figure 4. After confirming that there are enough tetrads, the slide is considered to be suitable for analysis and the other flower buds are discarded. If there are not enough tetrads, the procedure is repeated with another bud, and so on.

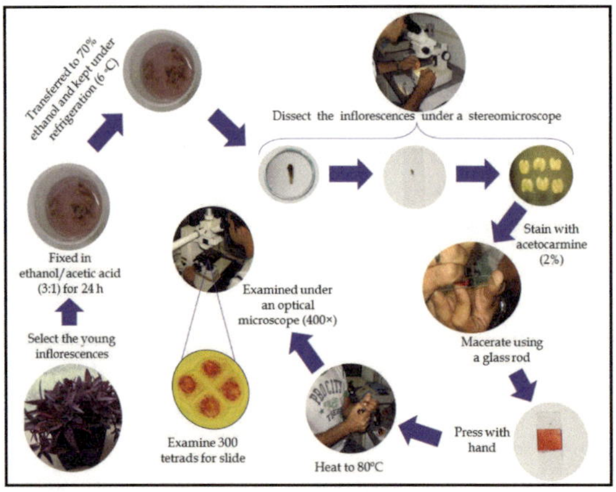

Fig. 4. Steps for preparing the slides

4.2.4 Cytogenetic analysis

The cytogenetic analysis for estimating the number of micronuclei is done under an optical microscope (400X). It should include counting 300 tetrads (Figure 5a) per slide and, for each exposure, five slides from individual inflorescences should be prepared, thus totaling 1,500

tetrads per treatment. It is important to code the slides so that the analysis can be done blindly. They should only be decoded after finishing the cytogenetic analysis.
Structures measuring 1/3 to 1/5 of the sized of the nucleus, with similar chromatin distribution and distinct separation from the nucleus, are taken to be micronuclei (Figure 5b).

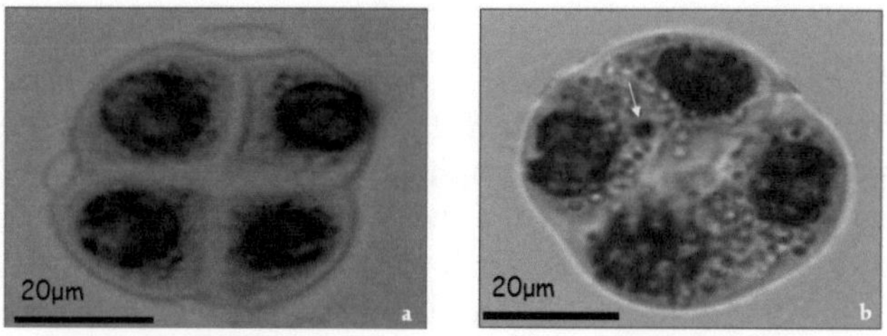

Fig. 5. Photomicrograph of tetrad of *Tradescantia pallida* with and without micronuclei (arrow)

4.2.5 Statistical analysis

The results obtained using Trad-MCN should always be subjected to adequate statistical analysis. Several tests can be used, for example Dunnet´s test, Student's *t*-test and Kruskal-Wallis test.

4.3 Advantages and disadvantages of Trad-MCN

Trad-MCN is a fast, simple and low-cost test that generates reliable results. Micronuclei are observed under an optical microscope and are easily identified. The plants are easy to cultivate, occupy little space and develop satisfactorily throughout the year, in the open air in subtropical regions and in greenhouses anywhere. The test is considered to be versatile, since it enables genotoxic evaluations on pollutants both *in situ* and in a laboratory, and because it is sensitive to contaminants of gaseous, liquid and radioactive nature, thus enabling biomonitoring of the soil, water and atmosphere.
The sterility of *Tradescantia* clones and the capacity of these plants for vegetative propagation eliminates the variation coming from meiotic recombination, thereby ensuring that genetic uniformity is maintained (which would be lost through this process). In addition, meiotic chromosomes are more sensitive to mutagens than mitotic chromosomes are (Rodrigues, 1999). Metabolism of pre-carcinogens into carcinogens occurs mostly in the plant itself, and enzyme activation is unnecessary.
Inflorescences of *Tradescantia* can be fixed and stored for subsequent cytological observation for long periods. The cytological preparations can also be stored for future studies.
The main limitation of Trad-MCN is that counting the micronuclei constitutes an underestimate of the real occurrence of genetic damage, given that chromosome rearrangements (translocations, inversions, etc.) are not detected. Because the test is highly sensitive, special care is required to ensure that the spontaneous micronucleus frequency is not altered by the action of agents other than those under investigation. In addition, the micronucleus frequency undergoes variations depending on climatic conditions. The mutagenic effects observed through the test cannot be directly associated with carcinogenicity, and the differences between plant and animal metabolism have to be taken into account, which limits the ability to extrapolate from the results (Ma, 1981).

4.4 Detection of atmospheric pollution using Trad-MCN

Plants and clones of *Tradescantia* are particularly sensitive to chemical mutagens, especially those that are present in the atmosphere. For this reason, they have been extensively used for biomonitoring of atmospheric pollution, in order to assess the genotoxic effects of gases and particles that are generated through the activity of factories, industrial plant, incinerators and landfills, and through burning fossil fuels.

4.4.1 *Tradescantia* clones

Ma et al. (1996) conducted a study in which the aim was to evaluate the genotoxic effects from atmospheric pollution generated through gas emissions consequent to the processes of dumping garbage in landfills and garbage incineration. They investigated occurrences of mutations in stamen hairs (Trad-SHM) and the micronucleus count in tetrads (Trad-MCN) in plants of clone 4430 that were exposed in the vicinity of a landfill and an incinerator (*in situ* biomonitoring) and in a laboratory. Four gases are emitted in such processes: toluene, ethylbenzene, trichloroethylene and ethyltoluene. For each test, two controls were analyzed: one kept in the laboratory and the other in the field. The observed frequency of stamen hair mutations, but not the frequency of micronuclei in tetrads, was significantly greater in the plants exposed in the landfill, in comparison with both of the controls. The average pink mutation rates and micronucleus frequencies in the incinerator tests were borderline positive. The authors considered that these results might be due to the length of exposure of the plants: not only was this short (5-7h), but also it did not include times of the day (early morning and late afternoon) when the wind died down. The results obtained from exposure of plants in the laboratory showed higher frequencies of micronuclei, but not of stamen hair mutations, in all the tests. This showed that both the landfill and the incinerator processes generated gases with relatively high genotoxicity.
The genotoxicity of the gases emitted by incinerators was also evaluated by Fomin & Hafner (1998) using Trad-MCN with three different approaches: a) direct fumigation of *Tradescantia* with diluted incinerator emissions; b) exposure of *Tradescantia* to smoke condensates; and c) in situ monitoring of genotoxicity of emissions near a municipal incinerator. The results obtained from direct fumigation of the plants and exposure to smoke condensates showed that the micronucleus frequencies in the exposed plants were significantly greater. The results of in situ monitoring of genotoxicity near a municipal waste incinerator showed that the *Tradescantia* micronucleus frequencies depended on the distance from the incinerator, and the direction of wind.
Misík et al. (2007) used plants of clone #03 to evaluate the genotoxicity of the urban pollution caused by an incinerator and a petrochemical plant in the city of Bratislava (Slovakia). The plants were placed at distances of 150m and 200m from the incinerator and petrochemical plant, respectively. During the period in which the first data were gathered (1997 to 2000), the micronucleus frequencies in the exposed plants were significantly higher than the frequencies observed in the control plants, but no difference in micronucleus occurrence was detected in the plants collected between 2003 and 2005.
These authors considered that the observed reduction in genotoxicity was due to the substantial reduction in atmospheric pollution that had been brought about through using new technologies. Reductions in the genotoxicity of the air caused by installation of protective measures had previously been observed by Fomin & Hafner (1998).
Rodrigues et al. (1996) showed that Trad-MCN, but not Trad-SHM, was effective for detecting the mutagenic effects of ozone, even at low concentrations. In fumigation

chambers, these authors subjected plants of clone 4430 to doses of 50 and 100 ppb for 6h, over periods of 1-3 days. Higher frequencies of micronuclei, but not of mutations in stamen cells, were observed in relation to the control group.

Monarca et al. (1999) evaluated the air quality in Brescia (Italy) in a study that included in *situ* biomonitoring and laboratory analysis on the genotoxic effects of particulate matter collected from the air of that city. In the laboratory, young *Tradescantia* inflorescences were kept for 24h in 1 ml of distilled water containing extracts of particulate matter dissolved in dimethyl sulfoxide (DMSO). These samples had been gathered from two streets in the city: one residential and the other, with heavy traffic. As a control, the plants were kept for the same length of time in distilled water and DMSO. The results from the laboratory exposure to particulate matter from the street with heavy traffic indicated that the micronucleus frequency was greater than in the control plants. In the *in situ* monitoring, the plants were placed in three areas of Brescia for 24h: area A, a link road in a heavily industrialized district; area B, a bypass around the town center; and area C, the square in front of the railway station. The results from the *in situ* biomonitoring did not, however, show any increased micronucleus frequencies.

Mutagenic effects from pollution originating from gasoline and diesel combustion in automotive vehicles were also investigated by these authors in a tunnel in the city of Perugia with heavy traffic movements and another in the city of Brescia with moderate traffic. Although the lengths of exposure of the plants in the tunnel in Perugia (1h and 5h) were shorter than those of the plants in the tunnel in Brescia (24h), higher micronucleus frequencies than in the control plants were only observed in the plants exposed in the Perugia tunnel. These authors concluded that *in situ* monitoring using Trad-MCN only seemed to be effective when the pollution levels were high.

Villarini et al. (2009) also conducted a monitoring study on atmospheric pollution in Perugia, using Trad-MCN and the Comet test on *Nicotiana tabacum* cv. Xanthi. These authors analyzed three urban sites in the city with different levels of pollution. The length of exposure was 24h and the sites were monitored five times: twice in the winter and once each in the spring, summer and autumn. The plants exposed in the area with greatest pollution were the ones that presented the highest micronucleus frequencies. Trad-MCN showed a better correlation with the pollution levels than did the Comet test.

The genotoxic effects from atmospheric pollution that are detected by Trad-MCN depend on the season of the year in which the plants are exposed, among other factors (Isidori et al., 2003). These authors exposed plants of clone 4430 at 17 sites in the city of Caserta (southern Italy), with a variety of traffic levels, during both the winter and the summer. Comparison with the control group revealed micronucleus frequencies that were significantly higher in the plants exposed during the winter, at all the sites, and at only two of the sites at which the plants had been exposed during the summer.

Within the framework of a European network for air quality assessment through using bioindicator plants, Trad-MCN was used to monitor urban areas in ten cities (Klumpp et al., 2006). The results obtained revealed that the genotoxic effects were higher in areas with heavier traffic. It was concluded that this test was a suitable tool for detecting local 'hot spots' of mutagenic air pollution in urban areas.

The correlation between micronucleus frequency in tetrads of *Tradescantia* and the adjusted mortality rates for cardiovascular and respiratory diseases and cancer was investigated by Mariani et al. (2009). Their study included biomonitoring at 28 sites with different degrees of atmospheric pollution in the city of São José dos Campos (Brazil) over a five-month period.

The higher micronucleus frequencies observed were in the plants that were exposed in areas of heavy traffic and in areas close to a petrochemical complex. Significant associations were detected between micronucleus frequency and the mortality rate due to cardiovascular diseases and cancer.

4.4.2 Tradescantia pallida (Rose) Hunt. cv. purpurea Boom

Biomonitoring studies on plants of the species *T. pallida* have mostly been conducted by Brazilian researchers. The results obtained have confirmed the results from pioneering studies that showed that naturally occurring plants of this species are as sensitive as clones (Suyama et al., 2002).

Batalha et al. (1999) used seedlings of *T. pallida* to evaluate the air quality in the city of São Paulo (Brazil), which is the largest urban center in the southern hemisphere. The seedlings were exposed to air samples collected from the city center that had been diluted in distilled water at concentrations of 15 and 30 mg/l. The micronucleus frequency in the plants exposed to the more concentrated solution was significantly greater than the frequency observed in the plants exposed to the lower concentration, thus demonstrating the genotoxic effects from the atmospheric pollution in this city. The air quality in the city of São Paulo was also evaluated by Guimarães et al. (2000) using this methodology and their results showed the effectiveness of *T. pallida* plants for detecting the genotoxic effects from atmospheric pollution.

Carvalho-Oliveira et al. (2005) also conducted a biomonitoring study using *T. pallida* plants in the city of São Paulo and demonstrated that diesel fuel burned by automotive vehicles induced genotoxicity. These authors evaluated occurrences of micronuclei over a three-day period during a strike by bus drivers and one week after the strike ended. The results obtained showed that the micronucleus frequency was significantly higher during the strike.

Carreras et al. (2006) investigated the genotoxicity of atmospheric pollution in the city of Córdoba (Argentina) between October 2004 and April 2005. *T. pallida* plants were exposed at three sites: one in the city center, characterized by heavy traffic of automotive vehicles; another on the university campus, located along a side road with heavy traffic of gasoline and diesel-powered vehicles (buses and trucks); and a third site in a residential area without significant pollution sources. Twenty young *T. pallida* inflorescences were collected from each sampling site in November, February and April. The lowest micronucleus frequencies were observed in the plants placed in the residential area. Similar results using the same methodology were obtained by Prajapati & Tripathi (2008) in a study on air quality biomonitoring in the city of Varanasi (India).

The results obtained through studies carried out by our team, in cities in the state of Bahia (Brazil) have systematically indicated the effectiveness of Trad-MCN using *T. pallida*, both for assessing the genotoxic effects consequent to air pollution originating from automotive vehicles and for assessing soil contaminants. Higher micronucleus frequencies associated with air contamination due to automotive vehicles were detected even in cities with low traffic flows.

The air quality in Senhor do Bonfim, a small city in the state of Bahia (Brazil), was evaluated by Andrade Júnior et at. (2008). This biomonitoring study involved exposure of *T. pallida* in two areas with different traffic levels: (1) on the Lomanto Junior highway, close to the ring road, where there is heavy traffic of cars, trucks and buses; and (2) in Praça Nova do Congresso, a public square located in the city center, where the traffic is less intense. The control consisted of plants that were kept in a traffic-free area. Young inflorescences were gathered every month over a 12-month period. The statistical analysis on the results showed that the micronucleus frequency was significantly higher in the exposed plants than in the controls. However, there was no statistically significant difference between the two exposure sites.

Air monitoring in the city of Feira de Santana, the second largest urban center in the state of Bahia (Brazil) was carried out by Meireles et al. (2009) using Trad-MCN with *T. pallida* between February 2006 and December 2007. Two types of monitoring were undertaken in this study: active and passive. Three localities with different traffic flows were chosen for testing: (1) the Campo Limpo traffic circle, which according to data that we collected had a traffic load of approximately 8200 vehicles/hour, mainly consisting of buses and trucks; (2) Getúlio Vargas Avenue which also had heavy traffic, although less than at Campo Limpo (3500 vehicles/hour according to data from the municipal authorities of Feira de Santana); and (3) an area with very low vehicle movement (about 10 vehicles/day, according to our observations during the study period).

Inflorescences were collected on a monthly basis from plants growing in these locations, under both passive and active monitoring regimes. Micronucleus occurrences were found to be proportional to vehicle flows under both monitoring regimes, while the plants under active monitoring showed greater sensitivity to atmospheric contamination.

5. Conclusion

From this presentation, it can be seen that, despite certain limitations, Trad-MCN carried out using plants or clones of *T. pallida* is an important tool for biomonitoring of atmospheric pollution.

6. Acknowledgement

We thank Feira de Santana State University, Bahia State University and FAPESB for their financial support for our research.

7. References

Alves, E.; Giusti, P.; Domingos, M.; Saldiva, P.; Guimarães, E. & Lobo D. (2001). Estudo anatômico foliar do clone híbrido 4430 de *Tradescantia*: alterações decorrentes da poluição aérea urbana. Revista Brasileira de Botânica, Vol. 24, No. 4, December 2001, pp. 567-576, ISSN 0100-8404

Andrade Júnior, S.; Santos Júnior, J.; Oliveira, J.; Cerqueira, E. & Meireles, J. (2008). Micronúcleos em tétrades de *Tradescantia pallida* (Rose) Hunt. cv. *purpurea* Boom: alterações genéticas decorrentes de poluição aérea urbana. Acta Scientiarum: Biological Sciences, Vol. 30, No. 3, July-September 2008, pp. 295-301, ISSN 1679-9283

Arutyunyan, R.; Pogosyan, V.; Simonyan, E.; Atoyants, A. & Djigardjian, E. (1999). In situ monitoring of the ambient air around the chloroprene rubber industrial plant using the Tradescantia-stamen-hair mutation assay. *Mutation Research/Fundamental and Molecular Mechanisms of Mutagenesis*, Vol. 426, No. 2, May 1999, pp. 117-120, ISSN 1386-1964

Batalha, J.; Guimarães, E.; Lobo, D.; Lichtenfels, A.; Deur, T.; Carvalho, H.; Alves, E.; Domingos, M.; Rodrigues, G. & Saldiva, P. (1999). Exploring the clastogenic effects of air pollutants in São Paulo (Brazil) using the Tradescantia micronuclei assay. *Mutation Research*, Vol. 426, No. 2, May 1999, pp. 229-232, ISSN 0027-5107

Beelen, R.; Hoek, G.; van den Brandt P.; Goldbohm, R.; Fischer, P.; Schouten, L.; Armstrong. B. & Brunekreef, B. (2008). Long-term exposure to traffic-related air pollution and lung cancer risk. Epidemiology, Vol. 19, No. 5, September 2008, pp. 702-710, ISSN 1531-5487

Brunekreef, B, Holgate, ST. (2002). Air pollution and health. *Lancet,* Vol. 360, No. 9341, October 2002, pp. 1233-1242, INSS 0140-6736

Carreras, H.; Pignata, M. & Saldiva P. (2006). *In situ* monitoring of urban air in Córdoba, Argentina using the *Tradescantia*-micronucleus (Trad-MCN) bioassay. *Atmospheric Environment*, Vol. 40, No. 40, December 2006, pp. 7824-7830, ISSN 1352-2310

Carvalho, H. (2005). A *Tradescantia* como bioindicador vegetal na monitoração dos efeitos clastogênicos das radiações ionizantes. *Radiologia Brasileira*, Vol. 38, No. 6, November-December 2005, pp. 459-462, ISSN 0100-3984

Carvalho-Oliveira, R.; Pozo, R.; Lobo, D.; Lichtenfels, A.; Martins-Junior, H.; Bustilho, J.; Saiki, M.; Sato, I. & Saldiva, PH. (2005) Diesel emissions significantly influence composition and mutagenicity of ambient particles: a case study in São Paulo, Brazil. *Environmental Research,* Vol. 98, No. 1, May 2005, pp. 1-7, ISSN 0013-9351

Cohen, J.; Carlson, G.; Charnley, G.; Coggon, D.; Delzell, E.; Graham, J.; Greim, H.; Krewski, D.; Medinsky, M.; Monson, R.; Paustenbach, D.; Petersen, B.; Rappaport, S.; Rhomberg, L.; Ryan, P. & Thompson, K. (2002). A comprehensive evaluation of the potential health risks associated with occupational and environmental exposure to styrene. *Journal of Toxicology and Environmental Health,* Part B: Critical Reviews, Vol. 5, No. 1, January-June 2002, pp. 1-265, ISSN 1521-6950

Coronas, M.; Horn, R.; Ducatti, A.; Rocha, J. & Varga, M. (2008). Mutagenic activity of airborne particulate matter in a petrochemical industrial area. *Mutation Research/Genetic Toxicology and Environmental Mutagenesis,* Vol. 650, No. 2, February 2008, pp. 196-201, ISSN 1383-5718

Eitan, O.; Yuval; Barchana, M.; Dubnov, J.; Linn, S.; Carmel, Y. & Broday, D. (2010). Spatial analysis of air pollution and cancer incidence rates in Haifa Bay, Israel. *Science of the Total Environment,* Vol. 408, No. 20, September 2010, pp. 4429-4439, ISSN 0048-9697

Emmerling-Thompson, M. & Nawrocky, M. (1982). Evidence of gametic mutation for flower color in *Tradescantia. Environmental and Experimental Botany,* Vol. 22, No. 4, November 1982, pp. 403-408, ISSN 0098-8472

Ennever, F.; Andreano, G. & Rosenkranz, H. (1988). The ability of plant genotoxicity assays to predict carcinogenicity. *Mutation Research,* Vol. 205, No. 1-4, May-August 1988, pp. 95-105, ISSN 0027-5107

Falistocco, E.; Torricelli, R.; Feretti, D.; Zerbini, I.; Zani, C. & Monarca S. (2000). Enhancement of micronuclei frequency in the Tradescantia/micronuclei test using a long recovery time. *Hereditas,* Vol. 133, No. 2, May 2000, pp. 171-174, ISSN 0018-0661

Fomin, A. & Hafner, C. (1998). Evaluation of genotoxicity of emissions from municipal waste incinerators with Tradescantia-micronucleus bioassay (Trad-MCN). *Mutation*

Research/Genetic Toxicology and Environmental Mutagenesis, Vol. 414, No. 1-3, May 1998, pp. 139-148, ISSN 1383-5718

Gichner, T. & Veleminský, J. (1999). Monitoring the genotoxicity of soil extracts from two heavily polluted sites in Prague using the Tradescantia stamen hair and micronucleus (MNC) assays. *Mutation Research/ Fundamental and Molecular Mechanisms of Mutagenesis*, Vol. 426, No. 2, May 1999, pp. 163-166, ISSN 1386-1964

Grant, W.; Lee, H.; Logan, D. & Salamone, M. (1992). The use of *Tradescantia* and *Vicia faba* bioassays for the *in situ* detection of mutagens in an aquatic environment. *Mutation Research/Fundamental and Molecular Mechanisms of Mutagenesis*, Vol. 270, No. 1, November 1992, pp. 53-64, ISSN 1386-1964

Guimarães, E.; Domingos, M.; Alves, E.; Caldini, N.; Lobo, D.; Lichtenfels A. & Saldiva, P. (2000). Detection of the genotoxicity of air pollutants in and around the city of São Paulo (Brazil) with the Tradescantia-micronucleus (Trad-MCN) assay. *Environmental and Experimental Botany*, Vol. 44, No. 1, August 2000, pp. 1-8, ISSN 0098-8472

Hanahan, D. & Weinberg, R. (2000). The hallmarks of cancer. *Cell*, Vol. 100, No. 1, January 2000, pp. 57-70, ISSN 0092-8674

Holland, N.; Bolognesi, C.; Kirsch-Volders, M.; Bonassi, S.; Zeiger, E.; Knasmueller, S. & Fenech, M. (2008). The micronucleus assay in human buccal cells as a tool for biomonitoring DNA damage: the HUMN project perspective on current status and knowledge gaps. *Mutation Research*, Vol. 659, No. 1-2, July-August 2008, pp. 93-108, INSS 0027-5107

Ianistcki, M.; Dallarosa, J.; Sauer, C.; Teixeira, C. & da Silva, J. (2009). Genotoxic effect of polycyclic aromatic hydrocarbons in the metropolitan area of Porto Alegre, Brazil, evaluated by Helix aspersa (Müller, 1774). *Environmental Pollution*, Vol. 157, No. 7, July 2009, pp. 2037-2042, ISSN 0269-7491

Isidori, M.; Ferrara, M.; Lavorgna, M.; Nardelli A. & Parrella, A. (2003). In situ monitoring of urban air in Southern Italy with the Tradescantia micronucleus bioassay and semipermeable membrane devices (SPMDs). *Chemosphere*,Vol. 52, No. 1, July 2003, pp. 121-126, ISSN 0045-6535

Joly, Aylthon Brandão. (1998). Botânica: introdução a taxonomia vegetal (12th edition), Nacional, ISBN 8504002314, São Paulo

Kapka, L.; Zemła, B.; Kozłowska, A.; Olewinska, E. & Pawlas, N. (2009). Jakość powietrza atmosferycznego a zapadalność na nowotwory płuc w wybranych miejscowościach i powiatach województwa śląskiego. *Przeglad Epidemiologiczny*, Vol. 63, No. 3, March 2009, pp. 439-444, ISSN 0033-2100

Kawanaka, Y.; Matsumoto, E.; Sakamoto, K. & Yun, S. (2011). Estimation of the contribution of ultrafine particles to lung deposition of particle-bound mutagens in the atmosphere. *Science of the Total Environment*, Vol. 409, No. 6, February 2011, pp. 1033-1038, ISSN 0048-9697

Kawanishi, M.; Watanabe, T.; Hagio, S.; Ogo, S.; Shimohara, C.; Jouchi, R.; Takayama, S.; Hasei, T.; Hirayama, T.; Oda, Y. & Yagi, T. (2009). Genotoxicity of 3,6-dinitrobenzo[e]pyrene, a novel mutagen in ambient air and surface soil, in mammalian cells in vitro and in vivo. *Mutagenesis*, Vol. 24, No. 3, May 2009, pp. 279-284, ISSN 1464-3804

Klumpp, A.; Ansel, W.; Fomin, A.; Schnirring, S. & Pickl, C. (2004). Influence of climatic conditions on the mutations in pollen mother cells of *Tradescantia* clone 4430 and implications for the Trad-MCN bioassay protocol. *Hereditas,* Vol. 141, No. 2, May 2004, pp. 142-148, ISSN 0018-0661

Klumpp, A.; Ansel, W.; Klumpp, G.; Calatayud, V.; Carrec, J.; He, S.; Peñuelas, J.; Ribas, A.; RoPoulsen, H.; Rasmussen, S.; Sanz, M. & Vergne, F. (2006). Tradescantia micronucleus test indicates genotoxic potential of trafic emissions in European cities. *Environmental Pollution,* Vol. 139, No. 3, February 2006, pp. 515-526, ISSN 0269-7491

Lorenzi, H. & Souza, HM. (2008). Plantas ornamentais no Brasil: arbustivas, herbáceas e trepadeiras (4th edition), Instituto Plantarum, ISBN 9788586714306, São Paulo

Ma, T.; Sparrow, A.; Schairer, L. & Nauman, A. (1978). Effect of 1,2-dibromoethane (DBE) on meiotic chromosomes of *Tradescantia. Mutation Research/Genetic Toxicology,* Vol. 58, No.2-3, November 1978, pp. 251-258, ISSN 1383-5718

Ma, T. (1981). Tradescantia micronucleus bioassay and pollen tube chromatid aberration test for in situ monitoring and mutagen screening. *Environmental Health Perspectives,* Vol. 37, No 1, January 1981, pp. 85–90, ISSN 0091-6765

Ma, T. & Grant, W. (1982) The *Tradescantia* - adventurous plants. *The herbarist,* Vol. 48, April 1982, pp. 36-44, ISSN 0740-5979

Ma, T.; Cabrera G.; Cebulska-Wasilewska, A.; Chen, R.; Loarca, F.; Vandenberg, A. & Salamone, M. (1994). Tradescantia stamen hair mutation bioassay. *Mutation Research,* Vol. 310, No. 2, October 1994, pp. 210-220, ISSN 0027-5107

Ma, T.; Xu, C.; Liao, S.; McConnell, H.; Jeong, B. & Won, C. (1996). In situ monitoring with the Tradescantia bioassays on the genotoxicity of gaseous emissions from a closed landfill site and an incinerator. *Mutation Research,* Vol. *359,* No. 1, January 1996, pp. 39-52, ISSN 0027-5107

Marcondes, Ayrton Cesar. (1993). Programas de saúde (4th edition), Atual, ISBN 85-7056-238-1, São Paulo

Mariani, R.; Jorge, M.; Pereira, S.; Melione, L.; Carvalho-Oliveira, R.; Ma, T. & Saldiva, P. (2009). Association between micronuclei frequency in pollen mother cells of *Tradescantia* and mortality due to cancer and cardiovascular diseases: a preliminary study in Sao José dos Campos, Brazil. *Environmental Pollution,* Vol. 157, No. 6, June 2009, pp. 1767-1770, ISSN 0269-7491

Meireles, J.; Rocha, R.; Costa-Neto, A. & Cerqueira, E. (2009). Genotoxic effects of vehicle traffic pollution as evaluated by micronuclei test in *tradescantia* (Trad-MCN). *Mutation Research/ Fundamental and Molecular Mechanisms of Mutagenesis,* Vol. 675, No. 1-2, April 2009, pp. 46-50, ISSN 1386-1964

Minouflet, M.; Ayrault, S.; Badot, P.; Cotelle, S. & Ferard, J. (2005). Assessment of the genotoxicity of 137Cs radiation using Vicia-micronucleus, Tradescantia-micronucleus and Tradescantia-stamen-hair mutation bioassays. *Journal of Environmental Radioactivity,* Vol. 81, No. 2-3, July 2005, pp. 143-153, ISSN 0265-931X

Misík, M.; Ma, T.; Nersesyan, A.; Monarca, S.; Kim, J. & Knasmueller, S. (2011) Micronucleus assays with Tradescantia pollen tetrads: an update. *Mutagenesis,* Vol. 26, No. 1, January 2011, pp. 215–221, ISSN 0267-8357

Misík, M.; Micieta, K.; Solenská, M.; Misíková, K.; Pisarcíková, H. & Knasmüller, S. (2007). In situ biomonitoring of the genotoxic effects of mixed industrial emissions using the *Tradescantia* micronucleus and pollen abortion tests with wild life plants: demonstration of the efficacy of emission controls in an eastern European city. *Environmental Pollution,* Vol. 145, No. 2, January 2007, pp. 459-466, ISSN 0269-7491

Monarca, S.; Ferreti, D.; Zanardini, A.; Falistocco, E., & Nardi, G. (1999). Monitoring of mutagens in urban air sample. *Mutation Research,* Vol. 426, No. 2, May 1999, pp. 189-192, ISSN 0027-5107

Peluso, M.; Srivatanakul, P.; Munnia, A.; Jedpiyawongse, A.; Meunier, A.; Sangrajrang S.; Piro. S.; Ceppi, M. & Boffetta P. (2008). DNA adduct formation among workers in a Thai industrial estate and nearby residents. *Science of The Total Environment,* Vol. 389, No. 2-3, February 2008, pp. 283-288, ISSN 0048-9697

Perera, F.; Hemminki, K.; Jedrychowski W.; Whyatt R.; Campbell U.; Hsu Y.; Santella,R.; Albertini, R.; & O'Neill, J. (2002). In utero DNA damage from environmental pollution is associated with somatic gene mutation in newborns. *Cancer Epidemiology Biomarkers & Prevention,* Vol. 11, No. 10, October 2002, pp. 1134-1137, ISSN 1538-7755

Prajapati, S. & Tripathi, B. (2008). Assessing the genotoxicity of urban air pollutants in Varanasi City using Tradescantia micronucleus (Trad-MCN) bioassay. *Environment International,* Vol. 34, No. 8, November 2008, pp. 1092-1096, ISSN 0160-4120

Rodrigues, G.; Madkour, A. & Weinstein, L. (1996). Genotoxic activity of ozone in *Tradescantia. Environmental and Experimental Botany,* Vol. 36, No. 1, May 1996, pp. 45-50, ISSN 0098-8472

Rodrigues, G.; Ma, T.; Pimentel, D. & Weinstein, LH. (1997). Tradescantia bioassays as monitoring systems for environmental mutagenesis: A review. *Critical Reviews in Plant Sciences,* Vol. 16, No. 4, July 1997, pp. 325-359, ISSN 0735-2689

Rodrigues, Geraldo Stachetti. (1999). Bioensaios de toxicidade genética com *Tradescantia* (1th edition), Embrapa Meio Ambiente, ISSN 1516-4691, Jaguariúna

Shima, N. & Ichikawa, S. (1994). Sinergism detected among methyl methanesulfonate, ethyl methanesulfonate and X-rays in inducing somatic mutations in the stamen hair of *Tradescantia* clone BNL 4430. *Environmental end Experimental Botany.*Vol. 34, No. 4, October 1994, pp. 393-406, INSS 0098-6472

Suyama, F.; Guimarães, E.; Lobo, D.; Rodrigues, G.; Domingos, M.; Alves, E.; Carvalho, H. & Saldiva, P. (2002). Pollen mother cells of Tradescantia clone 4430 and *Tradescantia pallida* var. purpurea are equally sensitive to the clastogenic effects of X-rays. *Brazilian Journal of Medical and Biological Research,* Vol. 35, No. 1, January 2002, pp. 127-129, ISSN 0100-879X

Traversi, D.; Degan, R.; De Marco, R.; Gilli, G.; Pignata, C.; Ponzio, M.; Rava, M.; Sessarego, F.; Villani, S. & Bono, R. (2008). Mutagenic properties of PM2.5 air pollution in the Padana Plain (Italy) before and in the course of XX Winter Olympic Games of "Torino 2006". *Environment International,* Vol. 34, No. 7, October 2008, pp. 966-970, ISSN 0160-4120

Traversi, D.; Degan, R.; De Marco, R.; Gilli, G.; Pignata, C.; Villani, S. & Bono, R. (2009). Mutagenic properties of PM2.5 urban pollution in the northern Italy: the nitro-compounds contribution. *Environment International,* Vol. 35, No. 6, April 2009, pp. 905-910, ISSN 0160-4120

Trédaniel, J.; Aarab-Terrisse, S.; Teixeira, L.; Savinelli, F.; Fraboulet, S.; Gossot, D. & Hennequin, C. (2009). Pollution atmosphérique et cancer bronchique : données épidémiologiques. *Revue des maladies respiratoires*, Vol. 26, No. 4, April 2009, pp. 437-445, ISSN 0761-8425

Underbrink, A.; Schairer, L. & Sparrow, A. (1973). The biophysical properties of 3.9-GeV nitrogen ions. V. Determinations of the relative biological effectiveness for somatic mutations in *Tradescantia*. *Radiation Research*. Vol. 55, No. 3, September 1973, pp. 437-446, ISSN 0033-7587

Villarini, M.; Fatigoni, C.;Dominici, L.;Maestri, S.;Ederli, L.; Pasqual, S.; Monarca, S. & Moretti, M. (2009). Assessing the genotoxicity of urban air pollutants using two in situ plant bioassays. *Environmental Pollution,* Vol. 157, No 12, December 2009, pp. 3354-3356, ISSN 0269-7491

Vineis, P.; Hoek, G.; Krzyzanowski, M.; Vigna-Taglianti, F.; Veglia, F.; Airoldi, L.; Overvad, K.; Raaschou-Nielsen, O.; Clavel-Chapelon, F.; Linseisen, J.; Boeing, H.; Trichopoulou, A.; Palli, D.; Krogh, V.; Tumino, R.; Panico, S.; Bueno-De-Mesquita, H.; Peeters, P.; Lund, E.; Agudo, A.; Martinez, C.; Dorronsoro, M.; Barricarte, A.; Cirera, L.; Quiros, J.; Berglund, G.; Manjer, J.; Forsberg, B.; Day, N.; Key, T.; Kaaks, R.; Saracci, R. & Riboli, E. (2007). Lung cancers attributable to environmental tobacco smoke and air pollution in non-smokers in different European countries: a prospective study. *Environmental Health*, Vol. 6, No. 7, February 2007, pp. 1-7 ISSN 1476-069X

Wada, M.; Kido, H.; Kishikawa, N.; Tou, T.; Tanaka, M.; Tsubokura, J.; Shironita, M.; Matsui, M.; Kuroda, N. & Nakashim, K. (2001). Assessment of air pollution in Nagasaki city: determination of polycyclic aromatic hydrocarbons and their nitrated derivatives, and some metals. *Environmental Pollution,* Vol. 115, No. 1, November 2001, pp. 139-147, ISSN 0269-7491

Watson, L. & Dallwitz, M. (1992) onwards. The families of flowering plants: descriptions, illustrations, identification, and information retrieval. Version: 4th February 2011. http://delta-intkey.com

Yauk, C.; Polyzos, A.; Rowan-Carroll, A.; Somers, C.; Godschalk, R.; Van Schooten, F.; Berndt, M.; Pogribny, I.; Koturbash, I.; Williams, A.; Douglas, G. & Kovalchuk, O. (2008). Germ-line mutations, DNA damage, and global hypermethylation in mice exposed to particulate air pollution in an urban/industrial location. *Proceedings of the National Academy of Sciences*. Vol. 105, No. 2, January 2008, pp. 605-610. ISSN 1091-6490

12

Silver Fir Decline in Mixed Old-Growth Forests in Slovenia: an Interaction of Air Pollution, Changing Forest Matrix and Climate

Jurij Diaci
Department of Forestry, Biotechnical faculty, University of Ljubljana
Slovenia

1. Introduction

Silver fir (*Abies alba* Mill.) is from ecological, economical and social point of view one of the most important conifer species in Europe (Kramer, 1992; Prpic et al., 2001). Many organisms are closely linked to silver fir dominated habitats. Silver fir (hereafter fir) can be very productive tree species in optimal site conditions. On siliceous, heavy soils in the lowlands it is more resistant to drought than Norway spruce (*Picea abies* (L.) Karst.). However, fir is extremely susceptible to environmental change (Schütt, 1978), including climate change (Brinar, 1964; Wick & Möhl, 2006; Anic et al., 2009), wild ungulate browsing (Motta, 1996; Senn & Suter, 2003), management regime (Mlinsek, 1964; Hockenjos, 2008) and air pollution (Eckstein et al., 1983; Krause et al. 1986; Elling et al., 2009). It is especially vulnerable to SO_2 emissions. This was recognized by analyses of needles composition, comparative bio-indication of sites with different emission loads and tree-ring studies (e.g. Wentzel, 1980; Elling, 1987; Elling et al., 2009). In Slovenia in 1980 the SO_2 emissions amounted to about 235,795 tonnes/year and have significantly decreased until 2007 to 14,245 tonnes/year, which is almost 94% (ARSO, 2010). The decrease of emissions in Slovenia started in late 1980s and influenced recovery of silver fir vitality and growth (Prelc et al., 1993; Ficko & Boncina, 2006). Similar processes were reported also from other countries (Dobrowolska, 1998).

Beside air pollution also the climatic conditions, especially warm summers and repeated droughts had a significant impact on fir health (Mlinsek, 1964; Leibundgut, 1974; Becker et al., 1989; Bert, 1993; Thomas et al., 2002; Ficko et al., 2011). The first reports of local and regional fir decline in mixed fir-beech forests of the Dinaric Mountains date from the end of the 1920s (Safar, 1951). Sun exposed, rockier, and drier slopes were especially affected. The fir decline was probably triggered by harsh winters and hot, dry summers (e.g. 1950) and it was accompanied by bark beetle calamities.

Overbrowsing is another factor that can cause fir density decrease (Gill, 1992; Motta, 1996; Klopcic et al., 2010). Namely, fir combines several features that make it vulnerable to browsing: it is one of the most palatable species, it grows very slowly in shaded old-growth conditions, and it recovers poorly from browsing damage. Old-growth forests are often game reserves and are thus characterised by excessive game densities (Korpel, 1995; Kenderes et al., 2008; Vrska et al., 2009; Diaci et al., 2010).

Again, other explanations of fir decline may include various ecological factors from complex pollutant interactions in forest soils (Ulrich, 1981), pathogen infections and changes in sapwood of unhealthy trees (Brill et al., 1981; Blaschke, 1982) to the impact of fir limited genetic variability in Central Europe, which may have caused lack of adaptability (Larsen, 1986). However, also direct influences of forest management on fir health seem important. They may include introduction of silvicultural systems not adapted to silver fir ecology (e.g. clear cut system), changing of forest climate by building of forest roads and racks as well as heavy fellings, favouring of other species (e.g. Becker et al., 1989; Kramer, 1992; Levanic, 1997; Hockenjos, 2008). Distinguishing between factors causing silver fir decline and mode of their operation (locally, regionally or globally) is important for several reasons. Firstly, understanding direct influences of environmental change may add to better environmental policies, for example overall forest die-back in Europe in 1980s led to significant reductions of SO_2 emissions. Moreover, clear relations between management practices and response of tree species are important for silvicultural prescriptions and future forest management.

The vast majority of research on silver fir decline was carried out in managed forests, where natural factors (e.g. exceptional weather situations), indirect human influences (e.g. climate change, air pollution) were confounded with direct human influences (e.g. favouring of selected species by felling regime, silvicultural operations and building of forest road infrastructure). Therefore, research on mortality process in old-growth forest and forest reserves is extremely important to account for influence of management (Leibundgut, 1982; Korpel, 1995; Peterken, 1996). However, destructive methods, such as tree-ring analysis are often not allowed in protected forests, therefore long term studies of tree population structure are beneficial here. The oldest data from Slovenian old-growth forests dates back to 1892.

Mixed fir and beech (*Fagus sylvatica* L.) mountain forests represent one of the major forest types in Southeastern Europe (Horvat et al., 1974). They are well preserved since they have been influenced by humans to a lesser extent than lowland or high mountain forests have (Ficko et al., 2011). The majority of European temperate old-growth forests lie in this region (Leibundgut, 1982), while small-scale, uneven-aged silvicultural systems are typical of managed forests (Mlinsek, 1972).

Beside exogenous factors also endogenous factors as for example forest structure, tree architecture and tree age significantly affect the sensitivity of trees to emissions stress (Brinar, 1964; Bert, 1993; Ficko et al., 2011). Studies of forest decline in managed forests pointed out positive interactions between stand age and fir decline (Thomas et al., 2002). Dense stands, high competition and poorly developed and suppressed trees influence tree vulnerability to air pollution (Becker, 1989). On the other hand dominant trees with crowns above average might well receive higher concentrations of air pollutants - due to so called edge-effect (Filipiak & Napierala-Filipiak, 2009). Therefore, different social statuses of trees (dominant, co-dominant and suppressed trees) seem important for the impact of air pollution and thus influence tree health condition.

The objectives of our study were to: (1) analyse temporal changes in fir density between four mixed old-growth forests in Slovenia, (2) analyse the relationship between tree social status and fir decline, (3) compare trends with managed forests and old-growth forests in neighbouring countries, and (3) discuss the relevance of natural and anthropogenic factors causing the decline and their spatiotemporal mode of operation.

2. Study site and methods

Four mixed Dinaric old-growth forests from Slovenia were selected for this study: Strmec (45°38'N, 14°49'E), Krokar (45°33'N, 14°47'E), Rajhenavski Rog (45°40'N, 15°01'E) and Pecka (45°46'N, 15°00'E). All of the forests were located in the mountain vegetation belt of the Dinaric Mountains (Figure 1, Table 1). The bedrock consists of Cretaceous limestone for which karstic macro- and micro-topography is typical (e.g. sinkholes, rock outcrops). Due to variable microrelief different soil types could be found ranging from rendzic leptosols to calcocambisols. Soil depth was generally between 10 and 70 cm. Forest sites in Pecka (hereafter PE) and Rajhenavski Rog (hereafter RR) were classified as the Dinaric fir-beech forest type (*Omphalodo-Fagetum*) according to the Braun-Blanquet typology (Puncer, 1980). In Strmec (hereafter ST) and Krokar (hereafter KR) 45% and 78% of forest sites were classified as *Omphalodo-Fagetum,* respectively. Other sites within ST and KR were classified as beech forests. In all old-growth forests the prevalent tree species was beech with the most recently recorded proportion in the growing stock ranging from 43% in RR to 93% in KR (Table 1). The second most important species was fir, whilst other species as sycamore maple (*Acer pseudoplatanus* L.), wych elm (*Ulmus glabra* Huds.), Norway spruce (*Picea abies* (L.) Karsten), common ash (*Fraxinus excelsior* L.) and large-leaved lime (*Tilia platyphyllos* Scop.) made up in general less than 1 % of the total stem volume. All forests were protected by law as total reserves and no regular fellings were recorded. Forest disturbance regime was assumed to be driven by small canopy gap dynamics with occasional intermediate disturbances, especially in form of windthrow (Hartman, 1987; Nagel et al., 2006).

Fig. 1. Map of Slovenia with locations of four old-growth forest reserves. All reserves are located within Dinaric Mountains

Diameter distribution data was obtained from full callipering of the whole reserves in 10-year intervals. In these inventories, the diameter at breast height (dbh) and species of each tree more than 10 cm in dbh in the entire reserve were recorded. Since the inventories from 1950s and 1960s for KR and PE didn't include the third dbh class (10 cm < dbh ≤ 15 cm), for calculation of indicators only trees thicker than 15 cm were taken into account, while the figures show all the original data. The oldest data for RR and PE (1892) came from the first forest management plan and it has been obtained through sampling transects (Boncina, 1999). The stem wood volume was calculated using the same volume tables for all years and old-growth forests (Cokl, 1992). For the purpose of studying the interaction between fir decline and tree social status three extended dbh classes were used (edbh1 = 15 cm < dbh ≤ 30 cm, edbh2 = 30 cm < dbh ≤ 50 cm, edbh3 = 30 cm < dbh ≤ 50 cm).

Research site	Stand key	Area (ha)	Altitude a.s.l. (m)	Average annual precipitation (mm)	Average annual temperature (C°)	Parent material	Beech / silver fir / Norway spruce ratio
Strmec	ST	15.6	840-940	~1770	6-7	Limestone	79/21/0
Krokar	KR	74.5	840-1170	~2000	6	Limestone	93/7/0
Rajhenavski Rog	RR	52.1	740-880	~1760	6-7	Limestone	43/57/0
Pecka	PE	59.5	795-910	~1500	6-7	Limestone	82/18/0

Table 1. General characteristics of the four old-growth forests. Tree species ratio refers to the share in growing stock

3. Results

At the first inventory the highest fir share in the total tree density was recorded in RR (61%) and the lowest in KR (29%; Table 1). The same situation regarding ranking of fir share in the total tree density between the reserves was recorded also at the last inventory. However, during the same period fir share significantly decreased in all reserves. The highest absolute decrease in the fir share was recorded in PE, followed by KR, RR and ST. The time periods between the first and the last inventories were different between the reserves. Therefore, for a more objective comparison the annual difference in the fir share was calculated (Table 2). The highest annual decrease in fir share was recorded in ST, followed by KR, PE and RR.

Old-growth forest	Year of first / last inventory	Fir share in density at first inventory	Fir share in density at last inventory	Difference in fir share	Difference in fir share / year
Strmec	1984 / 2004	45.7%	34.0%	-11.7%	-0.59%
Krokar	1961 / 2004	28.9%	12.0%	-16.9%	-0.39%
Rajhenavski Rog	1957 / 2007	61.0%	48.9%	-12.1%	-0.24%
Pecka	1953 / 2003	36.7%	19.0%	-17.7%	-0.35%

Table 2. Changes of fir share in the total tree density

Fir density (dbh > 15 cm) at the first inventory spanned from 168.2 trees/ha in ST to 94.8 trees/ha in KR (Table 3). During following years it decreased in all reserves and reached highest values during the last inventory in RR with 97.3 trees/ha and lowest values in KR with 33.0 trees/ha. Maximum difference between the first and the last inventory divided by year (annual rate of fir density change; hereafter ARC) was recorded in ST, followed by PE, KR and RR with -3.67, -2.16, -1.44 and -1.28 trees/ha/year, respectively. However, it seemed likely that old-growth forests with higher fir initial density would have had experienced a higher absolute fir decrease. Therefore, ratio between initial and observed fir density was introduced as a more objective indicator of fir decline intensity. The highest relative annual rate of change (hereafter RARC) was recorded in ST, followed by KR, PE and RR.

Old-growth forest	Fir density at first inventory (a)	Fir density at last inventory (b)	Difference in fir density (b-a)	Annual rate of fir density change (ARC) ((b-a)/year)	Ratio between initial and final fir density (b/a)	Relative annual rate of fir density change (RARC) (b/a/year)
Strmec	168.2	94.7	-73.4	-3.67	56.3%	-2.18%
Krokar	94.8	33.0	-61.8	-1.44	34.8%	-1.52%
Rajhenavski Rog	161.1	97.3	-63.8	-1.28	60.4%	-0.79%
Pecka	146.5	38.3	-108.2	-2.16	26.2%	-1.48%

Table 3. Changes in fir density between the first and last inventory. All data are per ha

For the old-growth forests RR and PE data on the fir density from the first forest management plan was available (Table 4). These data may serve only as a rough estimate of conditions at that time, since forest compartments were larger, inventory was based on sample transects, and only the trees up to 80 cm in the dbh were measured. Comparison of the fir density among the first measurements in 1892 and inventories in the 1950s showed relatively small differences in both old-growth forests. RARC was in the case of RR positive and amounted to only 0.01%, and it was negative in the case of PE and amounted to -0.06%.

Old-growth forest	Fir density in 1892 (a)	Fir density in 1950s (b)	Difference in fir density (b-a)	Annual rate of fir density change (ARC) ((b-a)/year)	Ratio between initial and final fir density (b/a)	Relative annual rate of fir density change (RARC) (b/a/year)
Rajhenavski Rog	160.0	161.1	1.1	0.02	100.7%	0.01%
Pecka	152.0	146.5	-5.5	-0.09	96.4%	-0.06%

Table 4. Changes in fir density between the first inventory in 1892 for both old-growth forests and inventory in 1953 for PE and 1957 for RR. Density data for all inventories is for trees with dbh > 15 cm. During the first inventory only trees up to dbh < 80 cm were measured

We calculated ARC and RARC for periods between individual inventories for all old-growth forests. On the Figure 2 the mean ARC and RARC values in different periods were plotted against mean year of the period. For example, in ST there were three inventories: 1984, 1994 and 2004. ARC for the first (1984-1994) and second (1994-2004) period were -1.92 and -5.42 trees/ha/year, respectively. The mean years of the first and second period used for plotting of ARC were 1989 and 1999, respectively. As with all parameters in Tables 2 and 3 all ARC and RARC values were negative. The highest ARC values were -5.42 trees/ha/year in ST between 1994-2004, and -2.86 trees/ha/year for PE between 1953-1973. While the lowest ARC values -0.29 trees/ha/year were recorded in RR between 1957-1967, and -0.44 in PE between 1994-2003. RARC values spanned in interval from -0.18% (RR 1957-1967) and -3.64% (ST 1994-2004). The development of RARC indicator showed a similar course for the three forest reserves, KR, RR and PE. At the first and last inventories the index values were higher, and in the interim the lowest values were achieved. All minima were reached during 1980s. However, ST with shorter interval of inventories showed a steep decline during the last period.

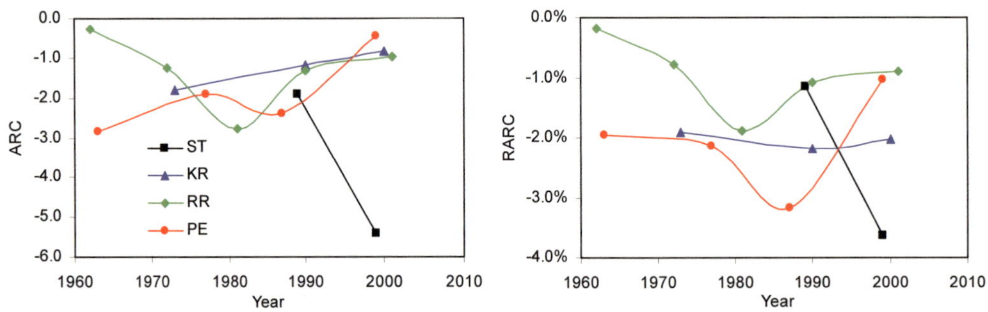

Fig. 2. Left: Annual rate of fir density change ARC (trees/ha/year) for four old-growth forests: ST = Strmec, KR = Krokar, RR = Rajhenavski Rog, PE = Pecka; Right: Relative annual rate of fir decline RARC (trees/ha/year/initial density of the period)

Further we were interested how fir decline affected trees with different social status (Table 5). As a surrogate for the tree social status, extended dbh classes were used (edbh). Tree diameter at breast height is namely closely related to tree height and social status. It was assumed that tress in edbh1 were mostly suppressed, tress in edbh2 were about to establish in the canopy layer, while trees in edbh3 formed the upper canopy layer and dominant individuals.

	Strmec			Rajhenavski Rog					
	1984-1994	1994-2004	**1984-2004**	1957-1967	1967-1976	1976-1985	1985-1995	1995-2007	**1957-2007**
edbh1	-0.58%	-3.54%	**-1.96%**	-0.37%	-1.02%	-3.50%	-0.94%	-1.34%	**-1.09%**
edbh2	-0.88%	-3.71%	**-2.13%**	-0.92%	-1.38%	-1.93%	-2.03%	-1.09%	**-1.09%**
edbh3	-3.23%	-3.91%	**-2.94%**	0.48%	-0.26%	-0.71%	-0.72%	-0.58%	**-0.34%**
	Krokar				Pecka				
	1961-1984	1984-1995	1995-2004	**1961-2004**	1953-1973	1973-1980	1980-1994	1994-2003	**1953-2003**
edbh1	-2.02%	-1.35%	-1.82%	**-1.44%**	-2.33%	-2.86%	-1.59%	0.99%	**-1.28%**
edbh2	-2.22%	-3.37%	-1.31%	**-1.69%**	-2.86%	-1.79%	-3.38%	-1.26%	**-1.65%**
edbh3	-0.69%	-2.48%	-3.56%	**-1.36%**	-1.06%	-1.82%	-3.94%	-2.84%	**-1.54%**

Table 5. Relative annual rate of fir density change RARC (trees/ha/year/initial density of the period) according to periods between inventories, extended dbh classes (edbh) and old-growth forests. Note: edbh1 = 15 cm < dbh ≤ 30 cm; edbh2 = 30 cm < dbh ≤ 50 cm; edbh3 = 30 cm < dbh ≤ 50 cm

We hypothesized that the dominant trees were most vulnerable to air pollution, since they were most directly exposed to its negative impacts. This was the case in ST for two periods, in PE for two periods, and in KR for one period. However, trees in edbh2 exhibited higher RARC during six periods (tree periods in RR, two in KR and one in PE), while the highest RARC in the suppressed trees of edbh1 was found during three periods (two periods in RR and one in PE). From this analysis we could not confirm the hypothesis about the greater sensitivity of the dominant trees to air pollution. Moreover, the lowest RARC was most frequently recorded in the sub-canopy tree layer (edbh2). This result is even more obvious if we compare RARC over the whole inventory periods (Table 4); it was the lowest in edbh2 for three out of four old-growth forests.

Figure 3 shows changes in empirical dbh distributions for selected years of inventories. A strong reduction in the number of fir trees is obvious throughout all periods. However, comparisons of dbh distributions also suggest differences in the structure of the fir regression. In the case of KR a substantial reduction within small and medium diameter trees was recorded, while large diameter trees were already rare at the first inventory. In PE and ST the fir reduction was more balanced across the whole dbh range. A special case was RR, where fir decrease was more pronounced within small and medium diameter tress, while large diameter trees have been shifted to even larger diameter classes.

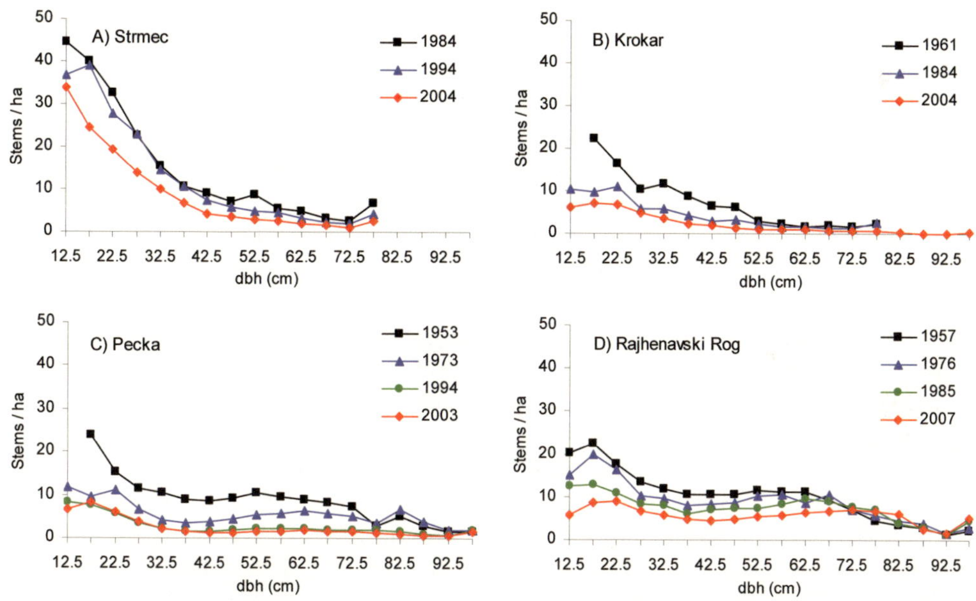

Fig. 3. Temporal changes of silver fir empirical diameter distributions for four mixed old-growth forests in Slovenian Dinaric Mountains

4. Discussion

First observations of fir decline in Slovenia were made in 1956 (Brinar, 1964; p. 111), they triggered research which focused on managed forests (Mlinsek, 1964). Fir decline in old-growth forests was underestimated, probably due to the abundance of old and non-vital trees, although old-growth forests may be more sensitive to air pollution than younger, managed forests. Because there was no awareness of fir decline in old-growth forests arguments emerged, that the fir decline in managed forests was triggered or influenced by heavy and unregulated post World War II. fellings. However, the first studies didn't confirm that hypothesis (Brinar, 1964). Our results also suggested a rather early start of the fir decline in old-growth forests. It is necessary to take into account the fact, that the studied old-growth forests were in the Dinaric part of Slovenia, which was not exposed to heavy direct emissions at the time. Therefore, long-range transport of pollutants might have been more important, while direct emissions became more intense later, especially during 1970s and 1980s. Thus, the results from our research don't support the primary hypothesis that forest management had a significant impact on silver fir decline in Slovenia. Similar results on early silver fir decline were reported also for mixed old-growth forests in other regions of Europe (e.g. Korpel, 1995; Diaci et al., 2008).

Our results indicate that the intensity of fir decline varied between the old-growth forests. The most significant decline was recorded in ST and KR. For both old-growth forests also some extreme, especially steeper slopes were characteristic, which were only partly overgrown by community *Omphalodo-Fagetum*. Both forests covered more pronounced mountain ridges compared to RR and PE, and their average altitude was slightly higher. ST was also characterized by dry sun exposed slopes. More intensive fir regression at the border of its natural distribution and on dryer sites was also reported by Ficko et al. (2011) for managed forests in Slovenia. Thomas et al. (2002) reported positive relationship between mean altitude and fir decline.

During last hundred years fir density in old-growth forests significantly changed. In the second half of the 19th century the extinction of large herbivores triggered abundant fir regeneration (Klopcic et al., 2010). Therefore, fir density gradually increased until the mid 20th century, when reintroduced ungulates reached excessive densities and overbrowsing begun. This was coupled with an increase of air pollution due to industrialisation. Our oldest data on fir density revealed relatively small changes in the fir density in RR and PE prior the mid 20th century, when compared with the second half of the same century. While old-growth inventories from 1950s indicated a gradual but significant decrease of fir density. The decrease peaked in KR, PE and RR during 1980s. Our data suggests that after this period the intensity of fir regression decreased again with the exception of ST. The synchronous trends of silver fir decline in old-growth forests can be closely linked to the records of SO_2 emissions in Slovenia (ARSO, 2010). In the reference years 1980 and 1990, SO_2 emissions in Slovenia amounted about 236,000 and 154,000 t/year, respectively. For Slovenia emissions peak is estimated even earlier around 1978, when two largest coal power plants were finished. According to the information presented in Stern (2006) in the period from 1975 to 1993 east Europe was the world's largest emitter of sulphur. However, the SO_2 emissions in Slovenia were not as high as in the northern Czech Republic, Germany, or south-western Poland ("Black Triangle"; Saltbones & Dovland, 1986). On the other hand, the forests of Slovenia were also affected by SO_2 emissions from Western and Northern Europe (Komlenovic, 1989).

In our research extended dbh classes (edbh) were used as proxies for assessing the social status of the trees. The results indicate variable fir decline rates among extended dbh classes. However, there was some evidence that overall fir decline rate was the highest in the edbh2 class ($30 < dbh \leq 55$ cm). When evaluating the impact of the extended dbh classes on silver fir decline, it should be noted that fir reduction in the edbh1 class was partly influenced by ungulate overbrowsing of fir regeneration, which started in the period after the WW II.

Moreover, when comparing the fir decline among extended classes, also the influence of the windthrow of 1983 in PE has to be considered. At that time especially large diameter firs were affected (Nagel et al., 2006). Trees in edbh2 class were on average between 20 and 35 m high and thus had been just beginning to enter the upper canopy layer. Considering high overall tree density in old-growth forests, trees in medium layer might be stressed due to crown and root competition. Similar results were reported from early research on fir decline in Slovenian managed forests (Brinar, 1964), where lowest fir vitality was found in firs with dbh between 30 and 40 cm. Our results are comparable with results of Becker (1989) who observed that dense stands and suppressed trees increased tree vulnerability to air pollution. On the other hand the hypothesis of large diameter trees being more susceptible to air pollution via wet deposition due to edge-effect (Filipiak & Napierala-Filipiak, 2009) can not be confirmed. In general the interactions between emissions impact and tree social status are complex. Ferlin (1990) found in evenaged spruce forests with high SO_2 emission loads that dominant trees were less damaged compared to co-dominant and suppressed trees. In the adjacent forest area with lower emission loads he found the opposite trend. Very likely, when studying the impact of emissions on trees of different social positions many different factors need to be taken into account: for example the distance to the main emission area, emissions mode (e.g. long or short range transport of emissions, wet or dry deposits) and the structure of the forest.

5. Conclusion

The literature outlines a number of possible causes for the acute fir decline during the second half of the last century. Very probably the reason was a complex interaction of many ecological factors. However, even more important than discussing possible influential factors is to highlight the predominant ones. Repeated inventories from old-growth forests in Slovenia and south-eastern Europe revealed a synchronous fir decline during the last fifty years, and a high association between SO_2 emission loads and intensity of fir decline (Diaci et al., 2008). Therefore, the results support the opinion that the first and the fundamental factor of fir decline was the polluted atmosphere – especially SO_2 emissions, while other factors including management, were of a lesser importance. Moreover, it may be likely that many other factors, like pathogens were more a consequence than a cause of the fir decline. However, fir density decrease in relatively remote old-growth forest areas, occasional fir vitality decline, as well as bark beetle calamities after extreme droughts (e.g. summer of 2003) confirm the impact of the climate change (warmer climate and incidence of summer draughts). Therefore, it seems that influences of air pollution and climate, which operate on different spatial scales, were related and they may have had a synergetic effect.

Our research indicated a greater sensitivity of the tree middle layer to air pollution, thus suggesting that the period of acute fir decline will be reflected in the mixed old-growth forest structure for centuries. Even more alarming is the fact that the fir regeneration overbrowsing by ungulates totally stopped fir recruitment. Thus, the fir density decrease will continue in spite of improved vitality. The results further suggest that for the future research of mixed old-growth forests more attention has to be devoted to the impact of polluted air, and indirect anthropogenic disturbances in general. Nevertheless, the reduced SO_2 emissions on a Europe-wide level represent an important example of best practice for environmental conservation, which should more often exposed to the public and replicated in other areas of human induced environmental change. To maintain viable populations of silver fir, a coordinated action by all the domains of forest management, including silviculture, game management, hunting, harvesting and marketing, is a requisite.

6. Acknowledgment

This research was supported by the Slovenian Research Agency (Research Programme P4-0059: Forest, forestry and renewable forest resources) and the Ministry of Agriculture, Forestry, and Food (Project CRP V4-0540: Conservation ecology and management of fir in Slovenia). For help on the manuscript, I wish to thank Tomaz Adamic.

7. References

Anic, I.; Vukelic, J.; Mikac, S.; Baksic, D. & Ugarkovic, D. (2009). Utjecaj globalnih klimatskih promjena na ekološku nišu obične jele (*Abies alba* Mill.) u Hrvatskoj. *Sumarski list*, Vol.133, No.3-4, pp. 135-144

ARSO, (2010). EIONET (Environmental Information and Observation Network), http://eionet-en.arso.gov.si/. ARSO Environmental Agency of the Republic of Slovenia

Becker, M.; Landmann, G. & Levy, G. (1989). Silver fir decline in the Vosges mountains (France): Role of climate and silviculture. *Water, Air, & Soil Pollution*, Vol.48, pp. 77-86

Bert, G.D. (1993). Impact of Ecological Factors, Climatic Stresses, and Pollution on Growth and Health of Silver Fir (*Abies-Alba* Mill) in the Jura Mountains - an Ecological and Dendrochronological Study. *Acta Oecologica*, Vol.14, pp. 229-246

Blaschke, H. (1982). Schadbild und Ätiologie des Tannensterbens: III. Das Vorkommen einer Phytophthora-Fäule an Feinwurzeln der Weißtanne (*Abies alba* Mill.), *European Journal of Forest Pathology*,Vol.12, pp. 232-238

Boncina, A. (1999). Stand dynamics of the virgin forest Rajhenavski Rog (Slovenia) during the past century, In: *Proceedings of the Invited Lecturers' Reports COST Action E4: Forest Reserves Research Network*, Diaci, J. (Ed.), pp. 95-110, Plesko, Ljubljana

Brill, H.; Bock, E. & Bauch, J. (1981). Über die Bedeutung von Mikroorganismen im Holz von *Abies alba* Mill. für das Tannensterben. *Forstwiss. Cbl.*, Vol.100, pp. 195-206

Brinar, M. (1964). Zivljenjska kriza jelke na slovenskem ozemlju v zvezi s klimaticnimi fluktuacijami. *Gozdarski vestnik*, Vol.22, pp. 97-144

Cokl, M. (1992). *Gozdarski prirocnik*, Biotehniska fakulteta, Oddelek za gozdarstvo, Ljubljana.

Diaci, J.; Rozenbergar, D.; Mikac, S.; Anic, I.; Hartman, T. & Boncina, A. (2008). Long-term changes in tree species composition in old-growth Dinaric beech-fir forest. *Glasnik za sumske pokuse*, Vol.42, pp. 13-27

Diaci, J.; Rozenbergar, D. & Boncina, A. (2010). Stand dynamics of Dinaric old-growth forest in Slovenia: Are indirect human influences relevant? *Plant Biosystems*, Vol.144, pp. 194-201

Dobrowolska, D. (1998). Structure of silver fir (*Abies alba* Mill.) natural regeneration in the 'Jata' reserve in Poland. *Forest Ecology and Management*, Vol.110, pp. 237-247

Eckstein, D.; Aniol, R.W. & Bauch, J. (1983). Dendroklimatologische Untersuchungen zum Tannensterben. *European Journal of Forest Pathology*, Vol.13, pp. 279-288

Elling, W. (1987). A procedure for the registration of the time course and degree of injury in conifer stands. *European Journal of Forest Pathology*, Vol.17, pp. 426-440

Elling, W.; Dittmar, C.; Pfaffelmoser, K. & Rötzer, T. (2009). Dendroecological assessment of the complex causes of decline and recovery of the growth of silver fir (*Abies alba* Mill.) in Southern Germany. *Forest Ecology and Management,* Vol.257, pp. 1175-1187

Ferlin, F. (1990). *Vpliv onesnazevanja ozracja na rastno zmogljivost odraslih smrekovih sestojev,* MSc Thesis, University of Ljubljana

Ficko, A. & Boncina, A. (2006). Silver fir (*Abies alba* Mill.) distribution in Slovenian forests. *Zbornik gozdarstva in lesarstva,* Vol.79, pp. 19-35

Ficko, A.; Poljanec, A. & Boncina, A. (2011). Do changes in spatial distribution, structure and abundance of silver fir (*Abies alba* Mill.) indicate its decline? *Forest Ecology and Management,* doi:10.1016/j.foreco.2010.12.014.

Filipiak, M. & Napierala-Filipiak, A. (2009). Effect of canopy density on the defoliation of the European silver fir (*Abies alba* Mill.) due to heavy industrial pollution. *Dendrobiology,* Vol.62, pp. 17-22

Gill, R.M.A. (1992). A review of damage by mammals in north temperate forests: 3. Impact on trees and forests. *Forestry,* Vol.65, pp. 363-388

Hartman, T. (1987). *Gozdni rezervati Slovenije - Pragozd Rajhenavski Rog.* Univerza v Ljubljani, Biotehniska fakulteta, Ljubljana

Hockenjos, W. (2008). *Tannenbäume - Eine Zukunft für* Abies alba, DRW-Verlag Weinbrenner, Leinfelden-Echterdingen

Horvat, I.; Glavac, V. & Ellenberg, H. (1974). *Vegetation Sudosteuropas,* G. Fischer, Stuttgart

Kenderes, K.; Mihok, B. & Standovar, T. (2008). Thirty years of gap dynamics in a central European beech forest reserve. *Forestry,* Vol.81, pp. 111-123

Klopcic, M.; Jerina, K. & Boncina, A. (2010). Long-term changes of structure and tree species composition in Dinaric uneven-aged forests: are red deer an important factor? European Journal of Forest Research, Vol.129, pp. 277-288

Komlenovic, N. (1989). Utjecaj SO_2 i nekih drugih polutanata na sumsko drvece s posebnim osvrtom na SR Hrvatsku. *Sumarski list,* Vol.113, pp. 243-260

Korpel, S. (1995)*: Die Urwälder der Westkarpaten.* Gustav Fischer Verlag, Stuttgart, Jena, New York

Kramer, W. (1992). Die Weißtanne (*Abies alba* Mill.) in Ost- und Südosteuropa. Gustav Fischer Verlag, Stuttgart, Jena, New York

Krause, G.; Arndt, U.; Brandt, C.J.; Bucher, J.; Kenk, G. & Matzner, E. (1986). Forest decline in Europe: Development and possible causes. *Water, Air and Soil Pollution,* Vol.31, pp. 647-668

Larsen, B.J. (1986). Das Tannensterben: Eine neue Hypothese zur Klärung des Hintergrundes dieser Rätselhaften Komplexkrankheit der Weisstanne (*A. alba* Mill.). *Forstw. Cbl.,* Vol.105, pp. 381-396

Leibundgut, H. (1974). Zum Problem des Tannensterbens. *Schweiz. Z. Forstw.,* Vol.125, pp. 476-484

Leibundgut, H. (1982). *Europäische Urwälder der Bergstufe,* Haupt, Bern

Levanic, T. (1997). Growth depression of silver fir (*Abies alba* Mill.) in the Dinaric phytogeographic region between 1960-1995. *Zbornik gozdarstva in lesarstva,* Vol.52, pp. 137-164

Mlinšek, D. (1964). Susenje jelke v Sloveniji - prvi izsledki. *Gozdarski vestnik,* Vol.22, pp. 145-159

Mlinsek, D. (1972). Ein Beitrag zur Entdeckung der Postojna Kontrollmethode in Slowenien. *Forstw. Cbl.*, Vol.91, pp. 291-296

Motta, R. (1996). Impact of wild ungulates on forest regeneration and tree composition of mountain forests in the Western Italian Alps. *Forest Ecology and Management*, Vol.88, pp. 93-98

Nagel, T.A.; Svoboda, M. & Diaci, J. (2006). Regeneration patterns after intermediate wind disturbance in an old-growth *Fagus-Abies* forest in southeastern Slovenia. *Forest Ecology and Management*, Vol.226, pp. 268-278

Peterken, G.F. (1996). Natural woodland: ecology and conservation in northern temperate regions, Cambridge University Press, Cambridge

Prelc, F., Veselič, Z., Jez, P. (1993). Rast jelke (*Abies alba* Mill.) se izboljšuje. *Gozdarski vestnik*, Vol.51, No.7-8, pp. 314-331

Prpic, B. (Ed.). (2001). *Obicna jela (*Abies alba *Mill.) u Hrvatskoj.* Akademija sumarskih znanosti, Zagreb

Puncer, I. (1980). Dinarski jelovo-bukovi gozdovi na Kocevskem. *Razprave*, Vol.22, No.6, pp. 161

Safar, J. (1951). Ugibanje i obnavljanje jele u prebornim sumama Gorskog Kotara. *Sumarski list*, Vol.75, pp. 299-303

Saltbones, J. & Dovland, H. (1986). Emissons of sulphur dioxide in Europe in 1980 and 1983, EMEP MSC-W Report 1/86, Norwegian Meteorological Institute, Oslo

Schütt, P. (1978). Die gegenwärtige Epidemic des Tannensterbens. *European Journal of Forest Pathology*, Vol.7, pp. 187-190

Senn, J. & Suter, W. (2003). Ungulate browsing on silver fir (Abies alba) in the Swiss Alps: beliefs in search of supporting data. *Forest Ecology and Managemet*, Vol.181, pp. 151-164

Stern, D.I. (2006). Reversal of the trend in global anthropogenic sulphur emissions, *Global Environmental Change*, Vol.16, pp. 207-220

Thomas, A.-L.; Gegout, J.-C.; Landmann, G.; Dambrine, E. & King, D. (2002). Relation between ecological conditions and fir decline in a sandstone region of the Vosges mountains (northeastern France). *Annals of Forest Science*, Vol. 59, pp. 265-273

Ulrich, B. (1981). Eine ökosystemare Hypothese über die Ursachen des Tannensterbens (*Abies alba* Mill.). *Forstwiss. Cbl.*, Vol.100, pp. 228-236

Vrska, T.; Adam, D.; Hort, L.; Kolár, T. & Janík, D. (2009). European beech (*Fagus sylvatica* L.) and silver fir (*Abies alba* Mill.) rotation in the Carpathians: A developmental cycle or a linear trend induced by man? *Forest Ecology and Management*, Vol.258, pp. 347-356

Wentzel, K.F. (1980). Weissitanne = immissionsempfindlichste einheimische Baumart. *Allgemeine Forstzeitschrift*, Vol. 35, pp. 373-374

Wick, L. & Möhl, A. (2006). The mid-Holocene extinction of silver fir (*Abies alba*) in the Southern Alps: a consequence of forest fires? Palaeobotanical records and forest simulations. *Vegetation History and Archaeobotany*, Vol.15, pp. 435-444

13

Removal Mechanisms in a Tropical Boundary Layer: Quantification of Air Pollutant Removal Rates Around a Heavily Afforested Power Plant

J. R. Picardo[1] and S. Ghosh[12]
[1]*School of Mechanical and Building Sciences, VIT University,*
[2]*ICAS Associate, School of Earth and Environment, University of Leeds*
[1]*India*
[2]*U.K.*

1. Introduction

Nature's astonishing biological resilience is closely linked to its self cleansing mechanisms. In a hugely populated developing world, economic growth comes with a heavy price. A multitude of industrial and transportation processes produce noxious gases which are released into the atmosphere. Fortunately many of these developing countries, like India, are blessed with a highly convective tropical boundary layer which dilutes pollution. In addition most of these developing Asian countries also receive intense monsoonal precipitations. Another common feature of these Asian nations is that their vegetative cover is dominated by evergreen plants, in contrast to many mid-latitude nations in the developed world where the number of evergreens are far limited. However, it is ironical that despite these natural propensities air pollution levels over Asian cities are much higher than their mid latitude counter parts. The purpose of this chapter is to describe, formulate and quantify the removal pathways of SO_2, a major air pollutant, through both dry and wet deposition whilst accounting for the endowments of nature enjoyed by tropical Asia. This is done via a detailed study on Asia's Largest Lignite based Power Plant- Neyveli Lignite Corporation (NLC) located in the Cuddalore district of Tamil Nadu in South India (Fig. 1). SO_2 is produced during the generation of power from lignite coal and released from elevated stacks (Table 1). An in depth modelling analysis of these emissions is presented addressing the dispersion of SO_2 in a tropical boundary layer using an atmospheric dispersion model developed as part of a consultancy with NLC and VIT University. The dry and wet deposition of the spatially distributed pollutant is then analyzed via suitable well established parameterizations. The quantification of removal mechanisms has a dual significance with respect to environmental studies. It is essential for calculating the atmospheric budget of trace gases as well as assessing the impact of emissions on local vegetation and structures due to acid deposition.

Two factors make NLC a particularly well suited study area for the purposes of this chapter; an extensive evergreen urban canopy and the North East (NE) monsoon which brings intense showers to the region from October to December. Regional climatology including year round high solar radiation and mild, almost non-existent, winters (temperature around

25 ^{0}C) make this a unique study. It is further set apart from investigations conducted in the mid-latitudes by the tropical vegetation and the intense monsoon showers which bear rain drops of larger mean diameter. All these factors are accounted for in the modelling analysis which is to follow, providing a basis for region specific environmental assessments. It will also serve as a roadmap for future studies in other Asian regions which have hitherto largely resorted to borrowing results from the mid-latitudes.

Fig. 1. Google Map showing the location of Neyveli Lignite Corporation (red A marker) in Tamil Nadu, India

Stack	Height (m)	SO_2 source strength (g s^{-1})
Thermal Power station-I		
1	60	227.82
2	60	271.35
3	60	153.23
4	120	305.99
Thermal Power station-I Expansion		
1	220	305.07
2	220	305.07
Thermal Power station-II		
1	170	359.38
2	170	359.38
3	170	359.38
4	220	317.45
5	220	317.45
6	220	317.45
7	220	317.45

Table 1. Details of stacks emitting SO_2 at NLC

2. Dry deposition

Deposition involves the transport of gaseous and particulate species from the atmosphere to physical surfaces at the ground where they are retained and thus removed from the atmosphere. In the absence of precipitation, they are brought to the surface by turbulent transport where they may be absorbed, adsorbed or chemically transformed. This process is called dry deposition. The deposition surface has a critical role to play and natural surfaces like vegetation, while difficult to describe in a study, promote dry deposition. This chapter is concerned with the mathematical modelling of SO_2 deposition. This section is devoted to the dry deposition of SO_2 onto a vegetative canopy while rain mediated removal is dealt with in the next section.

2.1 The urban canopy and the township at NLC

The founding fathers of NLC began a massive afforestation program which has resulted in the presence of 17 million tropical trees. The role of these trees in mitigating air pollution seems intuitive and a detailed quantitative investigation requires the application of dry deposition modelling techniques. Fig. 2 shows a map of NLC and an aerial view from Google Earth®. The region demarcated by a rectangle is the township of NLC which is home to 128,133 employees. Its proximity to the Thermal Power Station One (TPS1) makes it a particularly sensitive area which is likely to receive emissions from the stacks. Fortunately there is a considerable green cover over the township (Fig. 2) which promotes the deposition of pollutants and results in a cleaner atmosphere. The extent of this cleansing depends on the level of pollutant concentration and the environmental factors which modulate dry deposition. The quantification of the rate of dry deposition assumes greater significance in context of the health of the township's residents. The following sections are devoted towards this objective.

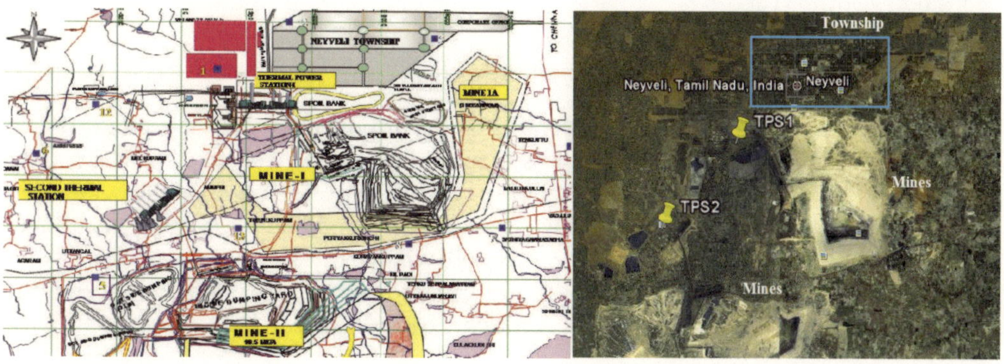

Fig. 2. Map of NLC and aerial view from Google Earth® of TPS1 and TPS2 and the Township (demarcated by a rectangle)

2.2 Parameterization of dry deposition to vegetation

The vegetative canopy is considered to be an irreversible sink for SO_2 and the flux of gas (F) to the ground is represented by a first order relationship. This flux is assumed to be uniform within the surface layer of the atmosphere (10-100 m) (Seinfeld and Pandis, 2006)

$$F = -V_d \times C_g \tag{1}$$

F has units of µg m^{-2}s^{-1}. C_g is the concentration of the gaseous species measured at a reference height within the surface layer (µg m^{-3}). V_d is a parameter called the deposition velocity and it has the units of m s^{-1} (hence the term velocity). Thus the problem of determining the flux of a species is transformed into the determination of its deposition velocity. The downward flux of the species is negative by convention and hence the deposition velocity is positive. Its value will depend on the reference height chosen and the assumption of uniform flux dictates that the concentration would also vary with height so as to keep the flux constant. The reference height in this study is taken to be 10 meters which is close to the vegetative surfaces (ensuring better agreement with the constant flux assumption) and is the height at which instrumentation is installed by NLC for meteorological and concentration measurements.

2.2.1 Deposition velocity- theory of resistances

The process of dry deposition is usually divided into three stages:

1. Transportation from the free atmosphere to the receptor surface (turbulent layer transport)
2. Transport through the quasi-laminar, stagnant air layer near the receptor surface (diffusive molecular transport)
3. Capture or absorption by the surface (in this case transport into the leaf stomata or cuticle or deposition onto the ground).

According to the universally adopted inferential resistance modelling approach, the dry deposition process is treated analogously to the flow of electrical current through a network of resistances in series. In this analogy, the aerodynamic resistance (R_a), the quasi-laminar resistance (R_b) and the surface resistance (R_c) refer to the aforementioned three stages of dry deposition respectively (all have units of s m^{-1}). The inverse of the total resistance is the dry deposition velocity (V_d).

$$V_d = \left(R_a + R_b + R_c\right)^{-1} \quad (2)$$

The aerodynamic component of the overall dry deposition resistance is typically based on gradient-transport theory and mass-transfer/momentum-transfer similarity. The resistance varies with the state of stability of the atmosphere. The quasi-laminar resistance depends on the diffusivity of the gas as well as the wind conditions. The aerodynamic and quasi-laminar resistances are computed by the following expressions (Seinfeld and Pandis, 2006). These expressions are valid only in the surface layer of the atmosphere where the species flux is constant. An approximate vertical extent is 100 m.

$$R_a = \begin{cases} \dfrac{1}{\kappa u_*}\left[\ln\left(\dfrac{z}{z_o}\right) + 4.7\left(\zeta - \zeta_0\right)\right] & (stable) \\ \dfrac{1}{\kappa u_*}\ln\left(\dfrac{z}{z_o}\right) & (neutral) \\ \dfrac{1}{\kappa u_*}\left[\ln\left(\dfrac{z}{z_o}\right) + \ln\left(\dfrac{\left(\eta_0^2+1\right)\left(\eta_0+1\right)^2}{\left(\eta_r^2+1\right)\left(\eta_r+1\right)^2}\right) + 2\left(\tan^{-1}\eta_r - \tan^{-1}\eta_0\right)\right] & (unstable) \end{cases} \quad (3)$$

Where $\eta_0 = (1-15\zeta_0)^{1/4}$ and $\eta_r = (1-15\zeta_r)^{1/4}, \zeta_0 = z_0/L$

The quasi-laminar resistance is calculated by the following expression

$$R_b = \frac{5Sc^{2/3}}{u_*} \qquad (4)$$

Sc is the Schmidt No; equal to the ratio of kinematic viscosity of air and the binary diffusivity of SO_2 and air. The friction velocity (u_*) can be calculated by (Xu and Carmichael, 1998)

$$u_* = \frac{\kappa u(z)}{\ln\left[(z-d)/z_0\right]} \qquad (5)$$

The Monin-Obhukov length (L), by definition, is the height at which turbulence produced by mechanical and buoyancy forces match. It characterizes atmospheric stability in the surface layer and is positive for a stable atmosphere and negative for an unstable atmosphere. It is determined from the Pasquill Stability classes by the method of Golder (1972) as detailed in Seinfeld and Pandis (2006). The roughness length (z_0) is taken as 1 m as recommended for urban locations by Voldner et al. (1985). The displacement length (d) is 70 - 80% of the height of the large roughness elements (Xu and Carmichael, 1998). The reference height (z) is taken as 10 m and the Von Karman constant (κ) as 0.4.

2.2.2 Surface resistance

The surface or canopy resistance is the most difficult to parameterize due to the complex nature of the processes involved in the absorption and retention of gases by vegetative surfaces. At the same time it is often the dominating resistance especially in the tropics where the atmosphere is highly convective. In recent studies conducted for this region, the parameterization of Wesely (1989) was used to calculate the surface resistance (Seth et al., 2010, Patra and Ghosh, 2010 and Picardo and Ghosh, 2011). It has been used in other studies for Asia as well (Xu and Carmichael, 1998 and Kumar et al., 2008). However, the several advancements made in the science of dry deposition and in the understanding of the dependence of surface resistance on environmental factors have rendered the parameterization of Wesely (1989) somewhat outdated. Many of these advancements are embodied in the work of Zhang et al. (2003b). These include a sunlit/shaded big leaf model for the calculation of the bulk canopy stomatal resistance from the individual leaf resistance via the Leaf Area Index (LAI). LAI is the ratio of leaf surface to ground surface and is around 1 for urban canopies and close to 6 for forests. Sunlit and shaded leaves are treated differently in this canopy -stomatal -resistance model. In Wesely's parameterization (1989), a base bulk stomatal resistance is provided and then modulated with radiation and temperature. This base bulk stomatal resistance value was specified for discrete seasonal categories and over various land types. The problem with this approach is that the seasonal categories considered by Wesely and the corresponding change in the canopy structure do not match the climate and vegetation characteristics in tropical Asia. Specifically, during the winter season at NLC, the temperature is around 25 °C and the vegetation is healthy. In contrast the winter seasonal category in Wesely (1989) describes conditions of subzero temperatures and snow covered ground! Moreover, the same land use type (e.g.

agricultural) can also have widely different vegetative characteristics depending on the geographical location. These short comings were realized by Gao and Wesely (1995) who introduced LAI into the stomatal resistance model of Wesely (1989). In Zhang et al. (2003b) all the resistances which are dependent on the canopy structure are related to LAI. This allows an accurate representation of the local vegetative characteristics and the seasonal dependence of green cover. The other improvements in Zhang et al. (2003b) include revised methods of accounting for wet surfaces and their effect on stomatal and non-stomatal resistances and a new parameterization of non-stomatal resistance which considers the effect of meteorological variations (Zhang et. al., 2003a).

For the above mentioned reasons, it was deemed necessary to adopt the parameterization of Zhang et al (2003b) for an accurate and region specific study of dry deposition. The surface resistance is represented as a combination of stomatal and non-stomatal resistances in parallel since stomatal uptake as well as cuticular absorption and deposition onto twigs and the ground occur simultaneously.

$$\frac{1}{R_c} = \frac{1 - W_{st}}{R_{st} + R_m} + \frac{1}{R_{ns}} \tag{6}$$

R_{st} is the canopy stomatal resistance. Stomatal uptake of gaseous species is controlled by the degree of stomatal opening. The major environmental factors which modulate stomatal opening are solar radiation, ambient air temperature, water vapor pressure deficit and leaf water stress. These are accounted for in the canopy resistance model. R_m is the mesophyll resistance which is treated as gas species dependent and specified as 0 for SO_2 since it is highly soluble in water (Zhang et al., 2002). Together they signify the total resistance to stomatal uptake. W_{st} accounts for the blocking of stomata during rains by the film of water which develops on the leaves. However, the net result of rain is a decrease in overall surface resistance since it greatly reduces the non-stomatal resistance (R_{ns}) of the surface in the case of a soluble gas like SO_2. The non-stomatal resistance is a combination of the in-canopy aerodynamic resistance (R_{ac}) and resistance to deposition to the ground (R_g), in series, along with canopy cuticular resistance in parallel (R_{cut}).

$$\frac{1}{R_{ns}} = \frac{1}{R_{ac} + R_g} + \frac{1}{R_{cut}} \tag{7}$$

The presence of wet surfaces due to rain or dew considerably decreases the cuticular and ground resistances. The friction velocity is included in the parameterization of in-canopy aerodynamic resistance, which is one of the advancements of this method. Apart from the canopy stomatal resistance, the LAI also has an effect on the in-canopy aerodynamic resistance and the canopy cuticular resistance. Hence, the LAI is quite an important parameter. The formulae for each of these terms and the parameters based on land use category are given in Zhang et al. (2003b) and Zhang et al. (2002). A large number of land use types are considered and the effect of environmental factors on stomatal conductance is accounted for via formulations which vary with the type of vegetation. Thus it is possible to include region specific information in the model and generate results which are far more compatible to the study area using the method of Zhang et al (2003b). In the present study, NLC is represented by an urban canopy land use type with tropical broadleaf vegetation.

2.2.3 Leaf area index for NLC

As described in the previous section, LAI is a key parameter which captures the local vegetative characteristics and its seasonal dependence. In this work, the LAI over NLC was obtained from MODIS satellite data (product MCD15A2). The data is available in 1 km resolution and is free to download. From the LAI data, which agrees with personal sampling of the vegetation, it was observed that the trees are at their lush best during October, the month of the onset of the NE monsoon and are at their leanest in May which is the peak of the dry summer season. NLC receives some showers from the South West (SW) monsoon as well -this brings relief from the summer heat and causes a rise in LAI from August onwards. Although there is a seasonal variation, there are no bare periods without any green cover-this is in sharp contrast to trees in the mid-latitudes. Moreover, it is comforting to note that the month of the least vegetative cover (i.e. May) is the hottest month when the boundary layer is at its most convective, leading to dilution of pollutants. Values of LAI are available with a spacing of 8 days. The averages for December'08, May'09 and October'09 were calculated as 0.91, 0.54 and 1.06 respectively. Images provided by the MODIS product (MCD15A2) of LAI over the peninsular part of the Indian subcontinent are displayed below (Fig. 3).

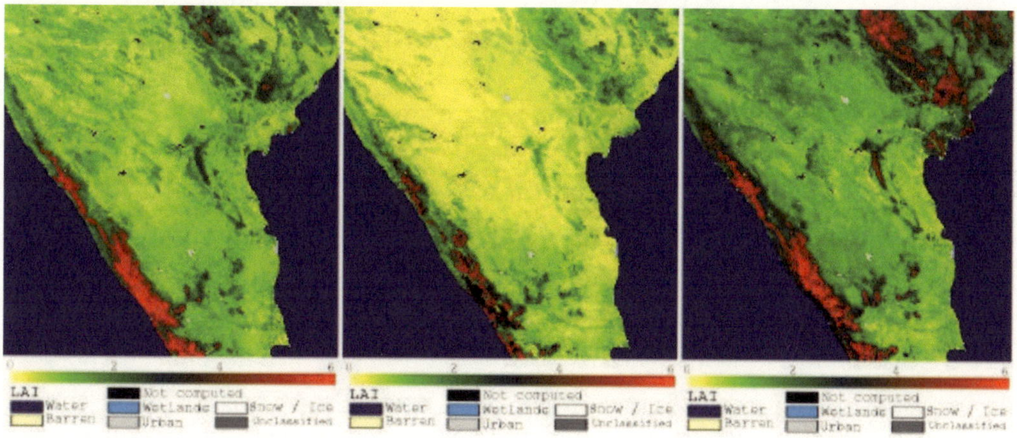

Fig. 3. LAI over the subcontinent for 10th Dec'08, 17th May'09 and 16th October'09 respectively (MODIS MCD15A2)

2.3 Deposition velocity of SO_2 at NLC

2.3.1 Calculation of deposition velocity for three seasons- summer, NE monsoon and mild winter

The average deposition velocity of SO_2 is computed by the method of Zhang et al. (2003b) for the months of December, May and October which represent the three major seasons experienced at NLC- mild winter, hot dry summer and wet North East Monsoon respectively. The statistical mean and standard deviation of the meteorological inputs, namely wind speed, solar radiation and relative humidity as well as the number of rainy days are given in Table 2. It is observed that the temperatures are generally above 25 °C even during December and that NLC receives considerable solar radiation throughout the year, including the month of October which experiences the maximum rainfall. The computed deposition velocities are presented in Table 3.

The seasonal modulation of the deposition velocity with LAI is apparent. In addition, the rains during October further increase deposition during the day and night due to the presence of wet surfaces on the leaves, twigs and ground. Matsuda et al. (2006) performed field experiments to determine the dry deposition velocity of SO_2 over a tropical forest in Northern Thailand. Although they studied a full fledged forest as opposed to an urban canopy, the vegetative characteristics of the region are similar to our study area. They observed much higher values of SO_2 deposition velocity during the rainy season as compared to the dry season with maximum values of 1.39 cm s^{-1} and 0.31 cm s^{-1} in the wet season and dry season respectively (daytime). They emphasize the importance of accounting for the effect of wet surfaces on non-stomatal resistance in order to accurately model the higher observed values of V_d during the rains. Moreover, they found that the value of deposition velocity predicted for tropical broadleaf trees by Zhang et al., (2003b) during the wet season was consistent with their experimental observations.

Month		Wind (m s^{-1})		Solar radiation (W m^{-2})		Temperature (°C)		Relative humidity (%)		Rainy Days
		MEAN	STD	MEAN	STD	MEAN	STD	MEAN	STD	
Dec'08	Day	1.37	0.46	346.12	146.41	28.00	2.00	63.16	2.57	4.00
	Night	0.55	0.39	NA	NA	22.00	0.96	58.77	2.26	
May'09	Day	1.84	1.81	475.45	92.38	36.65	2.36	55.57	5.43	0.00
	Night	1.19	0.65	NA	NA	27.79	1.62	52.91	4.95	
Oct'09	Day	1.64	0.88	416.00	111.69	32.93	1.92	52.57	0.95	13.00
	Night	0.51	0.70	NA	NA	25.20	1.05	50.88	0.35	

Table 2. Meteorological data used for computation of deposition velocity (NA: Not Applicable)

Season	LAI	V_d (cm s-1)	
		Day	Night
DEC 08- Mild winter	0.91	0.334	0.105
MAY 09- Hot Summer	0.54	0.215	0.145
OCT 09- NE Monsoon	1.06	0.478	0.214

Table 3. Deposition Velocity for the three major seasons at NLC

2.3.2 Variation of deposition velocity with environmental factors

All available parameterizations of deposition velocity are based on experimental observations from numerous field studies. The modulation of the various processes involved in dry deposition by environmental factors is quite complex and while formulations are provided for each individual process, it is difficult to comprehend the overall effect of any individual factor on deposition velocity. In this section, we focus on two important factors- wind speed and solar radiation and study their effect on deposition velocity as captured by the parameterization of Zhang et al. (2003b).

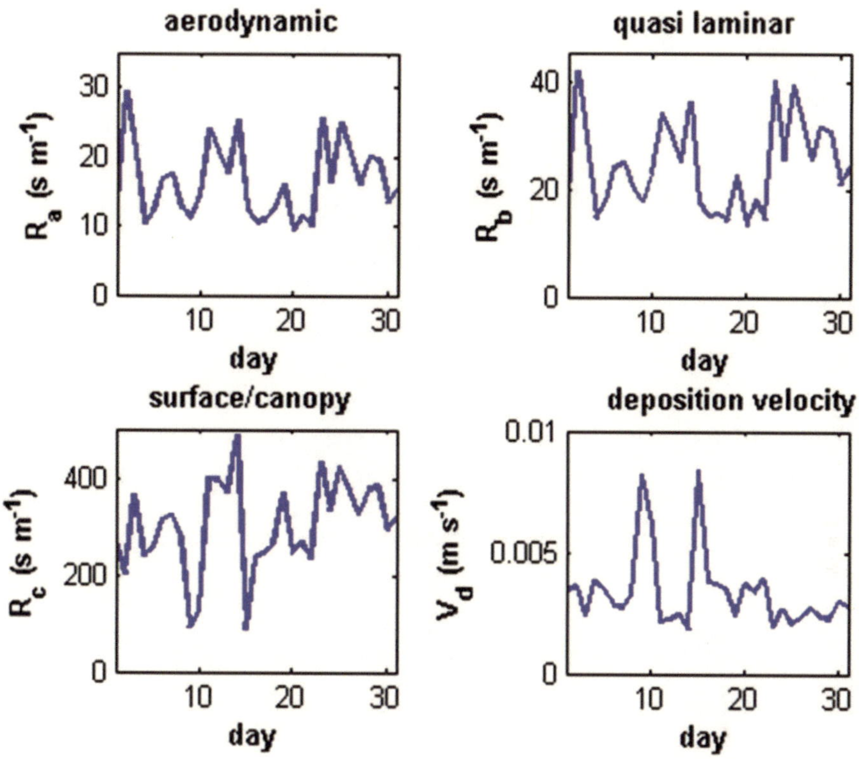

Fig. 4. Resistances and deposition velocity calculated for December 2008 at 14:30 IST

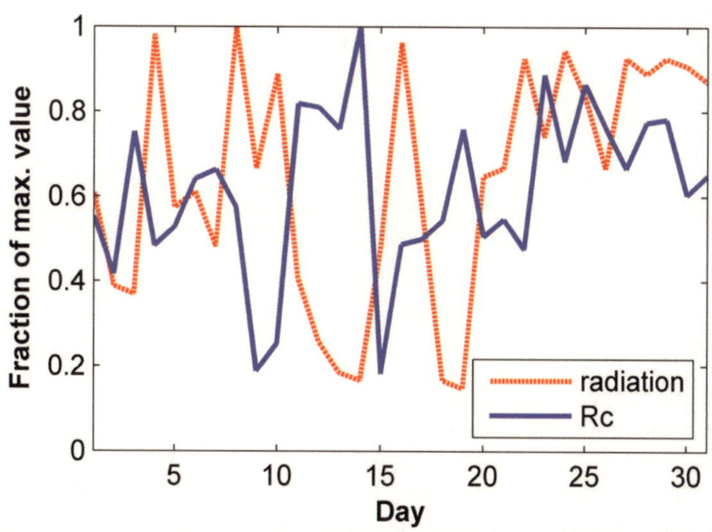

Fig. 5. Variation of canopy resistance with solar insolation (Dec 2008 14:30 IST)

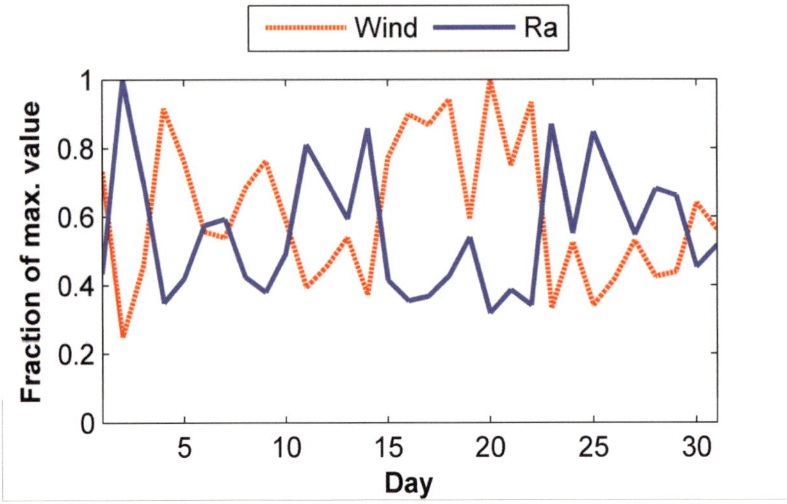

Fig. 6. Variation of aerodynamic resistance with wind speed (Dec 2008 14:30 IST)

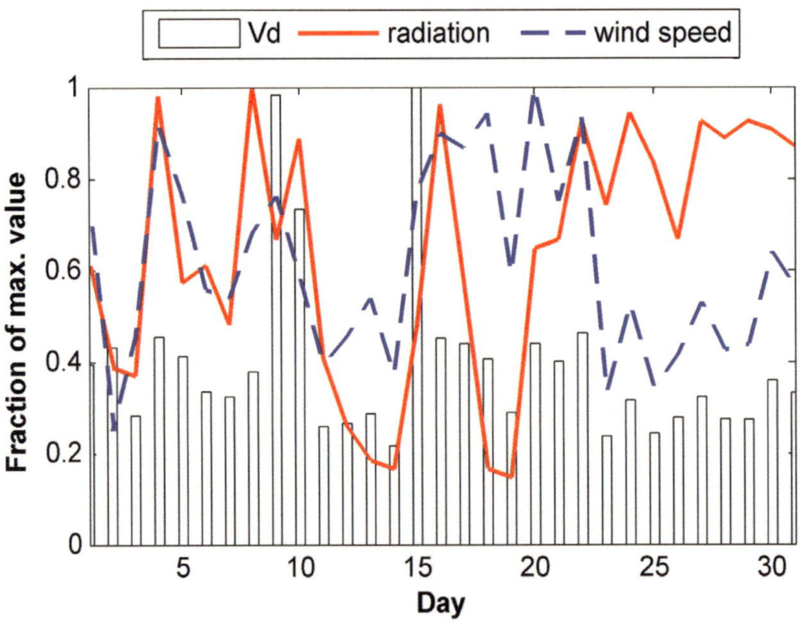

Fig. 7. Variation of deposition velocity with wind speed and solar insolation (Dec 2008 14:30 IST)

The values of the resistances and deposition velocity (V_d) computed for each day of December 2008 at 14:30 (daytime) is shown in Fig. 4. The canopy resistance (R_c) proves to be the controlling resistance in the dry deposition process, accounting for about 80% of the total resistance. Hence one would expect the solar insolation which strongly affects R_c to have a marked affect on V_d as well. This is indeed the case. In fact, the values of V_d are substantially lower at night since there is no solar radiation and the stomata are practically closed (Table 3).

The wind speed modulates R_a and R_b in the same manner. Normalized plots of R_c and R_a with solar radiation and wind speed respectively show an inverse dependence in both cases (Fig 5 and 6). However, while wind speed seems to be the sole important factor in controlling R_a, it is clear that there are significant contributing factors other than solar radiation which affect R_c. Fig. 7 shows the variation of V_d with both environmental factors. The tall standalone peaks correspond to rain and the effect of wet surfaces. A close look at days 12 to 14 seems to suggest that wind speed is the dominant overall factor. This is because, in addition to R_a, the wind also affects R_c by modulating the in-canopy aerodynamic resistance which a species experiences before deposition onto the ground or the lower parts of the trees (please refer Section 2.2.2).

2.4 Cleansing efficacy of the evergreen canopy

The flux of SO_2 to the ground at any location in NLC can be computed from Eq. (1) using the calculated deposition velocity if the concentration at the reference height is known. In the absence of concentration measurements, modelled values can be used to study the removal of SO_2 by the canopy. In this section we investigate the role of the urban canopy in improving air quality, especially in the township. On the 21st of May 2009, a southwest wind transported pollution directly over the township. The wind speed was 2.3 m s^{-1} and the solar radiation was 359 W m^{-2}.This situation provides an ideal setting for our study. However, before analyzing the deposition of SO_2, it is necessary to predict the concentration of SO_2 over the township.

2.4.1 Dispersion model for predicting SO2 concentration

In this study, a tailor made steady state atmospheric gaussian-dispersion model is used which was developed as part of a consultancy with NLC. This model is based on the gaussian plume formula which is applicable to the steady state emission of a gas from an elevated stack with a totally reflecting ground. Although this model does not account for deposition, the error in predicted SO_2 concentrations is relatively small especially since the township is close to the stacks and the source strengths are high. The accuracy is sufficient for the purposes of this study. The gaussian plume equation which predicts the concentration (µg m^{-3}) at any point around the stack is given by:

$$C_g(x,y,z) = \frac{q \times 10^6}{2\pi u \sigma_y \sigma_z} \exp\left(\frac{-y^2}{2\sigma_y^2}\right) \times \left[\exp\left(\frac{-(z-h)^2}{2\sigma_z^2}\right) - \exp\left(\frac{-(z+h)^2}{2\sigma_z^2}\right)\right] \tag{8}$$

The stack is taken to be at the origin with the x axis along the centerline of the plume which is in the mean direction of the wind. The y axis is along the horizontal and the z axis along the vertical. According to this model, as the plume travels with a mean speed u m s^{-1} along the wind direction, it disperses horizontally and vertically so that the average steady state concentration at any cross section of the plume follows the normal Gaussian probability distribution. σ_y and σ_z (in meters) are the standard deviations of the concentration in the y and z directions. q is the source strength (g s^{-1}) and h is the effective stack height (the vertical rise of the plume, before it bends over, added to the physical stack height- m). The dispersion parameters (σ_y and σ_z) depend on atmospheric stability and distance from the stack and are computed using the formulae recommended by Briggs (1973) based on the Pasquill atmospheric stability classes (Turner, 1969) as detailed in Seinfeld and Pandis

(2006). A detailed exposition of gaussian plume dispersion models can be found in Seinfeld and Pandis (2006) and Hanna et al. (1982). This model is also used in the study of wet deposition of pollutants in the latter part of this chapter.

2.4.2 Deposition from a plume

The ground level concentration computed from the dispersion model is shown in Fig. 8. The township is demarcated by a rectangle and the white markers represent the two power stations (TPS1 and TPS2) with TPS1 on the edge of the township. The text markers indicate the locations of air quality monitoring stations. From Fig. 8 it is clear that much of the township experiences concentrations above 10 μg m^{-3}. Areas closer to TPS1 receive higher amounts of the polluting gas and the concentration in the narrow region surrounding the plume centerline exceeds 100 μg m^{-3}.

Due to the presence of the evergreen canopy, there is a continuous deposition of material from the plume onto the trees. This flux is greater in regions of higher concentration and can be evaluated using Eq. (1). Material will be deposited as long as the plume remains aloft over the canopy and the wind direction does not change. Considerable amount of pollution is deposited and contours of the mass flux are shown in Fig. 9. Approximately 1.91 kg of SO_2 is deposited onto the canopy within the township, in an hour. However, this amount is insignificant when compared to the source strength of the emissions from the stacks which continuously pump pollution into the atmosphere (see Table 1). Since the township is very close to TPS 1 there is no perceivable change in the ambient air concentration while the plume remains aloft. However, areas surrounding NLC would benefit as the plume would have travelled a greater distance and much more SO_2 would have been deposited over the canopy which extends beyond the township. This study can be generalized for a wind driven plume, over any part of NLC with a canopy cover.

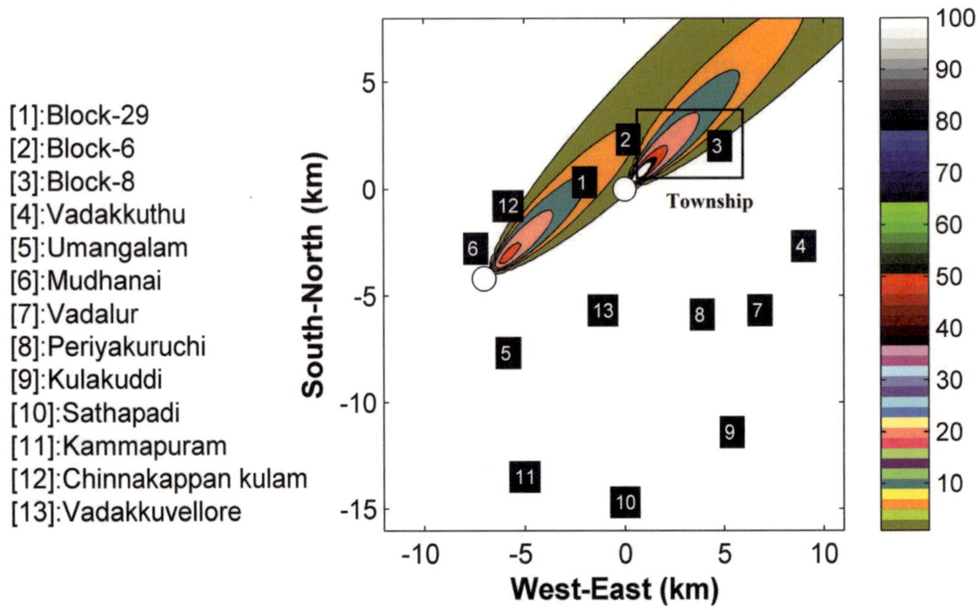

Fig. 8. The modelled ground level concentration over the township on 21st May 2009, 14:30 IST (μg m^{-3})

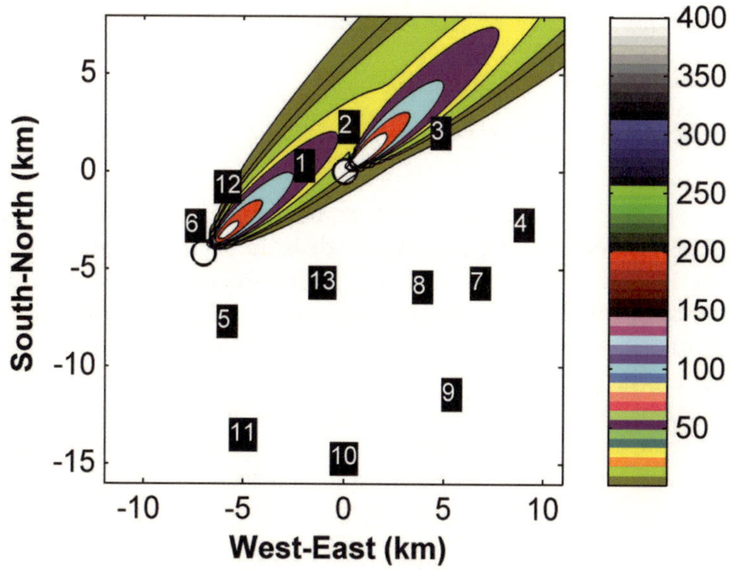

Fig. 9. Mass of SO_2 deposited per unit area per unit time on 21st May 2009, 14:30 IST ($\mu g\ m^{-2}\ hr^{-1}$)

2.4.3 Removal of residual pollution and improvement of air quality

In the previous section the deposition of SO_2 from a plume was analyzed. However, at any time only a small region of NLC can be directly affected by the plume. The other regions will experience residual pollution, left over from a previous visit of the plume or transported from other affected parts of NLC. The trees can reduce the residual SO_2 levels in these regions which do not receive a constant supply of SO_2 for a considerable period of time. These conditions are analogous to those prevalent in cities at night. There is a buildup of pollutants during the day when sources of gaseous pollutants are numerous. At night the emissions subside due to low levels of traffic and urban activity. Since there is no replenishment of pollution, dry deposition onto urban trees can result in improved air quality by the morning. In NLC, this situation is realized when a change in wind direction causes the plume to move away from the township or if a period of calm follows the plume's visit over the township. In either case, SO_2 will be left over the township and diluted by the convective motion of the atmosphere. The mixed layer of the atmosphere, within which this dilution is restricted, is the well mixed region of the atmosphere adjacent to the earth's surface. The height of the mixed layer varies with the time of day, the location of the site (latitude) and the atmospheric stability. The residual SO_2 confined within the mixed layer over the township will be gradually depleted due to dry deposition onto the trees. A first order removal of species from the bottom of a closed stirred tank is a simple way to model this process. A mass balance on SO_2 for a mixed layer of height H_{mix} yields:

$$\frac{dC_g}{dt} = -\frac{V_d C_g}{H_{mix}} \tag{9}$$

This equation when integrated yields the following expression for the time dependent concentration in the mixed layer of the atmosphere, where C_{g0} is the initial residual concentration.

$$C_g(t) = C_{g0} \exp\left(-\frac{V_d t}{H_{mix}}\right) \tag{10}$$

The initial residual pollutant concentration (C_{g0}) can be obtained by first estimating the total amount of pollution left over a given area and then distributing it uniformly throughout the mixed layer. This is done by integrating the plume concentration predicted by the dispersion model throughout the atmosphere for all points within a designated zone (in this case, the township) up to the mixing height and then dividing by the total volume of the region of integration. The present calculation requires an estimation of the height of the mixed layer. This height varies in the summer from 500 m in the morning up to 2-3 km in the late afternoon with an average of 1000 m (Seinfeld and Pandis, 2006). At night the inversion layer is much lower, sometimes only 100 m which can result in high pollution levels. At NLC, this is not much of a concern since the stacks are elevated and most of the pollution is transported above this height. A detailed method for estimating the mixed layer height and its day time variation is given by Luhar (1998). For the purpose of this work, representative values of 1000 m and 500 m are used. The cleansing of the atmosphere over the township due to the removal of SO_2 by the canopy is depicted in Fig. 10. At lower mixing heights, the concentration is much higher, but so is the intensity of the cleansing action. The deposition velocity is low in May and so is the removal of SO_2. However, it is nature's boon that the lowest values of deposition velocity coincide with the hottest month (May) when the day time mixing height is high and pollution episodes are unlikely. This is in contrast to mid-latitudes where the lowest deposition velocity values occur during winter when the mixing height is low. The above calculation is repeated using the deposition velocity of October (Table 3) and the results are depicted in Fig 11. A comparison with Fig. 10 clearly demonstrates the accelerated pace of atmospheric cleansing due to the higher deposition velocity of October.

As mentioned previously, the height of the mixed layer can be very low at night. This can be a problem in polluted cities of developing countries where the majority of emissions have ground sources such as moving vehicles and burning garbage. In such cases the pollutants can become highly concentrated and have a harmful impact on the health of the local population. The presence of pollutant tolerant vegetation can be a mitigating factor as the removal by deposition increases with increase in concentration of the deposition species. The ability of the vegetation to survive under daily pollutant stress is a matter which begs further investigation, especially in context of the continued urbanization in India and other developing countries of Asia. The response of the vegetation may be quite different from that of European and North American plants and calls for another region specific study. The trees at NLC have been able to survive the continuous exposure to the emissions of the power plants but this is, in part, due to the convective atmosphere and the elevated emission sources.

The visionary founders of NLC started an afforestation program several decades ago and now the thriving township is receiving the full benefits of this action. The presence of the green canopy provides a round-the-year removal mechanism for pollutants and thus contributes to better living conditions in terms of real air quality improvement apart from aesthetic benefits. Pollutants are also removed from the atmosphere by rain. This mechanism is much stronger than dry deposition but is operational for a much shorter time. Wet deposition is dealt with in the next section.

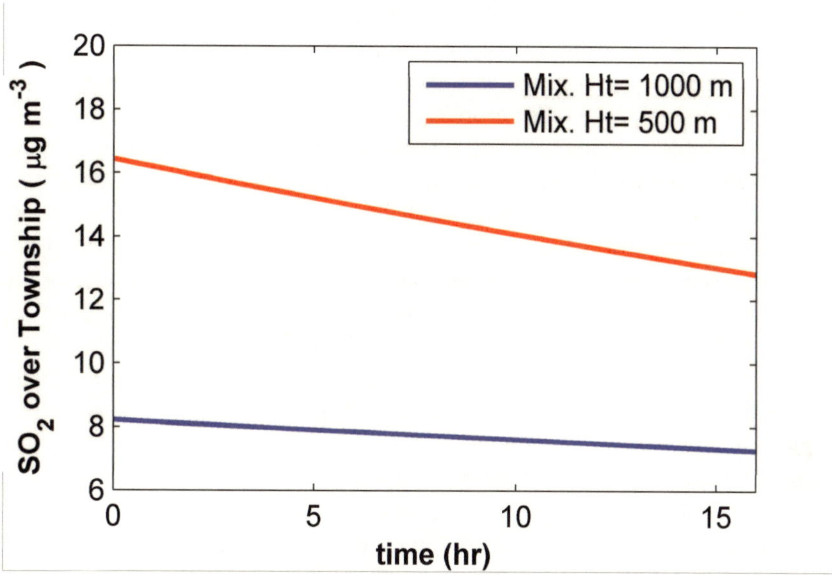

Fig. 10. Depletion of residual SO_2 concentration over the township on 21st May 2009

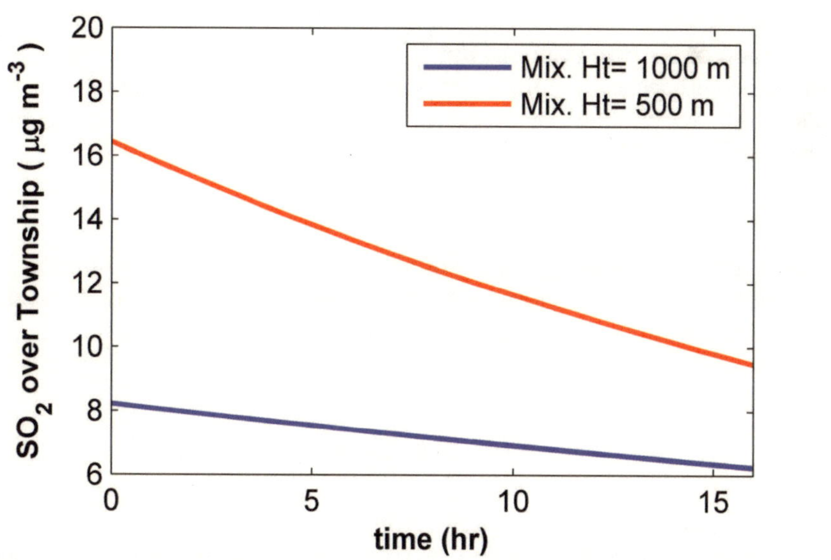

Fig. 11. Depletion of residual SO_2 concentration over the township considering the higher deposition velocity of October

3. Wet deposition- below cloud rain washout

Wet deposition is a broad term encompassing all the natural processes by which material is scavenged from the atmosphere by hydro-elements (cloud and fog drops, snow and rain) and brought to the Earth's surface. Wet deposition mediated by rain, can occur via in-cloud and below cloud scavenging. The former involves in-cloud processes occurring at the cloud base which incorporates substances into the cloud and ultimately transports them to the Earth's surface via rain. The latter, termed Washout refers to the removal of substances from the atmosphere directly by rain drops as they fall from the cloud base towards the ground. We are only concerned with the latter more significant effect.

SO_2 is a water soluble species. During precipitation events, soluble substances in the atmosphere are absorbed by the multitude of falling rain drops and brought to the ground. This process cleanses the atmosphere and leads to an 'after-rain freshness', often experienced by residents of urban and industrialized areas. In order to quantify this phenomenon it is necessary to analyze the mass transfer of a gaseous species into falling rain drops. It must be borne in mind that the drop size is distributed over a range of diameters. Moreover, the drop size distribution (DSD) varies with the geographical location (which inherently includes dependency on the cloud type, cloud base height, local environment etc.) and the intensity of the showers (rain rate). Finally, the results should be linked to the previously described dispersion model (Section 2.4.1) in order to predict the extent of atmospheric cleansing. Washout transports the polluting substance from the atmosphere to the ground. The pollutant is not rendered harmless by transformation or incorporation into a biological cycle (as is the case in dry deposition onto a canopy). This raises the question of pollution at the receptor surface in the form of acid rain. Acid rain is a recognized problem, especially around industrial areas which can cause damage to man-made structures as well as vegetation. Thus, it is important to estimate the pH of the rain water received over NLC.

3.1 Quantification of the rain washout flux

The flux of a gaseous species from the air into rain drops, per unit height of the atmosphere (W, g $m^{-3}s^{-1}$), due to washout by rain, can be approximated by a linear first order relationship (Seinfeld and Pandis, 2006).

$$W = \beta \times C_g \tag{11}$$

β is the scavenging coefficient with units of s^{-1} and C_g is the gas concentration. The flux due to washout (F_W) of a gaseous species to the ground from a column of the atmosphere, of unit cross-sectional area (where the concentration of the gas is horizontally constant), is given by:

$$F_W = \int_0^H (\beta \times C_g)\, dh \tag{12}$$

If the integrated concentration of the species, within a column of the atmosphere, from the ground to the upper reaches of the emissions (H, m) is represented as C_T (g m^{-2}) then for a given value of β(assumed constant at all spatial positions) the flux at any point (x,y) on the ground is given by:

$$F_W(x,y) = \beta \times C_T(x,y) \tag{13}$$

Given the spatially distributed concentration ($C_g(x,y,z)$) around an emission source, which can be obtained using the dispersion model (Section 2.4.1), the concentration at a point is integrated with height to evaluate C_T. With a suitable value of β the flux of species to the ground at that point due to washout by rain is computed.

The above formulation is a first order removal process which considers the rain drops to be a perfect sink, just as the trees were in the dry deposition analysis. However, this assumption is questionable since it is known that SO_2 is not irreversibly soluble and hence the water drop would eventually reach a saturation state after which no further dissolution would be possible. On the other hand, the drops do not fall indefinitely but rather have a finite distance to travel before reaching the ground. Therefore, if during its fall a rain drop remains sufficiently far from saturation, allowing it to absorb SO_2 continuously, then one may consider, for the purposes of this study, that SO_2 is indeed being irreversibly absorbed. A further investigation of this assumption and its validity is presented in the next section which leads on to additional interesting information regarding the relationship between rain drop size and rain water pH.

3.2 Interface mass transfer of SO_2 into a falling rain drop

3.2.1 Transient concentration of dissolved SO_2 inside a falling drop

For effective pollutant washout to occur, a rain drop should absorb pollution through most of its descent. As described in the previous section, it is vital to the description of washout via the scavenging coefficient. For a reversibly soluble gas like SO_2 it is therefore important to calculate the distance through which a drop would fall before saturation.

A rain drop experiences considerable shear at its surface as it falls through the atmosphere which induces internal circulations within the drop. This allows the assumption of a well mixed drop i.e. the concentration gradients within the drop are neglected. The resistance to mass transfer is assumed to exist only in the gaseous film surrounding the drop. It should be noted that if the well mixed assumption was dropped and the liquid phase resistance considered in addition to the gas phase resistance, then this would lead to slower mass transfer with a longer saturation time (Pruppacher and Klett, 1997). Once at the drop surface, the SO_2 dissolves into the drop and is transformed into HSO_3^- ions (bisulphite). Further reaction of these ions to sulphite and other ions is not considered since the dissociation constant is much smaller for any reaction following the initial dissociation to HSO_3^-. With the above assumptions in mind, the process is simplified to the transport of SO_2 in the atmosphere, through a gaseous film surrounding the drop, to the drop surface where SO_2 is absorbed and increases the concentration of HSO_3^- uniformly within the drop (well mixed assumption). This situation is described by the following differential equation (Pruppacher and Klett, 1997).

$$\frac{dC_L}{dt} = \left(\frac{12 f_g D_g^*}{D^2}\right)\left(C_g - \frac{C_L{}^2}{K_H K_1 RT}\right) \tag{14}$$

C_L is the HSO_3^- concentration within the drop (mol L^{-1}) and C_g is the SO_2 concentration in the atmosphere. In this equation the units of C_g are mol L^{-1}. However for convenience the ambient air concentration values are reported in μg m^{-3} as was done in section 2 of this chapter. D is the drop diameter (m), K_H is Henry's Law constant and K_1 is the dissociation constant of the reaction of formation of HSO_3^-. f_g is the ventilation coefficient which is the ratio of the mass transfer of the gas for a drop falling at its terminal velocity to that of a stationary drop. D_g^* is the modified diffusivity ($m^2\ s^{-1}$) which is obtained from the binary diffusivity of SO_2 (D_g, $m^2\ s^{-1}$) in air by the following expression:

$$D_g^* = D_g / \left[1 + \left(8D_g / D\alpha V_G\right)\right] \tag{15}$$

α, the mass accommodation coefficient, is assumed to be 0.5 for this study (Pruppacher and Klett, 1997) and V_G, the molecular thermal velocity can be computed by (Seinfeld and Pandis, 2006)

$$V_G = \left[8RT/\pi M\right]^{1/2} \tag{16}$$

R is the universal gas constant (8.314 J $K^{-1}mol^{-1}$), T is the temperature (K) and M is the molecular mass of the species (kg mol^{-1}). In this study we are concerned with wet scavenging over the NLC region (approx. 20 km^2) surrounding the source of emissions. Since these emissions are unlikely to reach the cloud base over this region, we can assume the concentration of HSO_3^- in the rain droplets, as they fall from the cloud base, to be negligible. Then the initial condition for Eq. (14) is $C_L = 0$. This first order differential equation can be solved to yield an expression for the transient concentration of HSO_3^- in the drop.

$$C_L(t) = C_{Lsat} \tanh\left(\frac{kC_g t}{C_{Lsat}}\right) \tag{17}$$

C_{Lsat} is the concentration in a saturated droplet which is in equilibrium with the ambient air concentration of SO_2 and

$$C_{Lsat} = \left(C_g K_H K_1 RT\right)^{1/2} \tag{18}$$

k is the average mass transfer coefficient (s^{-1}) which accounts for both collisional as well as diffusional uptake.

$$k = 12 f_g D_g^* / D^2 \tag{19}$$

The value of the ventilation coefficient can be determined from the following empirical relation (Pruppacher and Klett, 1997) in terms of the Reynolds (Re) and Schmidt Numbers (Sc).

$$f_g = 0.78 + 0.308 Sc^{1/3} \mathrm{Re}^{1/2} \tag{20}$$

These dimensionless numbers are defined as:

$$Sc = D_g/\nu, \mathrm{Re} = DU_t/\nu \tag{21}$$

ν is the kinematic viscosity of air. The following relation is used for calculating the terminal velocity (U_t, m s^{-1}) of a drop (Johnson, 1982) where Q is an empirical constant with a value of 8630 s^{-1}.

$$U_t = Q(D/2) \tag{22}$$

Concentration profiles of HSO_3^- in a drop falling through a SO_2 laden atmosphere, as given by Eq. (17), are plotted in Fig. 12 for various drop diameters. The time of fall, beginning with the first encounter of the drop with SO_2 pollution, is expressed in terms of the distance of fall after evaluating the terminal velocity for the drop from Eq. (22). The values of the parameters used in the above equations are given in Table 4 and are obtained from Pruppacher and Klett (1997). The atmospheric concentration of SO_2 is taken to be 100 μg m^{-3} which is the ambient concentration expected close to the NLC stacks (See Fig. 8).

D_g, Gas Phase diffusivity	14.1×10^{-6} m^2 s^{-1}
M, Molecular Mass SO_2	64×10^{-3} kg mol^{-1}
ν, Kinematic viscosity of air	14.1×10^{-6} m^2 s^{-1}
K_HRT, Henry's law constant product	30 (dimensionless)
K_1, Dissociation Constant	1.23×10^{-2} mol L^{-1}

Table 4. Values of the parameters used in Eq. (17)

It is observed that the distance through which a drop falls, before it becomes saturated, increases with the drop's diameter. While the small drops (D < 1 mm) saturate in less than 100 m, the moderate sized ones (D~2 mm) fall for 500 m before attaining saturation. It is observed that the average rain-drop diameter is higher for showers of higher precipitation rates. For the high intensity NE Monsoon showers there is a preponderance of moderate to large sized drops. Moreover, the SO_2 in the atmosphere will be concentrated within the plume. Thus, the distance through which the drop falls, while absorbing SO_2, is the plume width alone and not the entire distance from the cloud base to the ground. Further, in regions of high concentration close to the stacks, the plume width will be small. When the plume width is large (100 to 200 m and beyond) the concentrations will be much lower than 100 μg m^{-3} and so the drops will fall through larger distances than shown in Fig. 12 without attaining saturation. With the above considerations in mind it can be safely assumed that the rain drops do not get saturated as they fall through NLC plumes. Hence the rain will wash out SO_2 from the upper reaches of the emitted plume to the ground level and the process can be treated as a first order removal. NLC is one of the largest power plants in Asia-this reasoning prevails for most power plant emissions over the Indian Subcontinent as well as over other countries which receive intense, monsoon-type rains.

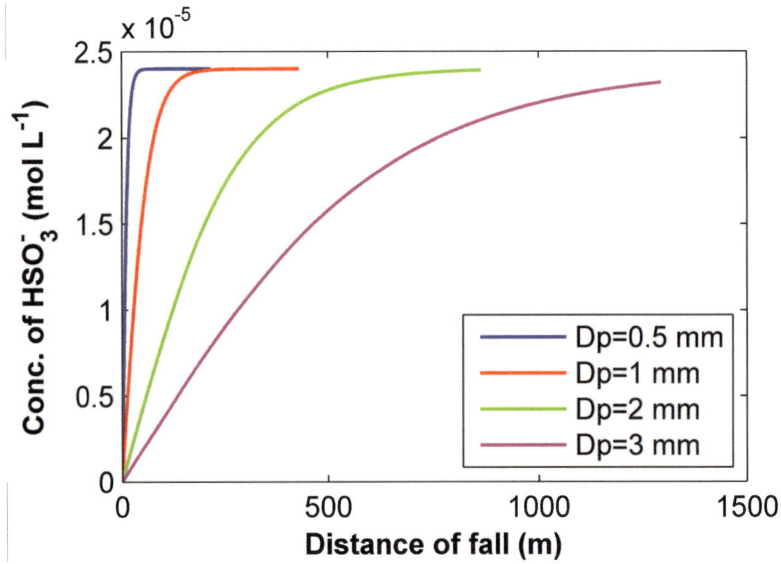

Fig. 12. Concentration of HSO_3^- inside a drop as it falls through SO_2 laden atmosphere with a concentration of 100 μg m^{-3}

3.2.2 Ground level pH of drops over NLC

Apart from the washout of emissions we are also concerned with the acidity of rain water. Thus, it is useful at this stage to compute the pH of the rain drops when they reach the ground. The pH of a rain drop leaving the cloud base (pH_{ini}) is assumed to be 5.6 due to absorption of CO_2 in the upper atmosphere. Applying the principle of electro-neutrality, the increase in H^+ ions due to the absorption of SO_2 and its subsequent dissociation to HSO_3^- ($[H^+]_{abs}$) can be computed. Finally the pH of the drops at the ground can be obtained from the total H^+ ion concentration (pH_{ground}).

$$\left[H^+\right]_{ini} = 10^{(-pH_{ini})}, \qquad \left|pH_{ini} = 5.6\right. \tag{23}$$

$$\left[H^+\right]_{abs} = \left[HSO_3^-\right]_{abs} \tag{24}$$

$$pH_{ground} = -\log_{10}\left(\left[H^+\right]_{ini} + \left[H^+\right]_{abs}\right) \tag{25}$$

The pH of drops of different sizes when they reach the ground is depicted in Fig. 13. The lower flat part of the curve represents drops which get saturated prior to their reaching the ground. They attain a minimum pH of 4.5 which corresponds to the saturation concentration of HSO_3^- in a drop surrounded by a gas phase concentration of 100 μg m^{-3}. The larger sized drops (diameter exceeding 1.5 mm) are unsaturated. The ground level pH of these drops increases with their diameter which suggests that a preponderance of large drops in rain showers will ensure a higher rainwater pH. This result is significant when one realizes that heavier rains have larger drops. However, heavier rains also scavenge more pollution. The dependence of rain water pH on rain rate will be further analyzed in section 3.4.3.

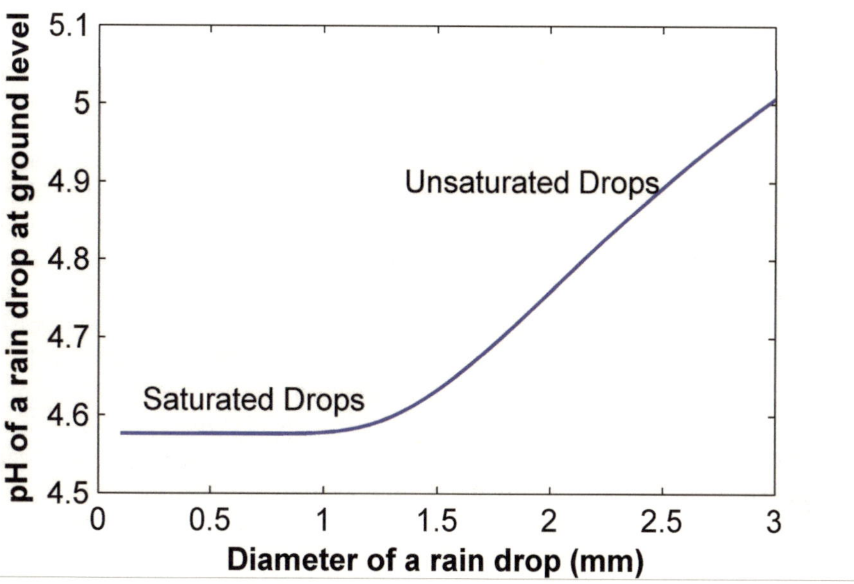

Fig. 13. pH of a drop as it reaches the ground after falling through a SO_2 laden atmosphere with a concentration of 100 μg m^{-3}

3.3 Formulation of the scavenging coefficient for NLC during the NE Monsoons

India receives heavy rainfall from two distinct monsoons. The South West Monsoon bearing moisture from the Indian Ocean, holds sway over most of the Subcontinent from June to September. The North East Monsoon (NE Monsoon) or Retreating Monsoon brings moisture from the Bay of Bengal and empties itself over the South Eastern coast of India during the months of October to December. It is the latter which forms the basis of this study. It is timely as some of the world's mega cities (i.e. Chennai and Kolkata) are along this coastline. The State of Tamil Nadu, located along the South East coastline, is most affected by the NE Monsoon and NLC receives rain rates in excess of 50 mm hr^{-1}.

In order to study washout by the NE monsoons, it is necessary to apply the previous mass transfer study which was aimed at an individual drop, to a multitude of drops of varying size which are present in a rain shower. This is made possible via the formulation of the scavenging coefficient (Section 3.1). For a particular size distributed spectra of rain drops (which varies with rain rate) the scavenging coefficient is given by (Seinfeld and Pandis, 2006):

$$\beta = \int \pi K_C D^2 N(D) d(D) \tag{26}$$

N(D) is the drop size distribution function which represents the number concentration of drops of a given size ($m^{-3}mm^{-1}$). K_c is the empirically determined mass transfer coefficient for the transfer of a gaseous species to falling drops. It can be evaluated by the following correlation (Bird et. al., 2002). Note that this empirical equation is a form of Eq. (20) and describes the same phenomena of mass transfer to a drop of size D.

$$K_c = (D_g/D)(2 + 0.6Sc^{1/3}\,\mathrm{Re}^{1/2}) \tag{27}$$

The DSD (drop size distribution) function ($N(D)$) is determined by fitting experimental data to mathematical distributions. Many different distributions have been used to represent the DSD of rain. These include the Marshal Palmer, Lognormal and the Modified Gamma Distributions. In the work of Konwar *et al.* (2006), it is shown that the Lognormal and Modified Gamma distributions are similar and provide a better fit than the Marshal Palmer distribution for DSD data of rain samples over Gadanki located in Tamil Nadu, India. Moreover, the Modified Gamma distribution is shown to give the best fit for the DSD. Roy *et al.* (2005) have studied the DSD of rain over the Cuddalore district of Tamil Nadu and have fitted the data to the same gamma distribution. Fortunately, NLC, the target area for the application of our work, is located in the district of Cuddalore itself. Assured by the work of Konwar *et al.* (2006) we adopted the gamma distribution of Roy *et al.* (2005) for our study. Curves of the gamma distribution for certain rain rates are shown in Fig. 14. It can be seen that the number of larger drops increases with rain rate. The modified gamma distribution function is given below. N_0, μ and λ are the intercept parameter, shape parameter and slope parameter of this distribution respectively. Roy et al. (2005) fitted DSD data for rain rates ranging from 0.5 mm hr^{-1} to 59 mm hr^{-1}.

$$N(D) = N_0 D^{\mu} \exp(-\lambda D) \tag{28}$$

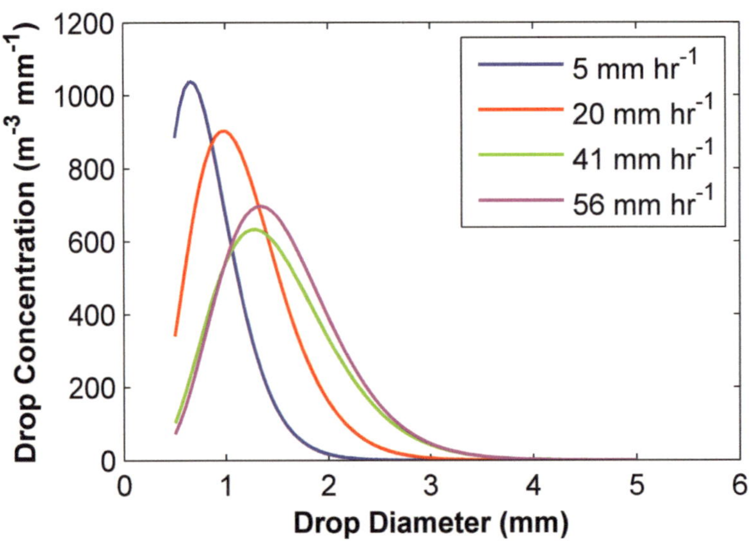

Fig. 14. Modified gamma distribution of rain drop sizes for various rain rates over Cuddalore (Roy et al., 2005)

Knowing the modified gamma distribution function (Eq. (28)), Eq. (26) can be numerically integrated to yield the value of the scavenging coefficient. This is done for all the rain rates studied by Roy et al. (2005) and the scatter plot shown in Fig. 15 is generated. On specific occasions NLC receives rain rates greater than those studied by Roy et al. (2005), sometimes up to 100mm hr^{-1}. In order to deal with such outlying cases as well as intermediate rain rates, a regression line (Eq. 29) is fitted to the scavenging coefficient results and is used to predict the scavenging coefficient for any rain rate (p, mm hr^{-1}).

$$\beta = \left(2.1961 \times 10^{-5}\right) p + 1.9244 \times 10^{-4} \tag{29}$$

Eq. (29) is valid for all non-zero rain rates. It fits the data better within the range of interest as compared to a regression line through the origin (please see Fig. 15).

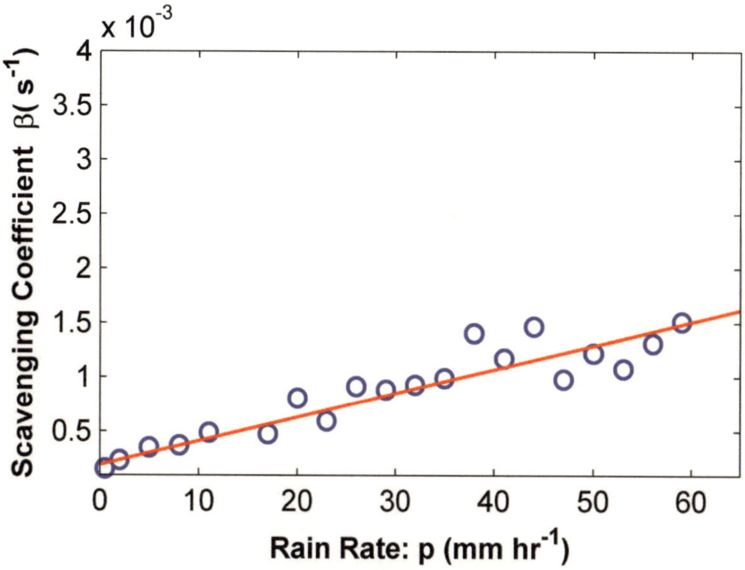

Fig. 15. Scavenging coefficient computed for various rain rates and fitted to a linear regression line

3.4 Atmospheric cleansing due to washout and rain water quality

3.4.1 Incorporation of washout into the dispersion model

In order to determine the reduction in ambient air SO_2 concentration during a rain shower, it is necessary to combine the results of the previous section with the dispersion model that is used to predict the spatially distributed SO_2 concentration around the stacks (section 2.4.1). In fact, the usefulness of applying the scavenging coefficient to analyze washout lies in the ease with which it can be incorporated into a dispersion model to develop a dispersion-deposition model.

During a rain event, a decrease in concentration will occur at all points in the atmosphere and the magnitude of decrease will be proportional to the concentration at that point. This decrease can be accounted for by a reduction in the source strength used in the gaussian dispersion formula (section 2.4.1). This reduction can be approximated by multiplication with an exponential factor which involves the product of the scavenging coefficient and the distance from the stack, along the prevailing wind direction. The washed out concentration is given by (Seinfeld and Pandis, 2006).

$$C_{g,washout}(x,y,z) = \exp\left(-\beta\frac{x}{u}\right) \times \frac{q \times 10^6}{2\pi u\sigma_y\sigma_z}\exp\left(\frac{-y^2}{2\sigma_y^{\ 2}}\right) \times \left[\exp\left(\frac{-(z-h)^2}{2\sigma_z^{\ 2}}\right) - \exp\left(\frac{-(z+h)^2}{2\sigma_z^{\ 2}}\right)\right] \quad (30)$$

Thus, as a parcel of air travels through rain and away from the stack, the concentration of the soluble gaseous species will be exponentially depleted due to scavenging by the rain. This exponential depletion is a result of the first order removal used to describe washout.
The rain water bearing dissolved SO_2 will reach the ground with an increased acid content. The rain water pH can be evaluated by considering the increase in H^+ ions due to the dissociation of dissolved SO_2 into HSO_3^-, as was done in the case of a single drop (Section 3.2.2). First, the amount of SO_2 brought to the ground per unit area per unit time by the rain is computed using Eq. (13). The volume of rain water received by that surface during a unit of time is simply the rain rate. Thus the concentration of HSO^{3-} ions can be estimated at any location on the ground, which in turn allows a computation of the pH at that point.

3.4.2 Wet deposition of SO_2 over NLC

The model developed in Section 3.4.1 was applied to a typical day in October (specifically 21st Oct 2007) which received heavy showers to the tune of 43.2 mm hr^{-1}. A horizontal wind of magnitude 1.11 m s^{-1} blew in from the north-north-east at the time of the shower. The scavenging coefficient at this rain rate is 1.14×10^{-3} (Eq. (29), Fig. 15). The ground level concentration of SO_2 surrounding the Thermal Power Stations in the absence of rain and the depleted levels due to washout are shown in Fig. 16. It is clear that the sharp NE monsoon showers rapidly cleanse the ambient air. Next, the rain water pH was calculated and contours of the same are displayed in Fig. 17. As may have been expected, the rain water is acidic in the region along the plume centerline (where the atmospheric concentration of SO_2 is maximum) with a pH of 4. However, the ph rises rapidly with distance from the centerline and the three receptors (station 5, 13 and 11 in Fig.17) within the plumes horizontal extent are affected by mild acid rain (pH above 4).

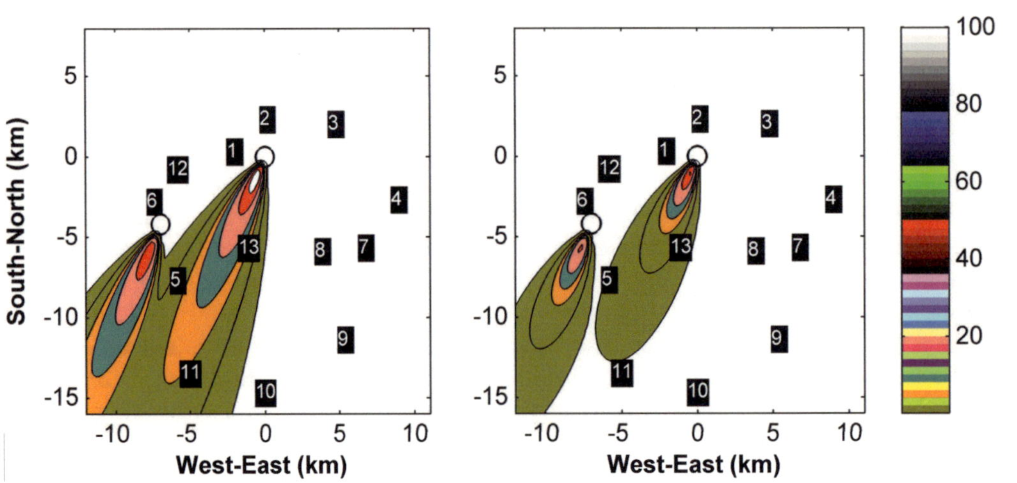

Fig. 16. The SO_2 concentration over NLC before and during the rain showers with rain rate of 43.2 mm hr^{-1}

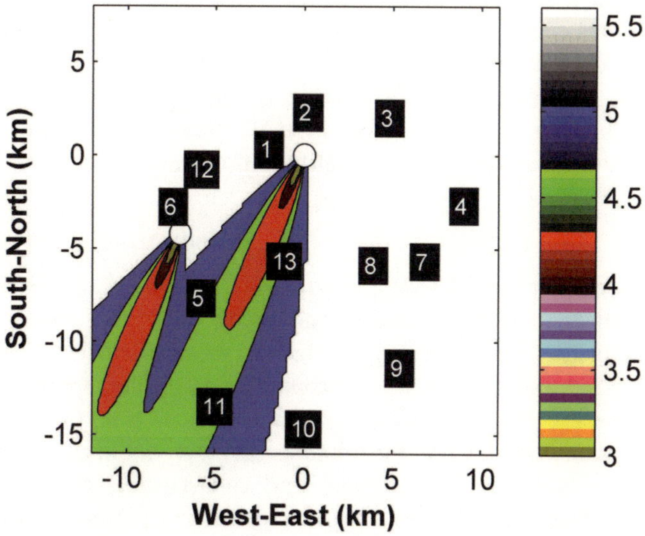

Fig. 17. Contours of the rain water pH at the ground for a rain rate of 43.2 mm hr^{-1}

3.4.3 Dependence of rain water pH on rain rate

It was discussed in section 3.2.2 that the pH of a rain drop increases with its diameter. This suggests that the presence of larger drops in rain showers will result in a higher pH. Further, it was observed in section 3.3 (Fig. 14) that the number concentration of larger drops increases with increase in rain rate. However, it must be borne in mind that higher rain intensities lead to increased scavenging as evidenced by the rising scavenging coefficient with rain rate in Fig. 15. Thus greater amounts of acidic pollutant will be present in the surface rain water. To analyze the relationship between rain rate and rain water pH, the previous computation of pH contours is repeated for an arbitrary small rain rate of 10 mm hr^{-1} and a large rain rate of 100 mm hr^{-1}. The scavenging coefficients at these rain rates are

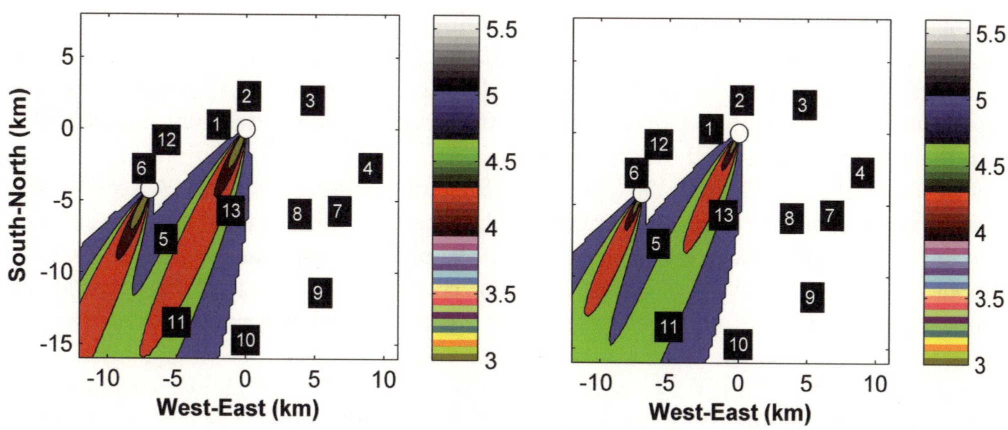

Fig. 18. Contours of the rain water pH for rain rates of 10 mm hr^{-1} (left) and 100 mm hr^{-1}

4.12×10^{-4} s^{-1} and 2.39×10^{-3} s^{-1} respectively (Eq. (29)). The degree of atmospheric cleansing will vary in proportion to the scavenging coefficients and these results are not displayed due to space constraints. The pH contours for rain rates of 10 mm hr^{-1} and 100 mm hr^{-1} are shown in Fig. 18. It is clear from a comparison of these images that the pH of the rain water does increase with rain rate. Although more SO_2 is scavenged at higher rain rates, it is diluted in larger amounts of water. Thus, the heavy rains of the NE monsoon will result in surface rain water of a higher pH as compared to mild mid-latitude precipitation.

4. Conclusion

At the end of the first decade of the twenty first century, the developing nations of the world are in a quandary. Constant industrialization and rising per capita energy consumption implies increasing fossil fuel derived energy usage. In most countries, renewable energy sources alone will not be able to meet the exorbitant energy demands in the foreseeable future. At the same time, the pollution caused by toxic emissions from thermal power plants and industries can no longer be ignored. Sustainable development implies the maintenance of a delicate balance between human progress and conservation/promotion of Nature. Environmental impact assessment studies of new and existing projects form an important component of the roadmap to sustainability. Such studies often turn to mathematical modelling methods for analyzing the impacts of polluting releases, especially gaseous emissions in the context of thermal power plants. Thus far, most studies in Asia have resorted to borrowing results from mid latitude analyses or adopting hasty adaptations of models which were developed for regions quite unlike their own. The widely different climatology, ecology and general environment of the Asian region demands region specific studies and analyses.

In this work a detailed modelling study of the removal mechanisms of gaseous pollutants is presented for the Neyveli Lignite Corporation (NLC), located in Tamil Nadu, India. Removal of SO_2 via dry deposition and rain scavenging is addressed with a particular emphasis on using local meteorological data and determining region specific model parameters. At various points in the analysis, it was observed that nature had provided the region with several advantages as far as mitigation of pollution was concerned. These include the fact that the leanest state of the vegetative canopy coincides with the hot summer when the convective boundary layer will ensure dilution of polluting emissions. The presence of two monsoonal seasons and the heavy rain rates (with larger rain drops) which result in higher pH is another boon to the region. This study is an example of the important inferences that can result from region specific studies that may be missed if unsuitably adapted, borrowed models are used.

While it was demonstrated that the canopy acts as a round-the-year sink for SO_2 and is able to improve air quality and living conditions of the residents, it is yet a matter of speculation as to the effect of continuous exposure to SO_2 on the plants. These plants will have different responses when compared to plants in other continents and should be investigated accordingly. It would be interesting to ascertain whether the monsoon rain received by the Indian sub-continent has larger drop sizes when compared to rain of similar intensity in other parts of the world e.g. Brazil. Most importantly, it is hoped that more such studies are carried out in developing countries where sustainable development is as yet only an ideal and Nature's resilience is constantly put to the test.

5. Acknowledgements

We thank the Neyveli Lignite Corporation (NLC) for funding this work and providing necessary meteorological data. We also thank the Director, School of Mechanical and Building Sciences, VIT University.

6. References

Bird, R. B.; Stewart, W. E. & Lightfoot, E. N. (2002). *Transport Phenomena, 2nd Ed.*, John Wiley (Asia), Singapore, pp. 681

Gao, W. & Wesely, M. L. (1995). Modeling Gaseous Dry Deposition over Regional Scales with Satellite Observations-I. Model Development, *Atmospheric Environment*, Vol. 29, No. 6, pp. 727-737

Hanna, S. R.; Briggs, G. A. & Hosker, Jr. R. P. (1982). *Handbook on Atmospheric Diffusion*, DOE TIC-11223, Technical Information Center, U.S. Dept. of Energy, USA, pp. 25-35

Johnson, D. (1982). The Role of Giant and Ultragiant Aerosol Particles in Warm Rain Initiation. *J. Atmos. Sci.*, Vol. 6, pp. 448-460

Konwar, M.; Sarma, D. K.; Das, J. & Sharma, S. (2006). Shape of the Rain Drop Size Distributions and Classification of Rain Type at Gadanki. *Indian J. Radio Space*, Vol. 35, pp. 360-367

Kumar, R.; Srivastava, S. S. & Kumari, K. M. (2008). Modeling Dry Deposition of S and N Compounds to Vegetation. *Indian J. Radio Space*, Vol. 37, pp. 272-278

Luhar, A. K. (1998). An Analytical Slab Model for the Growth of the Coastal Thermal Internal Boundary Layer under Near-Neutral Onshore Flow Conditions. *Boundary-Layer Meteorology*, Vol. 88, pp. 103-120

Matsuda, K.; Watanabe, I.; Wingpud, V.; Theramongkol, P. & Ohizumi, T. (2006) Deposition Velocity of O_3 and SO_2 in the Dry and Wet season above a Tropical Forest in Northern Thailand. *Atmospheric Environment*, Vol. 40, pp. 7557-7564

NASA, MODIS LAI/FPAR product, Available from: http://modis.gsfc.nasa.gov/

Patra, S. & Ghosh, S. (2010) Quantifying Trace Gas Uptake Rates by Passion Flower Draped Facades and Roofs. *Proc. World Green Roof Congress*, London, 2010

Picardo, J. R. & Ghosh, S. (2011). Establishing the Efficacy of the Cleansing Action of Tropical Evergreens: A Modeling Analysis of Asia's Largest Lignite based Power Plant. *Proc. 1st EnvironmentAsia International Conference*, Bangkok, Thailand, 2011.

Pruppacher, H. R. & Klett J. D. (1997). *Microphysics of Clouds and Precipitation, 2nd Ed.*, Dordrecht, The Netherlands, Kluwer Academic Publishers, pp. 770-772

Roy, S. S.; Datta, R. K.; Bhatia R. C. & Sharma A. K. (2005). Drop Size Distributions of Tropical Rain over South India. *Geofizika*, Vol. 22, pp. 105-130

Seinfeld, J. H. & Pandis, S. N. (2006). *Atmospheric Chemistry and Physics, 2nd ed.* John Wiley, New Jersey, USA, pp. 828-979.

Seth, U. K.; Sarkar, S.; Bardhan, R. & Ghosh, S. (2010). Asia's Largest Lignite based Power Plant's success story: Efficient Removal of SO_2 through a Manmade Forest Canopy. *Proc. World Congress of Engineering*, Imperial College, London, No. 2

Wesely, M. L. (1989). Parameterization of Surface Resistance to Gaseous Dry Deposition in Regional Scale Numerical Models. *Atmospheric Environment*, Vol. 23, pp. 1293-1304

Xu, Y. & Carmichael G. (1998). Modeling the Dry Deposition Velocity of Sulphur Dioxide and Sulphate in Asia. *Journal of Applied Meteorology*, Vol. 37, pp. 1084-99

Zhang, L.; Moran, M. D.; Markar, P. A.; Brook, J. R. & Gong, S. (2002). Modelling Gaseous Dry Deposition in AURAMS: a Unified Regional Air Quality Modelling System. *Atmospheric Environment*, Vol. 36, pp. 537-560

Zhang, L.; Brook, J. R. & Vet, R. (2003a). Evaluation of a Non-Stomatal Resistance Parameterization for SO_2 Dry Deposition. *Atmospheric Environment*, Vol. 37, pp. 2941-2947

Zhang, L.; Brook, J. R. & Vet, R. (2003b). A Revised Parameterization for Gaseous Dry Deposition in Air Quality Models. *Atmos. Chem. Phys.*, Vol. 3, pp. 2067-2082

14

Interaction of Urban Vegetation Cover to Sequester Air Pollutants from Ambient Air Environment

Sharda Dhadse, D. G. Gajghate, P.R. Chaudhari, D. R. Satapathy and S. R. Wate
National Environmental Engineering research Institute, Nagpur India

1. Introduction

Acute air pollution problem is being faced in urban agglomeration due to economic expansion, increase in population, increased industrial activities and exponential growth in automobiles. The air pollution from these sources is imposing threat to urban human health. The morbidity and mortality caused by air pollution result in long term reduction of productivity and ultimately in overall deterioration of economic condition (Dockery & Pope, 1994; Anderson et al., 1992; Schwartz et al., 1996). In India the particulate matter problem is very significant due to the huge number of vehicles plying on the road, number of power plants, combustion processes, dust storms and domestic emissions (Gurjara et al., 2004). In the recent studies, exceeding levels of PM10 are observed (TERI, 1997; Chelani et al., 2001).

The trees in urban environment are continuously exposed to air pollutants, which play an important role in maintaining ecological balance by actively participating in the nutrients cycle. Many trees are effective for trapping and absorbing air pollutants and acts as sink to several air pollutants (Allan & Krupa, 1986; Bell & Treshow, 2002; Farmer, 1993; Barker & Tingey, 1992; De Kok & Whaltey 1984; De Kok & Stulen, 1998; Treshaw & Anderson, 1989; Nowak *et al.* 1997; Shyam *et al.*, 2006). Hence it is more beneficial to see the impact of pollution on vegetation especially on roadside trees. (Gajghate & Hasan, 1999; Kotoh *et al.* 1989; Kozhauharov *et.al.* 1985; Ninave *et.al.*, 2001; Mellios *et al.*, 2006; Mutena, 2004; Tommervik *et al.*, 1995).

Nagpur City is the best place to study the interactions of atmospheric pollutants such as SO_2, NO_2 and suspended particulate matter (SPM) on vegetation, as it is high traffic zone with industrial area on the outskirts as well as have good vegetation cover in the city. Nagpur city is very well known as second Green City in India. Plantations are actively carried out every year in the city with the active participation of local administration and non-governmental organizations (NGOs). The city is also richly dotted by well maintained parks, plantations, forest patches and agricultural fields. This has given lush green aesthetic appearance to the Nagpur city.

Present research is carried out, to study the status of urban pollution in relation to biodiversity in the Nagpur city using ambient air quality monitoring, remote sensing for landuse cover, ground truth and anatomical and biochemical responses of the trees to air pollution.

2. Materials and methods

2.1 Sampling locations

The Nagpur city is situated in between $20^0 30'$ and $21^0 30'$ N latitude and $78^0 30'$ and $79^0 30'$ E longitude. The strategic situation of the Nagpur City in the central part of India has lead a rapid expansion of city and ever increasing environmental problems with reference to pollution of air, water and soil. Ambient air quality monitoring, vegetation survey and remote sensing study have been carried out in three different areas of the city having different activities namely industrial (MIDC), commercial (Itwari) and residential (NEERI Campus).

2.2 Remote sensing study

In order to strengthen the baseline information on existing land use pattern, the remote sensing data has been collected for Nagpur area lying between ($21^0 03'$ - $21^0 13'$) N longitude (79^0 $00'$ - 79^0 $10'$) E latitude. The satellite data was acquired from the IRS P6 (RESOURCESAT -1) LISS III Scene (Path 99, Row 57 dated 04 Nov-2004; CD format) and the collateral data were used from Reference map, Toposheets 55 o/4 and 55 k/16.

2.3 Biological survey

Biological survey was carried out at these three sampling sites. The species of plants, their abundance and diversity were recorded and correlated with land use pattern of vegetation obtained through remote sensing imagery.

2.4 Air pollution tolerance index (APTI)

The plants namely *Bougainvellia spectabilis, Azadirachta indica, Pongamia pinnata* and *Polyalthia longifolia* growing in these area were selected for studying their biochemical responses to the impact of air pollution. The leaf samples were analyzed for pH (P) of leaf extract (Singh and Rao, 1993), Chlorophyll (T) (Arnon, 1949), ascorbic acid (A) (Singh, 1977) and Relative water content (R). Air pollution Tolerance Index (APTI) which gives an empirical value representing tolerance level of a plant to air pollution was used to interpret the impact of pollution on the plants. The APTI is calculated by the formula as $A(T+P)+R/10$.

2.5 Air monitoring

SPM in ambient air was measured by using standard High volume (Hi-vol) sampling technique. Gaseous samples were collected in absorbing solutions by tapping air and collected samples were analyzed using standard wet chemical method (Katz, 1977).

3. Results and discussions

3.1 Vegetation cover in study area through remote sensing

The land use / land cover status in Nagpur urban area as per IRS P6 LISS-III is shown in figure. 1. Eight different classes are identified in the Nagpur namely settlement, water bodies, fallow land, scrubland, bare soil/sand, vegetation-1, vegetation-2 and vegetation-3. Normalized Differentiation Vegetation Index (NDVI) values were computed for Nagpur City. The entire area has been classified into five categories namely non-vegetation, vegetation-1, vegetation-2, vegetation-3, and vegetation-4. The percentage compositions of

these classes in Nagpur City were non-vegetation (58.79%), vegetation-1 (17.40%), vegetation-2 (14.86%), vegetation-3 (3.77%), and vegetation-4 (5.19%) respectively.
The percentage compositions of different classes of vegetation (Table 1, Figure 1) in residential area are non-vegetation (71.23%), vegetation-1 (8.87%), vegetation-2 (8.82%), vegetation-3 (4.89%) and vegetation-4 (6.19%). The percentage compositions of different classes in industrial area are non-vegetation (74.08%), vegetation-1 (10.18%), vegetation-2 (7.31%), vegetation-3 (4.51%), and vegetation-4 (3.92%). Similarly, the percentage compositions of different classes in commercial area are non-vegetation (89.38%), vegetation-1 (6.18%), vegetation-2 (3.34%), vegetation-3 (0.49%) and vegetation-4 (0.61%). Table 1 show that the vegetation cover is highest at residential area i.e. 28.77%, lowest in commercial area i.e. 10.62% and moderate in industrial area i.e. 25.92%. Similarly the density, abundance and diversity of plants are highest at residential area while it is lowest in commercial area. The density and diversity of plants is medium at industrial area.

Classification Category	Residential Area		Industrial Area		Commercial Area	
	(Area in Km^2)	Area in %	(Area in Km^2)	Area in %	(Area in Km^2)	Area in %
Non-vegetation	19.547	71.23	20.330	74.08	24.529	89.38
Vegetation - 1	2.434	8.87	2.793	10.18	1.695	6.18
Vegetation - 2	2.420	8.82	2.007	7.31	0.917	3.34
Vegetation - 3	1.342	4.89	1.238	4.51	0.133	0.49
Vegetation - 4	1.700	6.19	1.075	3.92	0.618	0.61
Total (Vegetation)	7.896	28.77	7.113	25.92	3.363	10.62
Total Area	27.443	100	27.443	100	27.443	100

Table 1. Inventory of vegetation cover at different air quality monitoring locations

3.2 Floral diversity and abundance

The field survey of plants in Nagpur City showed a total of 103 plant species consisting of 45 tree species, 13 shrub species, 30 herb species, 8 species of climbers, 1 species of bamboo and 5 species of grasses. The density and diversity of plants were not equally distributed throughout the Nagpur City. The numbers of plants are more in well-planned residential area with adequate space for maintenance of green cover. Density and diversity of the plants is more in residential areas. Around 100 plants species were observed in residential areas with higher abundance of most of the species (around 800 trees/ha). The dominant plants were *Azadirachta indica, Albizzia procera, Annona squamosa, Leucaena lecocephala, Ipomoea fistulosa, Lantana camara, Alternanthera sessilis, Ageratum conizoides, Calotropis gigantia, Parthenium hysterophorus. Cyanodon dactylon and Dactyloctenium.* Among trees *Azadirachta indica, Leucaena leucocephala* are most abundant species. The shrub *Calotropis procera and Lantana camara* are abundant species. (Table 2).

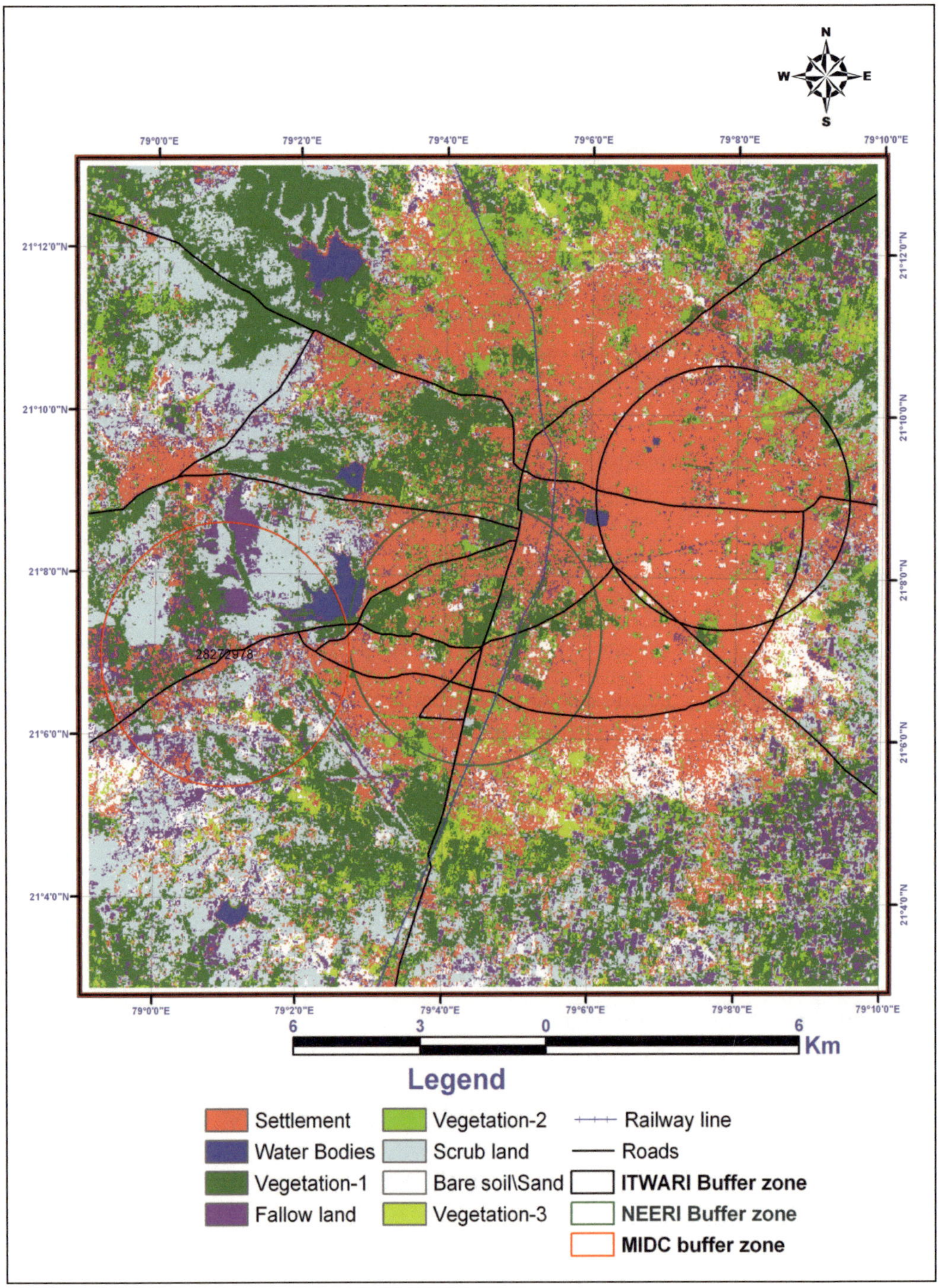

Fig. 1. Vegetation cover at Nagpur city

The industrial area has moderate amount of vegetation cover, which consist of avenue plantations, greenbelts and vegetation around industrial complexes. A total of 59 species of plants were recorded from this area. The density of trees is around 200 trees./ha. The dominant plants were *Azadirachta indica, Cassia fistula* and *Cassia siamea.* While among herbs *Parthenium hysterophorus, Ageratum conizoides* and *Alternanthera sessilis* were the most abundant species.

Sr. No.	Abundance Class	Abundance Scale (Species cover)	Species		
			Tree	Shrub	Herb
1	5	76-100%	*Azadirachta indica*	*Calotropis procera*	*Parthenium hysterophorus*
2	4	51-75%	*Leucaena leucocephala*	*Lantana camara*	*Ageratum conyzoides, Alternanthera sessile*
3	3	26-50%	*Albizzia procera*	*Ricinus communis*	*Malvastrum tricuspidatum*
4	2	6-25%	*Cassia siamea*	*Vitex negundo*	*Tridax procumbens, Ocimum sanctum*
5	1	up to 5%	*Psidium guajava*	*Indigofera tinctoria*	*Ricinus communis, Tinospora cordifolia*

Table 2. Abundance of major plant species in Nagpur urban area

The commercial area is the old part of the city with narrow lanes, closely set houses, shops and hectic commercial activity. Avenue plantation is negligible. Gardens and other plantation are almost absent. Thus, less number of plant species (32 species) were observed, the important species were *Azadirachta indica* and *Cassia siamea.*

3.3 Air quality status

Annual ambient air quality data is presented in Table 3. The annual average values of SPM varied ranged from 124, 134 and 195 μg/m^3 at residential, industrial and commercial sites respectively. Highest concentrations of SPM were recorded in commercial site. The reason for high SPM at commercial site is due to vehicular emissions, re-suspension of dust, commercial and domestic use of fuel etc. (NEERI Report, 2001).

The annual average values of NO_2 concentration were ranged from 18, 15 and 21 μg/m^3 at residential, industrial and commercial sites and those of SO_2 in air ranged from 6, 9 and 7 at residential, industrial and commercial sites respectively.

The residential area showed lowest levels of air pollutants as compared to those in industrial and commercial places. Though the values of SO_2 and NO_2 in residential, industrial and commercial areas are well below the standards, SPM values exceed the standards for commercial area i.e. 140 μg/m^3. Thus the ambient air quality is satisfactory in Nagpur city. However the rapid expansion of city, increasing number of automobiles and proposed big industrial area indicate that Nagpur would also be classified as polluted city if preventive environmental management measures are not undertaken.

Air Pollutant	Residential Site			Industrial Site			Commercial Site			Co-relation Coefficient
	N	AM	SD	N	AM	SD	N	AM	SD	
SPM	70	124	77	54	134	101	51	195	107	0.00278
SO_2	92	6	2	83	9	9	82	7	4	0.65
NO2	92	18	18	83	15	14	82	21	21	0.709

N: Sampling Days
AM: Arithmetic Mean
SD: Standard Deviation

Table 3. Annual Arithmetic Mean of SPM, SO_2 & NO_2

3.4 Sequestration of air pollutants by vegetation cover

The concentration of air pollutants was observed to be highest in industrial area followed by commercial area and residential area in decreasing order. It appears that ambient air quality in these microhabitats of Nagpur city is governed by sources of pollution as well as vegetation cover in that area. Table 3 shows that the biodiversity is highest in residential area followed by industrial area and commercial area in decreasing order.

Considering the air pollutant concentration in industrial area as 100%, the air pollutants SPM and SO_2 showed decrease in their concentration by 31.28% and 36.39% for SPM and 22.22% and 33.33% for SO_2 in commercial and residential area respectively. Thus filtering capacity of the plants is well known and these are utilized in greenbelt and in avenue plantation for filtration and reduction of dust concentration in air (Rao *et al*, 1993; Olszyk, 1984). The trees are also well known for acting as sink for SO_2 (Butterbach *et al.*, 1997; Shen *et al.*, 1995). The densities of trees (No/hector) were observed to be co-relating with reduction in ambient SPM and SO_2 concentration (Fig. 2 and 3).

Similarly the biodiversity of plants is also observed to be positively co-related with reduction in ambient concentrations of SO_2 and SPM. It is thus evident that increase in number of trees and increase in biodiversity helps in effective sequestration of air pollutants especially SO_2 and SPM. The ambient NO_2 concentration was observed to vary from 15 to 21 $\mu g/m^3$ (Fig 4). Each distribution pattern did not co-relate with number of trees or biodiversity of the area.

It has been reported that plant has high preference for SO_2 and extremely law preference for NO_2 to be absorbed and metabolized in the plant tissue. The preference for the plants for air pollutants has been investigated by Hill (1971) and Bennet and Hill (1973, 1975) that appear to follow the following order.

$$HF > SO_2 > Cl > NO_2 > O_3 > PAN > NO > CO$$

Smith (1981) assess pollutant removal efficiency of a model forest hector developed by USEPA which consisted of 95% year old plants will observed that, this forest patch annually removes 748 tonnes of SO_2 while only 0.38 tonnes of NO_2 is removed from atmosphere.

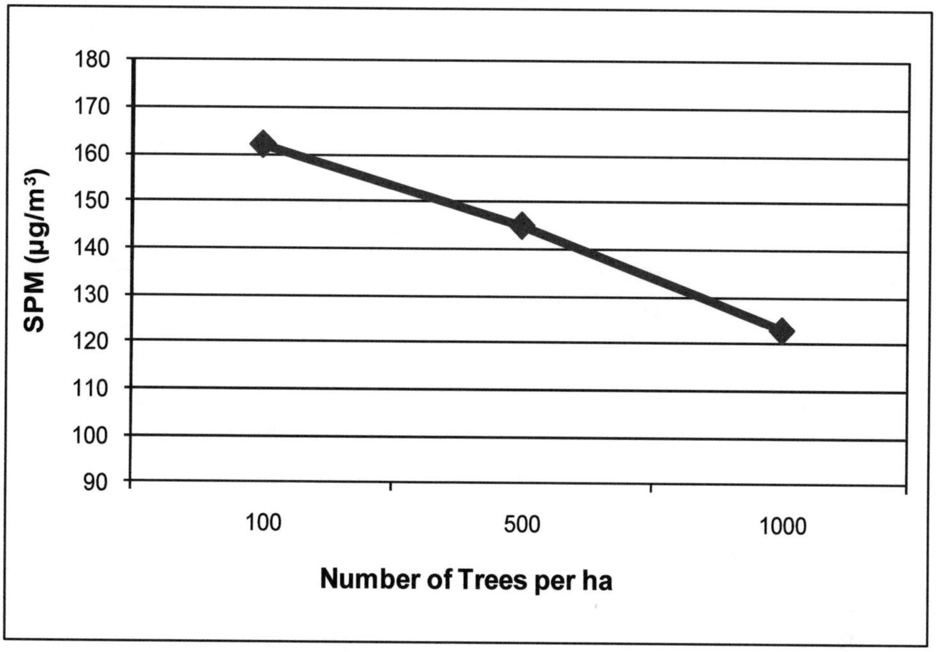

Fig. 2. Relationship between number of trees (per ha) and ambient SPM concentration (Slope: 0.04 and Intercept: 166.17)

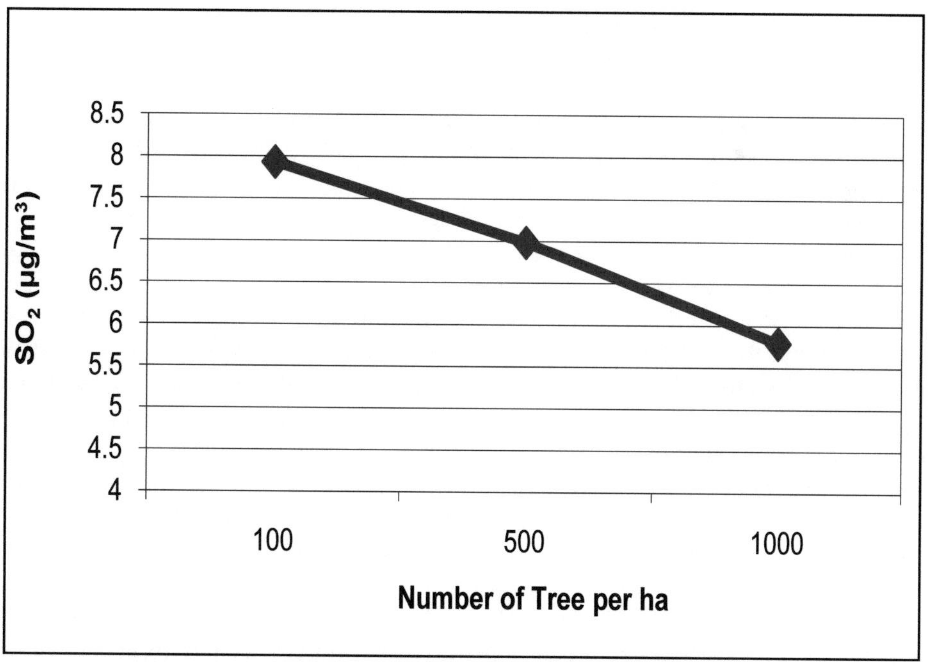

Fig. 3. Relationship between number of trees (per ha) and ambient SO_2 concentration (Slope: 0.00238; Intercept: 8.17)

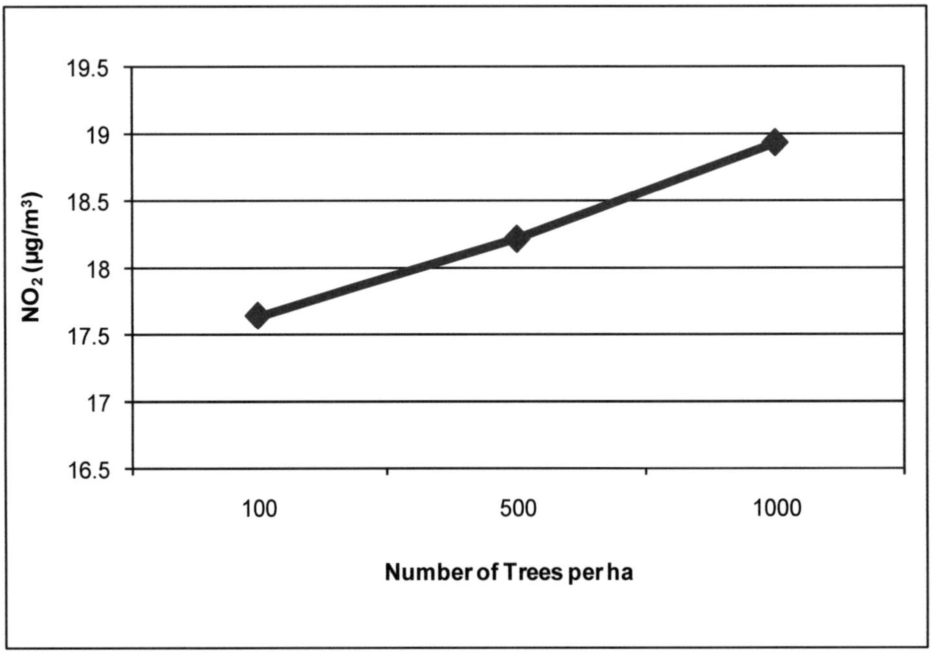

Fig. 4. Relationship between number of trees (per ha) and ambient NO_2 concentration (Slope: 0.00238; Intercept: 8.17)

3.5 Air pollution tolerance index (APTI)

The forgoing discussion shows that present level of pollutants are lower than the prescribed limits except SPM. However these chronic levels are prevailing for a long time i.e. more than 20 years. It is the vegetation cover that is acting as sinks to the air pollutants and playing important role in keeping the levels of pollution well below the standards. However the vegetation cover is relatively less in the commercial area and shows impact of air pollution. The vegetation of Nagpur city is exposed to dust pollution, which may affect the biochemical make up and tolerance capability of plants to the air pollution. The Air Pollution Tolerance Index for the plants from residential, commercial and industrial area is shown in figure III respectively. The APTI of the plants from residential area ranged from 13.09 to 61.39 and from 16.69 to 31.44 in commercial area whereas from 22.02 to 83.15 for plants in industrial area. APTI values of the four common trees in polluted area were found to vary from 16.69 to 83.15 and the plants are listed according to their tolerance or sensitivity to pollution. The four species selected for APTI studies are tolerant to air pollutants in the geographical area of Nagpur city. Next to *Azadiracta sp.* with APTI value (61.36) *Baugainvellia sp.* was found to have APTI index values of 16.87 at residential area, 27.44 at commercial area and 27.83 in industrial area. *Polyalthia* sp. which ranks next showed APTI values of 18.40 at residential area, 21.64 and 22.12 at commercial and industrial area respectively (Fig. 5). Though these species showed different sensitivity, they all appear to be tolerant to air pollution as their APTI values showed to be increased in polluted environment. This may probably be the reason that the plants develop tolerance and detoxify the absorbed pollutants and as a result other biochemical constituents are not affected in such plants. This can be taken as an indication of the development of detoxification mechanism in the plant necessary for the tolerance.

This work has indicated the suitability of *Azadirachta indica* as the most tolerant species, suitable as sink for air pollution. It can be utilized for urban plantation and greenbelt development in industrial area to reduce the level of air pollution. However more research is necessary on a wide variety of trees, shrubs and herbs to prepare a biological sensitivity map of flora. The vast database would be useful in identifying tolerant plants, sink plants and management program.

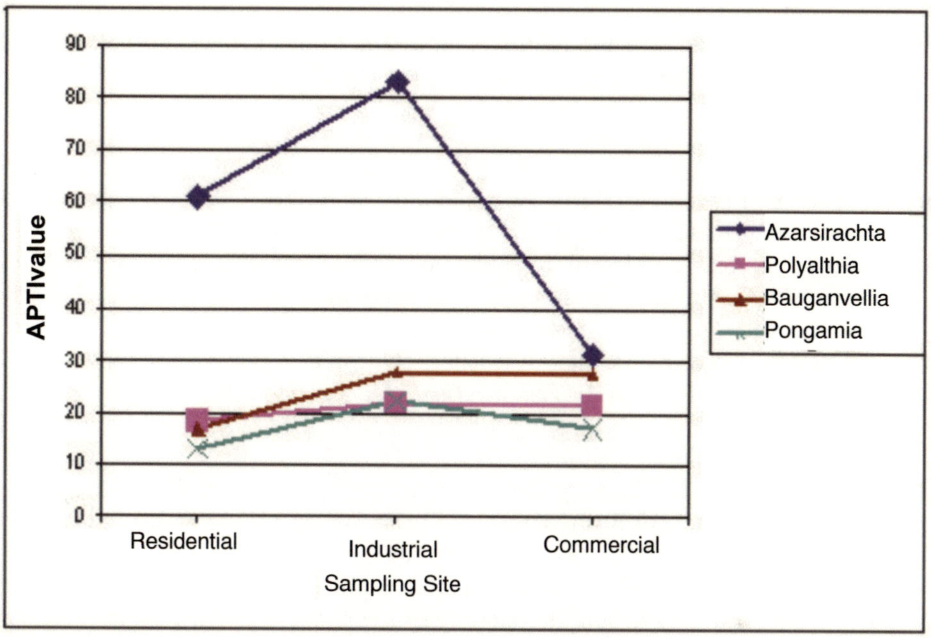

Fig. 5. APTI values of four tree species in different sampling sites

4. Conclusion

The consequences of complex effects of combinations of several atmospheric pollution and climate change in particular may threaten vegetation in ecosystem of urban atmosphere. The levels of pollutants SPM, NO_2 and SO_2 were highest at commercial site followed by industrial and then residential site. Status of air quality with respect to NO_2 and SO_2 showed the values within the range while SPM level is higher than the standard value except at commercial place. In response to these changing conditions plants adopt to their changed environment by showing different air pollution tolerance index. Out of four species namely *Azadirachta, Polyalthia, Baugainvellia* and *Pongamia,* only the *Azadirachta* is having the best air pollution tolerance index. This study can be helpful to identify the plant species for greenbelt development as one of the control measures for reduction of ambient concentration of air pollution. This tolerant species may be used for avenue plantation and beautification of the city. The vegetation is denser at residential site, while it is moderate at industrial site. Though the industrial area shows moderate air pollution but the small green belts in and around the industries do not allow the air pollutants to raise upto higher limits. Similarly at commercial location the lack of vegetation causes increase in the levels of NO_2, SO_2 and SPM. This proves that the plants are the best sinks for all types of air pollutants. More work in this area is needed for more precise information on expected changes in vegetation exposed to air pollutants. Similar studies can be extended to identify air pollution tolerant species at different environmental pollution level in different climatic conditions, which can be used to maintain the air quality level within the permissible level.

5. Acknowledgment

Authors are grateful to the Director, NEERI Nagpur and Council of Scientific and Industrial Research (CSIR) for providing the infrastructure and financial support.

6. References

Allan, H., L & Krupa, S. V. (1986). Air pollutants and their effects on the terrestrial Ecosystem. *Series in advances in environmental Sci & Tech.*

Anderson, K. R., Avol, E. L., Edward, S.A., Shamoo, D.A., Peng, R.C., Linn, W.S. & Hackny, J.S. (1992). Controlled exposures of volunteers to respirable carbon and sulphuric acid aerosols. *J Air Waste Manag Assoc, 42,* 770-776.

Arnon, D.F. (1983). Copper enzymes in isolated chloroplasts polyphenol oxidase in *Beta vulgaris. in: Proc Symp on Air Pollution Control held at IIT, Delhi,* 17-18.

Barker, J. R. & Tingey D. T. (1992). *Air pollution effects on biodiversity.* pub. Van Nostrand Reinhold, New York.

Bell J.N.B. & Treshow, M. (2002). *Air pollution and plant life.* pub. John Wiley & Sons Ltd.

Benett, J.H. and Hill, A.C. (1973). Absorption of gaseous air pollutants by a standardized plan canopy. *J. Air. Pollut. Contr. Ass., 28,* 203-206.

Bennett, J.H. and Hill, A.C. (1975). Interaction of air pollutants with canopies of vegetation. J.S. Mudd and T.T. Koslowski (eds). *Response of plants to Air Pollution,* Academic Press, New York, 273-306.

Butterbach -Bahl K. R., Gasche, L. Breuer & H. Papen (1997). Fluxes of NO and N_2O from temperate forest soils: impact of forest type, N deposition and of liming on the NO and N_2O emissions. *Impacts of air pollutants on vegetation in developing countries,* 48, 1-2, 79-90.

Chelani, A.B., Gajghate, D.G. & Hasan, M.Z.. (2001). Airborne toxic metals in air of Mumbai city, India. *Bull Environ Contam Toxicol, 66,* 196-205.

De KoK Luit, J. & Stulen, B. I. (1998). *Response of plant metabolism, Air Pollution and Global change.* Publishers Leiden.

De KoK, M.L. & Whaltey, F.R. (1984). *Gaseous air pollutants and plants metabolism,* Butterworth London: 480.

Dockery, D.W. & Pope, C.A. (1994). Acute respiratory effects of particulate air pollution. *Ann Rev Public health, 15,* 107-132.

Farmer, A.M. (1993). The Effect of dust on vegetation - A review. *Environment Pollution, 79*: 63-75.

Gajghate, D.G. & Hasan, M.Z. (1999). Ambient lead level in urban Area. *Environmental contamination and Toxicology, 62,* 403-408.

Gurjara, B.R., Aardennea, J.A., Lelievelda, J. & Mohanb, M. (2004). Emission estimates and trends (1990-2000) for mega city Delhi and implications. *Atmos Environ, 38*, 5663-5681.

Hill, A.C. (1971). Vegetation: A sink for atmospheric pollutants. *J. Air Pollution Control Assoc., 21:* 341-346.

Katz, M. (1977). *Standard methods for air sampling and analysis.* 2nd edition, APHA Press Inc., Spring Field VA.

Kotoh, T., Kasuya, M., Kagamimork, S., Kazuka, S. & Kawano, S. (1989). Effects For air pollution on Tannin biosynthesis and predation damage in *Cryptomeria Japonica. Phytochem, 28*, 439-445.

Kozhauharov, S.I., Petrova, A.V. & Veroukova, L. (1985). Some aspects of using plants as indicators of environmental pollution. *In symposium on biomonitoring state of environment, Indian National Science Academy,* New Dilhi-110002, 98-105.

Mellios, G., Aalst, R. V & Samaras Z. (2006). Validation of road urban emission inventories by means of concentration data measured at air quality monitoring station in Europe. *Atmospheric Environment,* 40, 7362-7377.

Mutena, F. (2004). Measuring air quality over large urban area: development and application of an air pollution index at the urban area of Naples. *Atmospheric Environment,* 38, 6195-6202.

NEERI Report, (2001). Ambient air quality status for ten cities of India. No.11. submitted to Central Pollution Control Board, Government of India, New Delhi.

Ninave, S.Y., Chaudhari, P.R., Gajghate, D.G. & Tarar, J.L. (2001). Foliar Biochemical features of plants as indicators of air Pollution. *Bull. Environ. Contam. Toxicolol,. 67,* 133-140.

Nowak, D.L., Michale, P.J., Ibarra, M., Crane, D., Stevens, J. C. and Luley C.J. (1997). The effect of urban vegetation on air pollution in 22nd Nato/CCMS International Technical Meeting on air Pollution modeling and its application, 276-282 Nato/Csms, Brussels.

Olszyk, D.M. and Tingey, D.T. (1984). Phytotoxicity of air pollutants evidence for Phyto Detoxification of SO_2 but not O_3. *Plant Physiol, 74,* 999-1005.

Rao, M. V., Khujneri, S, Dubey, P. S. & Kumawa, D. M. (1993). Response of eight tropical plants to enhanced ammonia deposition under field conditions prevalent with SO_2 and NH_3. *J. Water, Air & Soil Pollution,* 71, 331-345.

Schwartz, J., Dockery, D.W. & Neas, L.M. (1996). Is daily mortality associated specifically with fine particles?. *J Air Waste Manag Assoc.* 46,927–939.

Shen, J., Zhao, Q., Tang, H., Zhang, F., Feng, Z., Okita, T., Ogura, N. & Totsuka, T. (1995). Concentrations and depositions of SO_2, SO_4^{2-} etc. in a Changing suburban forested area. *Water, Air & Soil Pollution,* 85, 3, 1299-1304.

Shyamshilpa, H. N., Verma, S. & Bhangana. (2006). Air pollution & its impact on plant Growth. New India Publishing Agency.

Singh, A. (1977). Practical Plant Physiology. Kalyani Publishers, New Delhi, p. 266.

Singh, S.K. & Rao, D.N. (1993). Evaluation of plants for their tolerance to air pollution . *In: Proc. Symp on Air Pollution control held at IIT Delhi, Nov, 1983,* 218-224.

Smith, W.H. (1981). Air pollution and forests interactions between Air Contaminants and Forest Ecosystems Springer – Verlag, New York.

TERI. (1997). Environmental aspects of energy use in large Indian metropolitan cities. Tata Energy Research Institute (TERI), New Delhi.

Tommervik, H., Johansen, B.E. & Pedersen, J.P. (1995). Monitoring the effects of air pollution on terrestrial ecosystem in Varanger (Norway) and Nikel-Pechenga (Russia) using remote sensing. *The Science of Total Environment*. 160,161, 753-767.

Treshaw, M. & Anderson, F. K. (1989). *Plant stress from air Pollution*. John Wiley and Sons.

15

Method OF INAA for Critical Evaluation Pollution of Ecosystem

Blanka Maňkovská and Július Oszlányi
Institute of Landscape Ecology, Slovak Academy of Sciences, Bratislava
Slovakia

1. Introduction

The heavy metals in mosses biomonitoring network was originally established as a Swedish initiative (Růhling, Tyler, 1968, 1971). It is assumed that in Slovakia (SK) a large gradient of atmospheric deposition load of elements exists, because part of the SK territory belongs to one of the most polluted areas in central Europe known as the 'Black Triangle II'. In order to recognise the distribution of element deposition in SK, the moss monitoring technique, also known as bryomonitoring, was applied to the whole territory in 1990, 1995, 1996, 1997, 2000 and 2005 (Maňkovská, Oszlányi, 2008). Bryomonitoring is a suitable technique using moss analysis to determine the levels of atmospheric deposition of the elements. The technique has been highly standardised and international bryomonitoring programs coordinated by Nordic countries have a pan-European character (Harmens et al.2008; Maňkovská, 1997; Maňkovská et al. 2003; 2008a b; Schröder et al.2008; Suchara et al. 2007; Zechmeister et al. 2003). These are characterized by a high concentration of toxic elements such as As, Cd, Cr, Cu, Hg, Fe, Mn, Ni, Pb, V and Zn. The aim of this paper is to present actual data of the first survey of 9 elements in mosses (*P. schreberi, H. splendens* and *Dicranum* sp.) in five Slovak sites: National parks (Vysoké Tatry, Nízke Tatry, Západné Tatry, Slovenský raj) and Landscape protection area (Veľká Fatra) and Báb Research Sites. An additional aim of this report is to summarize changes in heavy metal concentrations in mosses in Slovakia between 1990 and 2005 and to summarize concentration of Ag, Al, As, Au, Ba, Br, Ca, Ce, Cl, Co, Cr, Cs, Fe, Hf, I, In, K, La, Mg, Mn, Mo, Na, Ni, Rb, Sb, Sc, Se, Sm, Sr, Ta, Tb, Th, Ti, U, V, W, Yb, Zn, Zr in mosses in Slovakia for 2000.

2. Material and methods

The mosses *P. schreberi, H. splendens* and *Dicranum* sp. have been taken in compliance with the international methods (ICP, 1994) in permanent areas situated at the intersection of a 16 x 16 km pan-European network. Moss samples were collected according to the procedures used in deposition surveys in the Scandinavian countries. The collection of samples was performed during the first half of August 1990, 1995, 1996, 1997, 2000 and 2005. The samples consisted of the last three years' annual segments. Separately, the Vysoké Tatry, Nízke Tatry, Západné Tatry-Jelenec and Slovenský raj National parks, Veľká Fatra in Landscape protection area and the Báb Research Sites were evaluated.

Neutron activation analysis (NAA) was performed in 2000 in the Frank Laboratory of Neutron Physics, Dubna, Russia for 39 elements (Ag, Al, As, Au, Ba, Br, Ca, Ce, Cl, Co, Cr, Cs, Fe, Hf, I, In, K, La, Mg, Mn, Mo, Na, Ni, Rb, Sb, Sc, Se, Sm, Sr, Ta, Tb, Th, Ti, U, V, W, Yb, Zn, Zr). In the laboratory, the samples were carefully cleaned from needles, leaves, soil particles and only the green and green-brown shoots representing the last three years growth were analyzed, after being air-dried to constant weight at 30–40 °C for 48 hours. The samples were neither washed nor homogenised. For short-term irradiation samples of about 300 mg were pelletized in simple press forms and heat-sealed in polyethylene foil. For epithermal neutron activation analysis samples prepared in the same manner were packed in aluminium cups for long-term irradiation. The samples were irradiated in the IBR-2 fast-pulsed reactor, in channels equipped with a pneumatic system. The neutron flux characteristics are shown in table 1. Two kinds of analysis were performed: to determine short-lived radionuclides the samples were irradiated for 3 minutes in the second channel (Ch2), and to determine elements associated with long-lived radionuclides samples were irradiated for 100 hours in the cadmium screened Ch1. After irradiation gamma-ray spectra were recorded twice for each irradiation using a high-purity Ge detector: the first one after decay periods of 2–3 minutes for 5 minutes, and the second one for 20 minutes, 9–10 minutes after the short irradiation. For long irradiation, samples were repacked into clean containers and measured after 4–5 days for 45 minutes, and after 20–23 days for 3 hours (Frontasyeva, Pavlov, 2000).

Irradiation position	Neutron flux density, $[n\times cm^{-2}\times s^{-1}]\times1012$		
	thermal (E = 0÷0.55 eV)	resonance (E = 0.55÷105 eV)	fast (E = 105÷25.106 eV)
Ch1 (Cd-screened)	0.023	3.3	4.2
Ch2	1.23	2.9	4.1

Table 1. Flux parameters of irradiation positions

The atomic absorption spectrometer Varian Techtron was used to determine concentrations of Cd, Cr, Cu, Hg, Ni, Pb and Zn. The elemental analyser LECO SC 132 was applied to determine the concentration of sulphur, and the elemental analyser LECO SP 228 was used to determine the total concentration of nitrogen.

The valid equation [concentration in moss] $mg.kg^{-1}$ = [4x atmospheric deposition] $mg.m^{-2}.year^{-1}$was used (Steinnes et al. 2001). The analysis results were interpreted in the form of contamination factors K_F as the rates median value of element in Slovak mosses C_{iSl} vs. Norway mosses C_{iN} ($K_F = C_{iSl}/ C_{iN}$). Median Norway value C_{iN} were taken from Steinnes et al. (2001).

The accuracy of data was verified by analysis of standard plant samples and by comparison with the results obtained in 109 laboratories within the IUFRO working group for quality assurance (Hunter, 1994). The QC of NAA results were ensured by analysis of reference materials: trace and minor elements in lichen IAEA-336 (International Atomic Energy Agency), IAEA-SL-1 (Trace elements in lake sediment) and SRM-1633b (Constituent elements in coal fly ash, US NIST-National Institute of Standards and Technology), SRM-

2709 (Trace elements in soil). For an assessment of vegetation we used current statistical methods, factor and correlation analysis.

3. Results and discussion

The results of analysing the concentration of 45 elements in the mosses (*P. schreberi, H. splendens, Dicranum* sp.) are given in Table 2. We present separate loading for Báb Research Sites, National Parks (NP) - Vysoké Tatry; Nízke Tatry; Západné Tatry - Jelenec valley; Slovenský raj, the Landscape protection area (LPA) of Veľká Fatra and for Slovakia. For comparison with a pristine territory the corresponding data for northern Norway (Steinnes et al., 2001) is shown in the left-hand column. Comparison with the limit values from Norway (Table 2) shows strong pollution of the examined areas of Slovakia with most of the elements. However, for Au, N, S and Zr data from Norway was not available.

Excesses of concentrations of elements in mosses in comparison with Norway were expressed by the coefficient of loading by air pollutants K_F and classified it into 5 classes; class 1 - elements are in normal standard concentrations and the coefficient does not exceed the value 1; class 2 - light loading (coefficient of loading ranges from 1 to 10); class 3 - moderate loading (coefficient ranges from 10 to 50); class 4 - heavy loading (coefficient ranges from 50 to 100) and class 5 - toxic (coefficient is higher than 100). As shown in Table 3, the coefficient of loading by air pollutants K_F for almost all elements is higher than one, excerpt for Au, Br, In, Mg, N, S, Se (Báb); Au, Br, Ca, I, Se (Vysoké Tatry); Au, Br, I, Mg, S, Se, Sm, Ti (Nízke Tatry); Au, Br, Ca, Hg, I, In, Mg, S, Se, Sm (Západné Tatry-Jelenec); Au, Br, In Sm, Se (Slovenský raj) and Au, Br,In Sm (Veľká Fatra).

Spatial trends of element concentrations in mosses were metal-specific. However in general the lowest concentrations were observed in the Nízke Tatry National park, and 10 times higher concentrations occurred in the Vysoké Tatry (Al, Cr, Hf, Sb, Ta, Yb, Zr); Západné Tatry (Cr, Hf, Sb, Yb, Ta, Zr); Slovenský raj (Ag, Hg, Hf, Mo, Pb, Ta, Tb, Yb, Zr); Veľká Fatra (Cr, Hf, Sb, Ta, Tb, Th, Yb, Zr), Báb (Al, Cd, Ce, La, Hf, Mo, Pb, Sb, Sc, Ta, Tb, Th, Yb, Zr) in comparison to the Norway values. The coefficient of loading by air pollutants K_F ranged from 4.2 in the Nízke Tatry to 11.8 in Slovenský raj. Since 1990, the metal concentration in mosses has declined for Cd, Cr, Cu, Fe, Hg, Ni, Pb, Zn (Fig. 1, 2, 3).

Ele-ment	**Báb** (n=40)		**Vysoké Tatry** (n=3)		**Nízke Tatry** (n=4)		**Západné Tatry** (n=14)		**Slovenský Raj** (n=5)		**Veľká Fatra** (n=6)		**Slovakia** (n=86)		**Norway**	
	Aver.	Exc.	Aver.	Exc.	Aver.	Exc.	Aver.	Exc.	Aver.	Exc.	Aver.	Exc.	Aver.	Exc.	Aver.	Exc.
Ag	0.021	4.2	0.031	6.2	0.027	5.4	0.033	6.6	0.072	14.4	0.032	6.4	0.038	7.6	0.005	1
Al	968	11.0	888	10.1	345	3.9	735	8.4	708	8.0	769	8.7	966	11.0	88	1
As	0.21	6.8	0.2	6.7	0.15	5.0	0.18	6.0	0.28	9.3	0.19	6.3	0.2	6.7	0.03	1
Au**	0.001	0.2	0.001	0.5	0.001	0.5	0.001	0.5	0.001	0.5	0.001	0.5	0.001	0.5	0.002	-
Ba	11.3	2.4	8.8	1.8	8.8	1.8	12.5	2.6	21.5	4.5	11	2.3	15.4	3.2	4.8	1
Br	1.16	0.9	1.03	0.8	0.78	0.6	0.85	0.7	0.81	0.6	0.95	0.8	0.91	0.7	1.25	1
Ca	1998	2.6	722	0.9	1006	1.3	113	0.1	1036	1.3	1485	1.9	1322	1.7	780	1
Cd	0.20	10.1	0.1	5.0	0.12	6.0	0.15	7.5	0.14	7.0	0.16	8.0	0.16	8.0	0.02	1
Ce	1.04	12.1	0.45	5.2	0.463	5.4	0.61	7.1	0.68	7.9	0.77	8.9	0.98	11.4	0.086	1
Cl	53	1.1	68	1.4	99	2.0	80	1.6	74	1.5	52	1.0	70	1.4	50	1

Co	0.28	6.5	0.28	6.6	0.13	3.1	0.34	8.0	0.41	9.4	0.32	7.3	0.38	8.8	0.043	1
Cr	1.64	9.7	1.83	10.8	0.8	4.7	1.8	10.6	1.41	8.3	2.33	13.7	2.18	12.8	0.17	1
Cs	0.10	3.3	0.13	4.3	0.2	6.7	0.12	4.0	0.11	3.7	0.11	3.7	0.13	4.3	0.03	1
Cu	2.29	2.1	2.1	1.9	1.6	1.5	5.5	5.0	4.1	3.7	2.1	1.9	2.5	2.3	1.1	1
Fe	437	4.8	425	4.7	186	2.0	377	4.1	455	5.0	497	5.5	555	6.1	91	1
Hf	0.163	81.5	0.128	64.0	0.055	27.5	0.131	65.5	0.125	62.5	0.125	62.5	0.168	84.0	0.002	1
Hg	0.047	3.6	0.034	2.6	0.032	2.5	0.01	0.8	0.147	11.3	0.039	3.0	0.102	7.8	0.013	1
I	0.7	1.5	0.3	0.6	0.4	0.8	0.4	0.8	0.7	1.4	0.5	1.0	0.5	1.0	0.50	1
In	0.02	0.4	0.05	1.0	0.12	2.4	0.04	0.8	0.04	0.8	0.02	0.4	0.04	0.8	0.05	1
K	1908	2.5	2362	3.1	2269	3.0	1936	2.6	1753	2.3	1674	2.2	1770	2.4	750	1
La	0.77	11.0	0.23	3.3	0.32	4.6	0.35	5.0	0.51	7.3	0.58	8.3	0.62	8.9	0.07	1
Mg	374	1.0	373	1.0	262	0.7	354	0.9	393	1.0	387	1.0	436	1.1	386	1
Mn	96	1.2	114	1.4	89	1.1	135	1.6	141	1.7	92	1.1	110	1.3	83	1
Mo	0.31	10.2	0.17	5.7	0.24	8.0	0.28	9.3	0.3	10.0	0.26	8.7	0.27	9.0	0.03	1
N*	5500	1.0	6642	1.2	5494	1.0	6268	1.1	5520	1.0	5407	1.0	5927	1.1	5638	-
Na	120	2.4	81	1.6	107	2.1	82	1.6	109	2.2	90	1.8	129	2.6	50	1
Ni	0.88	3.1	0.78	2.8	0.38	1.4	0.84	3.0	0.53	1.9	0.98	3.5	0.99	3.5	0.28	1
Pb	8.6	12.3	5.1	7.3	4	5.7	5.3	7.6	12.9	18.4	6.3	9.0	8.3	11.9	0.7	1
Rb	2.85	1.2	7.5	3.0	9	3.6	5.1	2.1	3.15	1.3	4.13	1.7	4.26	1.7	2.48	1
S*	464	0.9	492	1.0	404	0.8	473	0.9	596	1.2	511	1.0	502	1.0	508	-
Sb	0.19	19.3	0.17	17.0	0.13	13.0	0.2	20.0	2.06	206.0	0.24	24.0	0.38	38.0	0.01	1
Sc	0.165	11.0	0.118	7.9	0.048	3.2	0.105	7.0	0.11	7.3	0.143	9.5	0.153	10.2	0.015	1
Se	0.087	0.9	0.08	0.9	0.055	0.6	0.08	0.9	0.07	0.8	0.098	1.1	0.095	1.0	0.093	1
Sm	0.115	1.3	0.038	0.4	0.053	0.6	0.051	0.6	0.058	0.7	0.07	0.8	0.088	1.0	0.086	1
Sr	22.5	7.8	13.5	4.7	7	2.4	15.8	5.4	14	4.8	14.5	5.0	21.6	7.4	2.9	1
Ta	0.024	24.0	0.017	17.0	0.008	8.0	0.017	17.0	0.016	16.0	0.02	20.0	0.023	23.0	0.001	1
Tb	0.030	30.0	0.007	7.0	0.005	5.0	0.01	10.0	0.013	13.0	0.017	17.0	0.02	20.0	0.001	1
Th	0.133	13.3	0.05	5.0	0.05	5.0	0.08	8.0	0.09	9.0	0.11	11.0	0.13	13.0	0.010	1
Ti	9.5	1.6	16	2.7	3.8	0.6	11.6	2.0	9.3	1.6	10.3	1.7	14.3	2.4	5.875	1
U	0.028	7.0	0.018	4.5	0.033	8.3	0.025	6.3	0.025	6.3	0.028	7.0	0.035	8.8	0.004	1
V	1.575	4.6	1.13	3.3	0.71	2.1	1.48	4.4	1.24	3.6	1.88	5.5	1.85	5.4	0.34	1
W	0.078	2.6	0.05	1.7	0.08	2.7	0.07	2.3	0.08	2.7	0.08	2.7	0.07	2.3	0.030	1
Yb	0.093	31.0	0.038	12.7	0.016	5.3	0.038	12.7	0.043	14.3	0.05	16.7	0.063	21.0	0.003	1
Zn	15	2.0	16.3	2.2	9.5	1.3	15	2.0	19	2.6	11.5	1.6	15.4	2.1	7.4	1
Zr**	15	30.0	24	48.0	7.5	15.0	20	40.0	16	32.0	17	34.0	23.1	46.2	0.50	-
KF		**8.8**		**6.7**		**4.2**		**7.0**		**11.8**		**7.6**		**9.5**		**1**

Note: * Slovak median 2000 (Suchara et al. 2007); **Macedonia (Barandovski et al. 2006); Median Norway value (Steinnes et al. (2001); Aver. -Average; Exc. – exceedance of element concentrations in mosses comparison with Norway (Central Norway belongs to the least polluted regions in Europe); K_F - Coefficient of loading by air pollutants; n- number of Permanent monitoring plots.

Table 2. Atmospheric deposition of elements (average in mg.m^{-2}.rok^{-1}) in Báb Research Sites, Vysoké Tatry, Nízke Tatry, Západné Tatry, Slovenský Raj,Veľká Fatra , Slovakia and Norway calculated from concentration of elements in 3 year old segments of *P. schreberi; H. splendens* and *Dicranum* sp. in the year 2000

The examined Slovak territory shows that many regions have intense mining activity. These are characterized by a high concentration of toxic elements such as As, Al, Mn, Cd, Cr, Cu, Hg, Pb, and Sb. The most significant anthropogenic sources are fossil fuels combustion (electric power stations) located in Upper Nitra, and Vojany. Of the other industrial activities, metallurgy, nonferrous ores processing, and cement factories are also important, such as in Central Spiš, Central Pohronie, and Orava. In Slovakia many pollutant sources overlap which causes difficulty in source identification (Maňkovská, 1996; Suchara et al. 2007; Florek et al. 2008).

Sites	Contamination factor K_F					K_F
	< 1	1 -10	10-50	50- 100	>100	
Báb	Au, Br, In, Mg, N, S, Se	Ag, As, Ba,Ca, Cl, Co, Cr, Cs,Cu, Fe,Hg, I, K, Mn, Na, Ni, Rb, Sm, Sr, Ti, U, V, W, Zn	Al, Cd, Ce, La, Mo, Pb, Sb,Sc, Ta, Tb, Th,Yb, Zr	Hf		8.8
Vysoké Tatry	Au, Br, Ca, I, Se	Ag, As, Ba, Cd,Ce, Cl, Co, Cs, Cu, Fe, Hg, In, K, La, Mg, Mn, Mo, N,Na, Ni, Pb, Rb, S, Sc, Se, Sm, Sr, Tb, Th, Ti, U, V, W, Zn	Al, Cr, Sb, Ta, Yb,Zr	Hf		6.7
Nízke Tatry	Au, Br, I, Mg, S, Se, Sm,Ti	Ag, Al, As, Ba, Ca, Cd,Ce, Cl, Co, Cr, Cs, Cu, Fe, Hg, In, K, La, Mn, Mo, N,Na, Ni, Pb Rb, Sb, Sc, Sr, Ta, Tb, Th, U, V, W, Yb, Zn, Zr	Hf			4.2
Západné Tatry	Au, Br,Ca, Hg, I, In, Mg, S,Se, Sm	Ag, Al, As, Ba, Cd,Ce, Cl, Co, Cs, Cu, Fe, K, La, Mn, Mo, N,Na, Ni, Pb Rb, Sc, Sr, Tb, Th, Ti, U, V, W, Zn	Cr, Sb, Yb, Ta, Zr	Hf		7
Slovenský raj	Au, Br, In Sm, Se	Al, As, Ba, Ca, Cd,Ce, Cl, Co, Cr, Cs, Cu, Fe, I, K, La, Mg, Mn, N,Na, Ni, Rb, S, Sc, Sr, Th, Ti, U, V, W, Zn	Ag, Hg, Mo, Pb, Ta, Tb, Yb, Zr	Hf	Sb	11.8
Veľká Fatra	Au, Br,In Sm	Ag, Al, As, Au, Ba, Ca, Cd,Ce, Cl, Co, Cs, Cu, Fe,Hg, I, K, La, Mg, Mn, Mo, N,Na, Ni, Pb Rb, S, Sc, Se, Sr, Ti, U, V, W, Zn	Cr, Sb, Ta, Tb, Th, Yb, Zr	Hf		7.6
Slovakia	Au, Br, In	Ag, As, Ba, Ca, Cd, Cl, Co, Cs, Cu, Fe, Hg, K, La, Mg, Mn, Mo, N,Na, Ni, Rb, S,Sb, Sc, Sm, Sr, Ti, U, V, W, Zn	Al, Ce, Cr, Hf, Pb, Sb, Se, Ta, Tb, Th, Yb, Zr			9.5

Table 3. Coefficient of loading by air pollutants K_F in the year 2000

In comparison with the 1990 survey (Maňkovská, 1997), the average values in 2005 (Fig.1, 2, 3) for Cd, Cr, Cu, Fe, Hg, Ni, Pb, and Zn were reduced. Decreasing concentrations in Slovakia are connected with the decrease in production of steel and non-ferrous metals, and with the fazing out of leaded gasoline.

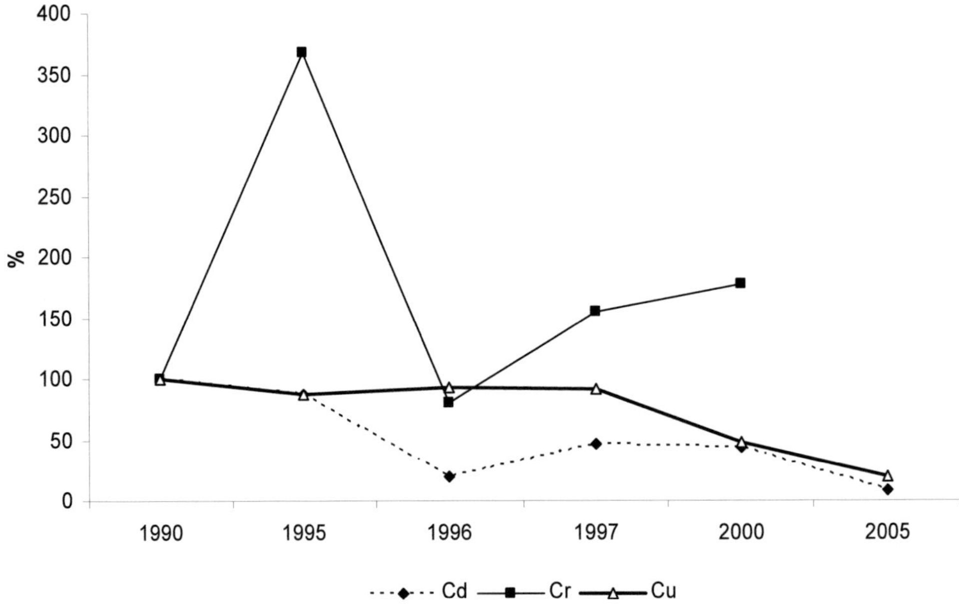

Fig. 1. Concentration of Cd, Cr and Cu (in % of average) in mosses for Slovakia in all survey years
Note: Year (number of PMP): 1990 (58); 1995 (79); 1996 (69); 1997 (74); 2000 (86); 2005 (82)
PMP- Permanent monitoring plots

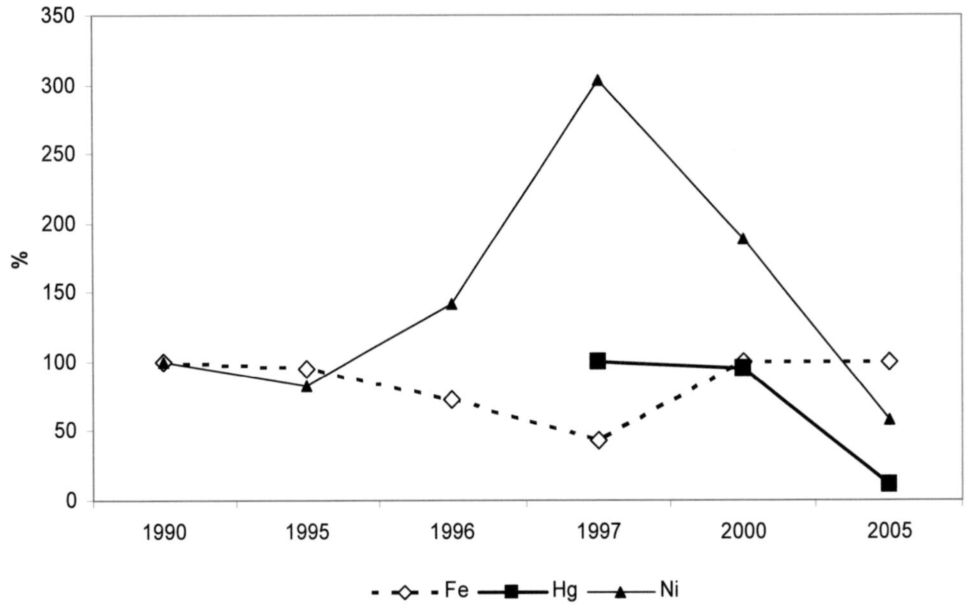

Fig. 2. Concentration of Fe, Hg and Ni (% of average) in mosses for Slovakia in all survey years
Note: Year (number of PMP): 1990 (58); 1995 (79); 1996 (69); 1997 (74); 2000 (86); 2005 (82),
PMP- permanent monitoring plots

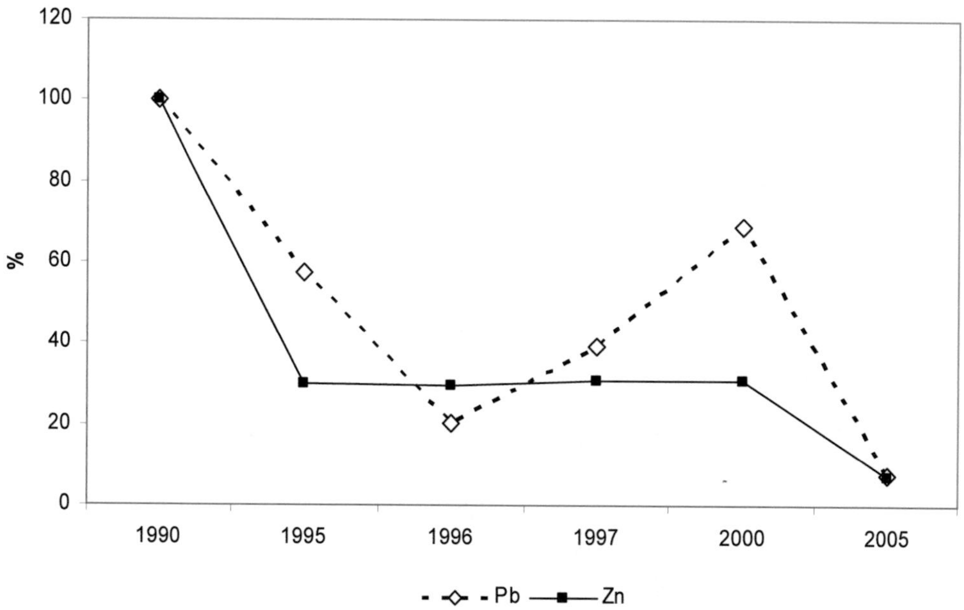

Fig. 3. Concentration of Pb, and Zn (% of average) in mosses for Slovakia in all survey years
Note: Year (number of PMP): 1990 (58); 1995 (79); 1996 (69); 1997 (74); 2000 (86); 2005 (82),
PMP- permanent monitoring plots

This Slovak moss data was subjected to principal component factor analysis (Varimax with Kaiser Normalization). The result of analysis is recorded in Table 4. The 8 factors explain 80% of the total variance in the data set. From the knowledge of the element composition of each factor and the values of factor loadings, the major sources can be identified. Interpretation of 8 factors in the sequence of their significance is as follows:

- Factor 1 is responsible for 40% of the total variance and it is characterized by the presence of all typical crustal elements. It can be explained by elements associated with mineral particles, mainly windblown dust and it includes 13 elements with Al, Sc, Ti, V, Fe, Zr, Re, Hf, Ta, and Th having loadings higher than 0,9, and the next ten elements (Na, Mg, Cr Co, Ni, Se, Sr, Cs, Ba, and U) have loading factors between 0.70-0.89.
- Factor 2 is the industrial component, with very high loadings for Cu, Zn, Ag, Sb, Pb and it is responsible for 10% of the total variance with a factor loading from 0.52 to 0.87. The maximum value is in area of Krompachy - Smolnícka Huta - Zlatá Idka. This factor reflects the impact of metallurgical plants.
- Factor 3 mainly includes Cl, Mn, In and Factor 6 contains K, Rb, and Cs. These elements are likely to be mainly of natural origin.
- Factor 4 includes As, Cd, Pb and S. The pollution with these metals is most likely caused by the lengthy transport. High levels of precipitation are strongly correlated with the heavy metal deposition, and this seems to be main source of heavy metal fallout at higher altitudes.
- Factor 5 has Mo and W elements as its main constituents. Their loading constitutes 0.77 of the factor. The factor components originate from engineering and instrument industry located in the towns of: Brezno, Martin, Dubnica, and Košice and in the triangle Stará Turá-Piešťany-Nové Mesto nad Váhom triangle.

- The major elements in Factor 7 are Ca, I, Br and In. In Slovakia there are 1626 registered mineral water springs of different chemical composition, and a portion of these contain I and Br in adequate quantity.
- Finally, Factor 8 explains 3% of the total variance, where the dominant element is mercury. Sources of contamination with Hg are related to metal processing industries, combustion fossil fuels and municipal solid wastes and trans-boundary contamination in the NW wind directions.

Factor	F1	F2	F3	F4	F5	F6	F7	F8
% of cumulative variability	43,9	53,3	59,8	64,5	68,2	71,4	74,4	76,9
Element								
Ag	0.14	**0.68**	0.15	0.04	0.12	–0.02	–0.20	–0.15
Al	**0.94**	0.06	0.06	0.10	–0.02	0.06	0.18	–0.06
As	0.17	0.36	0.26	**0.61**	0.34	–0.04	0.10	0.16
Au	0.35	0.32	0.43	–0.05	0.24	0.16	0.02	–0.02
Ba	**0.74**	0.35	0.29	–0.01	–0.06	0.16	–0.13	0.01
Br	0.42	–0.12	0.05	0.29	0.24	**0.45**	0.33	0.23
Ca	0.28	–0.03	0.08	0.20	0.07	–0.11	**0.65**	0.11
Cd	0.13	0.21	–0.04	**0.76**	0.14	–0.08	0.12	0.04
Ce	**0.95**	0.08	0.08	–0.06	0.10	0.16	0.01	0.06
Cl	0.08	0.08	**0.82**	0.29	–010	0.20	–0.03	0.14
Co	**0.84**	0.14	0.13	0.01	0.11	–0.04	0.09	0.16
Cr	**0.75**	0.00	0.24	0.26	0.15	–0.12	–0.03	–0.10
Cs	0.59	0.21	0.36	–0.11	–0.10	**0.52**	0.00	0.00
Cu	0.15	0.76	–0.03	0.24	0.17	–0.05	0.22	0.18
Fe	**0.93**	0.14	0.14	0.08	0.04	0.17	0.07	0.02
Hf	**0.92**	0.06	0.11	0.04	–0.03	0.11	–0.03	0.00
Hg	0.01	0.03	0.11	0.05	0.14	–0.06	0.09	**0.85**
I	0.30	0.13	0.10	0.11	0.23	0.05	**0.75**	0.09
In	0.02	0.06	**0.59**	–0.14	–0.14	0.06	0.36	–0.11
K	0.14	–0.03	0.52	0.11	0.21	**0.62**	–0.11	–0.09
La	**0.88**	0.07	0.03	–0.11	0.23	0.22	0.14	0.07
Mg	**0.78**	0.13	0.12	0.22	0.01	0.03	0.37	–0.16
Mn	0.38	0.12	**0.57**	–0.21	0.10	–0.05	0.07	0.20
Mo	0.21	0.21	–0.07	0.26	**0.77**	0.13	0.13	–0.06
Na	**0.80**	0.03	0.03	0.00	0.20	0.13	0.08	–0.17
Ni	**0.84**	0.02	0.09	0.11	0.03	–0.18	0.09	0.12
Pb	0.12	**0.55**	–0.22	**0.51**	0.15	–0.11	0.27	–0.17
Rb	0.28	–0.03	0.08	–0.18	0.10	**0.77**	–0.04	–0.03
S	0.06	0.33	–0.13	0.42	–0.33	0.27	0.18	0.46
Sb	0.09	**0.87**	0.09	0.05	–0.05	0.06	–0.02	0.06
Sc	**0.94**	0.08	0.06	0.01	0.08	0.15	0.16	0.01
Se	**0.85**	0.13	–0.03	0.28	0.17	0.10	0.02	0.04
Sm	**0.88**	0.01	–0.02	–0.06	0.25	0.22	0.10	0.01

Sr	**0.84**	-0.03	0.11	0.20	-0.10	-0.08	-0.06	0.13
Ta	**0.95**	0.08	0.09	0.04	0.04	0.11	0.05	0.01
Tb	**0.94**	0.11	0.05	-0.03	0.12	0.11	0.15	0.09
Th	**0.94**	0.12	0.11	-0.02	0.10	0.19	0.06	0.03
Ti	**0.90**	-0.01	0.03	0.11	0.02	0.00	0.22	-0.02
U	**0.78**	0.10	0.01	-0.05	0.27	0.32	0.24	-0.03
V	**0.90**	0.03	-0.06	0.17	0.09	0.00	0.25	-0.03
W	0.23	0.19	-0.01	0.11	**0.77**	0.13	0.15	0.22
Yb	**0.94**	0.10	0.02	-0.02	0.11	0.09	0.03	0.11
Zn	-0.04	**0.52**	0.22	0.33	0.28	-0.02	0.20	0.14
Zr	**0.93**	0.03	0.10	0.06	0.00	0.09	0.06	-0.05

Note: Eight main source types were identified. Characteristic elements for the sources types are marked in bold type

Table 4. WARIMAX rotated PC analysis of the first eight factors on 86 moss samples collected in the territory of Slovakia in 2000

4. Conclusion

- Mosses provide an effective method for monitoring trends in heavy metals pollution in Slovakia at a high resolution;
- Spatial trends of heavy metal concentrations in mosses were metal-specific. Since 1990, the metal concentration in mosses has declined for cadmium, chromium, cooper, iron, lead, mercury, nickel, and zinc.
- The coefficient of loading by air pollutants K_F for almost all elements is higher than one, excerpt for Au, Br, In, Mg, N, S, Se (Báb); Au, Br, Ca, I, Se (Vysoké Tatry); Au, Br, I, Mg, S, Se, Sm, Ti (Nízke Tatry); Au, Br, Ca, Hg, I, In, Mg, S, Se, Sm (Západné Tatry-Jelenec); Au, Br, In Sm, Se (Slovenský raj) and Au, Br,In Sm (Veľká Fatra), and this is an unfavourable outcome compared to the Norway values. Only the concentration of Au, Br, In, Mg, N, S, Se is lower in Báb Research Sites than relevant Norwegan vakues.
- This obtained data is useful as a reference level for comparison with the future measurements of air pollution in the examined area and also for biodiversity study. Finally, the significance of transboundary atmospheric transport in this region still remains a subject for future study.

5. References

Barandovski, L., Cekova, M., Frontasyeva, M., V., Pavlov, S., S., Stafilov, T., Steinnes, E., Urumov, V., 2006: Air Pollution Studies in Macedonia Using The Moss Biomonitoring Technique, NAA, AAS and GIS Technology. *JINR Preprint* E18-2006-160, Dubna: 5-15.

Florek M., Maňkovská B., Oszlányi J., Frontasyeva V., M., Ermakova, E., Pavlov S., S., 2007: The Slovak heavy metals survey by means the bryophyte technique. *Ekológia*, Bratislava, 26, 1: 99–114.

Frontasyeva, M., V., Pavlov, S., S., 2000: Analytical investigation at the IBR-2 Reactor in Dubna. *JINR Preprint* E14-2000-177, Dubna: 5–32.

Harmens, H., Norris, D., A., Koerber, G., R., Buse, A., Steinnes, E., Rühling, A., 2008: Temporal trends (1990–2000) in the concentration of cadmium, lead and mercury in mosses across Europe. *Environ. Pollut.*, 151: 368–376.

Hunter, I., R., 1994: Results from the Interlaboratory sample exchange. *IUFRO, Working Group* S1.02 -08 Foliar Analysis. Natural Resources Institute, Kent, 18 pp.

ICP, 1994: Manual on Methods and Criteria for Harmonized Sampling, Assessment, Monitoring and Analysis of the Effects of Air Pollution on Forest. *3rd edition Programme Coordinating Centre West*, BHF, Hamburg, 45 pp.

Maňkovská, B., 1996: Geochemical atlas of Slovakia – forest biomass (in Slovak, English). *Geologická služba slovenskej republiky*, Bratislava, 87 pp.

Maňkovská, B., 1997: Deposition of heavy metals in Slovakia - assessment on the basis of moss and humus analyses. *Ekológia*, Bratislava, 16: 433–442.

Maňkovská, B., Florek, M., Frontasyeva, M.V., Ermakova, E., Oprea, K., Pavlov, S.S., 2003: Atmospheric deposition of heavy metals in Slovakia studied by the moss biomonitoring technique. *Ekológia*, Bratislava, 22, 1: 211–217.

Maňkovská, B., Oszlányi, J., 2008 a: Mosses and foliage of forest tree species as biomonitors of nitrogen pollution. *International Journal of Environmental Studies*, 65, 3: 377–388.

Maňkovská, B., Oszlányi, J., Barančok, P., 2008 b: Measurement of the atmosphere loading of the Slovak Carpathians using bryophyte technique. *Ekológia*, Bratislava, 7, 4: 339-350.

Rühling, A., Tyler, G., 1968: An ecological approach to the lead problem. *Bot. Not.*, 121: 321–342.

Rühling, A, Tyler, G., 1971: Regional differences in the deposition of heavy metals over Scandinavia. *J. Appl. Ecol.*, 71, 8: 497–507.

Schröder, W., Pesch, R., Englert, C., Harmens, H., Suchara, I., Zechmeister, H., G., Thöni, L., Maňkovská, B., Jeran, Z., Grodzinska, K., Alber, R., 2008: Metal accumulation in mosses across national boundaries: Uncovering and ranking causes of spatial variation. *Environ. Pollut.*, 151: 377–388.

Steinnes, E., Berg, T., Sjobak, T., E., Uggerud, H., Vadset, M., 2001: Atmospheric deposition of heavy metals in Norway (in Norwegian). *Nation-wide survey* 2000, Report 838/01, State Pollution Control Authority, Oslo, 28 pp.

Suchara, I., Florek, M., Godzik, B., Maňkovská, B., Rabnecz, G., Sucharova, J., Tuba, Z., Kapusta, P., 2007: Mapping of Main Sources of Pollutants and their Transport in Visegrad Space. Silvia Taroucy Institute for Landscape and Ornamental Gardering Průhonice, CZ, ISBN 978-80-85116-55-7,78 pp.

Zechmeister, H., G., Grodzińska, K., Szarek-Lukaszewska, G., 2003: Bryophytes. In: Markert, B., A., Breure, A., M., Zechmeister, H., G., (eds.), *Bioindicators and biomonitors.* Elsevier, p. 3 29–375.